Field Extensions
and Galois Theory

ENCYCLOPEDIA OF MATHEMATICS
and Its Applications

GIAN-CARLO ROTA, Editor
Department of Mathematics
Massachusetts Institute of Technology
Cambridge, Massachusetts

Editorial Board

GIAN-CARLO ROTA, *Editor*
ENCYCLOPEDIA OF MATHEMATICS AND ITS APPLICATIONS

GIAN-CARLO ROTA, *Editor*
ENCYCLOPEDIA OF MATHEMATICS AND ITS APPLICATIONS

GIAN-CARLO ROTA, *Editor*
ENCYCLOPEDIA OF MATHEMATICS AND ITS APPLICATIONS

Other volumes in preparation

GIAN-CARLO ROTA, *Editor*

ENCYCLOPEDIA OF MATHEMATICS AND ITS APPLICATIONS

Volume 22

Section: Algebra

P. M. Cohn and Roger Lyndon, *Section Editors*

Field Extensions and Galois Theory

Julio R. Bastida
Department of Mathematics
Florida Atlantic University
Boca Raton, Florida

With a Foreword by

Roger Lyndon
The University of Michigan
Ann Arbor, Michigan

CAMBRIDGE
UNIVERSITY PRESS

CAMBRIDGE UNIVERSITY PRESS
Cambridge, New York, Melbourne, Madrid, Cape Town, Singapore,
São Paulo, Delhi, Dubai, Tokyo, Mexico City

Cambridge University Press
The Edinburgh Building, Cambridge CB2 8RU, UK

Published in the United States of America by Cambridge University Press, New York

www.cambridge.org
Information on this title: www.cambridge.org/9780521302425

© Cambridge University Press 1984

First published 1984

A catalogue record for this publication is available from the British Library

Library of Congress Cataloguing in Publication data

Bastida, Julio R.
   Field extensions and Galois theory.

   (Encyclopedia of mathematics and its applications;
v. 22)
   Bibliography: p.
   Includes index.
   1. Field extensions (Mathematics).   2. Galois theory.
1. Title. 11. Series.
QA247.B37 1984   512'.32   83-7160

ISBN 978-0-521-30242-5 Hardback
ISBN 978-0-521-17396-4 Paperback

A mi hijo,
Ricardo Antonio

Contents

Editor's Statement

A large body of mathematics consists of facts that can be presented and described much like any other natural phenomenon. These facts, at times explicitly brought out as theorems, at other times concealed within a proof, make up most of the applications of mathematics, and are the most likely to survive change of style and of interest.

This ENCYCLOPEDIA will attempt to present the factual body of all mathematics. Clarity of exposition, accessibility to the non-specialist, and a thorough bibliography are required of each author. Volumes will appear in no particular order, but will be organized into sections, each one comprising a recognizable branch of present-day mathematics. Numbers of volumes and sections will be reconsidered as times and needs change.

It is hoped that this enterprise will make mathematics more widely used where it is needed, and more accessible in fields in which it can be applied but where it has not yet penetrated because of insufficient information.

GIAN-CARLO ROTA

Foreword

Galois theory is often cited as the beginning of modern "abstract" algebra. The ancient problem of the algebraic solution of polynomial equations culminated, through the work of Ruffini, Abel, and others, in the ideas of Galois, who set forth systematically the connection between polynomial equations and their associated groups. This was the beginning of the systematic study of group theory, nurtured by Cauchy and Jordan to its flowering at the end of the last century. It can also be viewed as the beginning of algebraic number theory (although here other forces were also clearly at work), developed later in the century by Dedekind, Kronecker, Kummer, and others. It is primarily this number-theoretic line of development that is pursued in this book, where the emphasis is on fields, and only secondarily on their groups.

In addition to these two specific outgrowths of Galois's ideas, there came something much broader, perhaps the essence of Galois theory: the systematically developed connection between two seemingly unrelated subjects, here the theory of fields and that of groups. More specifically, but in the same line, is the idea of studying a mathematical object by its group of automorphisms, an idea emphasized especially in Klein's Erlanger Program, which has since been accepted as a powerful tool in a great variety of mathematical disciplines.

Apart from the historical importance of the Galois theory of fields, its intrinsic interest and beauty, and its more or less direct applications to number theory, these many generalizations and their important applications give further compelling reasons for seeking an understanding of the theory in its classical form, as presented in this volume. The Galois theory of field extensions combines the esthetic appeal of a theory of nearly perfect beauty with the technical development and difficulty that reveal the depth of the theory and that make possible its great usefulness, primarily in algebraic number theory and related parts of algebraic geometry.

In this book Professor Bastida has set forth this classical theory, of field extensions and their Galois groups, with meticulous care and clarity. The treatment is self-contained, at a level accessible to a sufficiently well-motivated beginning graduate student, starting with the most elementary facts about fields and polynomials and proceeding painstakingly, never omitting precise definitions and illustrative examples and problems. The qualified reader will be able to progress rapidly, while securing a firm grasp of the fundamental concepts and of the important phenomena that arise in the theory of fields. Ultimately, the study of this book will provide an intuitively clear and logically exact familiarity with the basic facts of a comprehensive area in the theory of fields. The author has judiciously stopped short (except in exercises) of developing specialized topics important to the various applications of the theory, but we believe he has realized his aim of providing the reader with a sound foundation from which to embark on the study of these more specialized subjects.

This book, then, should serve first as an easily accessible and fully detailed exposition of the classical Galois theory of field extensions in its simplest and purest form; and second, as a solid foundation for and introduction to the study of more advanced topics involving the same concepts, especially in algebraic number theory and algebraic geometry.

We believe that Professor Bastida has offered the reader, for a minimum of effort, a direct path into an enchantingly beautiful and exceptionally useful subject.

ROGER LYNDON

Preface

Since its inception at the beginning of the nineteenth century, the theory of field extensions has been a very active area of algebra. Its vitality stems not only from the interesting problems generated by the theory itself, but also from its connections with number theory and algebraic geometry. In writing this book, our principal objective has been to make the general theory of field extensions accessible to any reader with a modest background in groups, rings, and vector spaces.

The book is divided into four chapters. In order to give a precise idea of the background that the reader is expected to possess, we have preceded the text by a section on prerequisites. Except for the initial remarks, in which we indicate the restrictions that will be imposed on the rings considered throughout our presentation, the reader should not be concerned with the contents of this section until explicit reference is made to them. The first chapter is devoted to the general facts on fields and polynomials required in the study of field extensions. Although most of these facts can be found in one or another of the references given in the section on prerequisites, we have attempted to facilitate the reader's task by having them collected and stated in a manner suitably adapted to our purposes.

The theory of field extensions is presented in the subsequent three chapters, which deal, respectively, with algebraic extensions, Galois theory, and transcendental extensions. The chapter on algebraic extensions is of basic importance for the entire theory, and has to be thoroughly understood before proceeding further. The last two chapters, on the other hand, can be read independently of each other.

Chapters are divided into sections, and each section ends with a set of problems. The problems include routine exercises, suggest alternative proofs of various results, or develop topics not discussed in the text. We have refrained from identifying the more difficult, and as a rule, no hints are given for the solutions. A result stated in a problem is not used in the text, but it may be required for the solution of a later problem.

The choice of material was dictated by the dual objective of providing thorough coverage of each topic treated and of keeping the length of the book within reasonable bounds. We decided to include in the text the results that constitute the core of the general theory of field extensions. Those parts of the theory sufficiently developed to merit a book of their own have been left out entirely, and several specialized topics of considerable interest have been relegated to the problems. We have not attempted to discuss any serious applications of our subject to number theory or algebraic geometry, since doing this would have required the introduction of additional background material. However, as the reader cannot fail to notice, connections with number theory manifest themselves occasionally in the presentation.

We have included bibliographical notes at the end of each chapter. These will provide the reader with references to the works in which important contributions were first published, with easily available references on topics presented as problems and on alternative treatments of topics covered in the text, and with suggestions for further reading.

The reference list at the end of the book comprises mainly the works cited in the text and notes. The vast literature on field extensions and Galois theory and on their applications to number theory and algebraic geometry cannot be surveyed, even superficially, within the confines of a few pages. To get a good idea of the present state of the literature, the reader may consult the pertinent sections of *Mathematical Reviews*, the review journal of the American Mathematical Society.

It is with the deepest gratitude and respect that we acknowledge the help given to us by Professor Harley Flanders, without which this book could not have been written. He read the manuscript and made very substantive suggestions on both content and style; offered us unrestricted access to his notes on field extensions; discussed proofs, examples, and problems with us; and never betrayed the slightest impatience in dealing with us during the four-year period that we worked on this book.

We would also like to express our sincere appreciation to Professor Gian-Carlo Rota, for his kind invitation to write a volume for the *Encyclo-*

pedia; to Professors Paul M. Cohn and Roger C. Lyndon, for their valuable suggestions; to Professors Tomás P. Schonbek and Scott H. Demsky, for their help with the bibliographical material; to my students Lynn Garrett and Jaleh Owliaei, for their comments; to Ruth Ebel and especially Rita Pelava, for their efficient typing; and to my colleagues at Florida Atlantic University, for their constant encouragement.

JULIO R. BASTIDA
Boca Raton, Florida

Historical Introduction

Problems of geometric construction appeared early in the history of mathematics. They were first considered by the Greek mathematicians of the fifth century B.C. Only two instruments—an unmarked ruler and a compass—were permitted in these constructions. Although many such constructions could be performed, others eluded the efforts of these mathematicians. Four famous problems from the period that remained unsolved for a long time are the following: doubling the cube, which consists of constructing a cube whose volume is twice that of a given cube; trisecting the angle; squaring the circle, which consists of constructing a square whose area is that of a given circle; and constructing regular polygons.

At the end of the eighteenth century, when it was observed that questions on geometric constructions can be translated into questions on fields, a breakthrough finally occurred. The 19-year-old Gauss [2: art. 365] proved in 1796 that the regular 17-sided polygon is constructible. A few years later, Gauss [2: art. 365, 366] stated necessary and sufficient conditions for the constructibility of the regular n-sided polygon. He gave a proof only of the sufficiency, and claimed to have a proof of the necessity; the latter was first given by Wantzel [1] in 1837. In his investigations, Gauss introduced and used a number of concepts that became of central importance in subsequent developments. A by-product of the works of Gauss and Wantzel on regular polygons was a proof that an arbitrary angle cannot be trisected. The proof of the impossibility of doubling the cube is more elementary, but its discovery is difficult to trace. As to the remaining problem, it was realized that the proof of the impossibility of squaring the circle depended on knowing that the number π is transcendental; this missing ingredient was supplied in 1882 by Lindemann [1], who used analytic techniques to settle one of the more fascinating questions in this area of mathematics.

The general theory of fields evolved during the last half of the nineteenth century, when the algebraists made significant advances in the study of algebraic numbers and algebraic functions. The first systematic exposition of the theory of algebraic numbers was given in 1871 by Dedekind [4]; in this work, Dedekind introduced the basic notions on fields, but restricted the field elements to complex numbers. As regards transcendental numbers, the early contributions were made by analysts. The most notable of these contributions were that by Liouville [1] in 1851, devoted to the construction of classes of transcendental numbers, and those by Hermite [2] in 1873 and Lindemann [1] in 1882, in which proofs are given of the transcendence of the numbers e and π, respectively. But it was not until 1882 that transcendentals made their appearance in the theory of fields, when Kronecker [2] succeeded in using the adjunction of indeterminates as the basis for a formulation of the theory of algebraic numbers. It was also in 1882 that fields of algebraic functions of complex variables were introduced by Dedekind and Weber [1] in order to lay the foundations of the arithmetical theory of algebraic functions. This work, in which a purely algebraic treatment of Riemann surfaces is given, marks the beginning of what was to become a very fruitful interplay between commutative algebra and algebraic geometry. It was next discovered in 1887 by Kronecker [3] that every algebraic number field can be obtained as the quotient of the polynomial domain $\mathbf{Q}[X]$ by the principal ideal generated by an irreducible polynomial, showing in effect that the theory of algebraic numbers does not require the use of complex numbers. Finally, the abstract definition of a field as we know it today was given in 1893 by Weber [1] in an article on the foundations of Galois theory. Weber also observed in this work that Kronecker's construction can be applied to arbitrary fields, and in particular to every field of integers modulo a prime; and that as a result, we recover the theory of higher congruences previously developed by Galois [2], Serret [1: 343–370], and Dedekind [2].

The final step toward the axiomatic foundations of the theory of fields was taken by Steinitz [1] in 1910. Spurred on by both the earlier contributions and the discovery by Hensel [1] of the p-adic fields, Steinitz set out to derive the consequences of Weber's axioms. His work, in which field extensions were first studied in full generality and in which normality, separability, and pure inseparability were introduced in order to give a detailed analysis of the structure of algebraic extensions, became the cornerstone in the development of abstract algebra. In the words of Artin and Schreier [1]: "E. Steinitz hat durch seine 'Algebraische Theorie der Körper' weite Gebiete der Algebra einer abstrakten Behandlungsweise erschlossen; seiner bahnbrechenden Untersuchung ist zum grossen Teil die starke Entwicklung zu danken, die seither die moderne Algebra genommen hat". It is in the closing pages of Steinitz's article that the theory of transcendental extensions was first presented. However, before this theory could be brought to its present state, two significant additions were yet to be made, both partially motivated by questions in algebraic geometry. In 1939, MacLane [1] introduced the notion of separability for transcendental extensions. This was then followed in 1946 by the treatise on the foundations of algebraic geometry by Weil [1], in which the abstract notion of derivation is introduced in the study of separability.

Galois theory is generally regarded as one of the central and most beautiful parts of algebra. Its creation marked the culmination of investigations by generations of mathematicians into one of the oldest problems in algebra, the solvability of polynomial equations by radicals. The familiar formula for the roots of the quadratic equation was essentially known to the Babylonian mathematicians of the twentieth century B.C. No significant progress was made on polynomial equations of higher degree until the sixteenth century, when del Ferro and Ferrari discovered the formulas for the cubic and quartic equations, respectively. These results were first published by Cardano [1] in 1545; it is probably for this reason that Cardano's name has been traditionally associated with the formulas for the cubic equation.

These formulas express the roots of the equations in terms of the coefficients, using exclusively the field operations and the extraction of roots. Attempts to find such formulas for polynomial equations of higher degree were unsuccessful; and partly as a consequence of the work of Lagrange [2; 3] in 1770–1772, the algebraists of the period came to believe that it was impossible to derive them. This was proved to be the case at the beginning of the nineteenth century. Several proofs were published by Ruffini [1] between 1799 and 1813, but they were incomplete. The first satisfactory proof was given by Abel [2] in 1826, three years before his tragic death before the age of 27; between 1826 and 1829 he obtained further results on the solvability of polynomial equations by radicals, which were published in Abel [3; 1: II, 217–243, 269–270, 271–279].

The contributions of Ruffini and Abel were followed by the decisive results of Galois [1: 25–61] in 1832. Galois proved that the solvability of a polynomial equation by radicals is equivalent to a special property of a group naturally associated with the equation. Galois made this discovery before the age of 20, at a time when abstract algebra virtually did not exist!

Although Galois's result on the solvability of polynomial equations by radicals settled a problem that had eluded the efforts of some of the greatest mathematicians of earlier generations, later developments have shown that the ideas introduced by Galois in his solution surpass by far the importance of the problem that he originally set out to solve. First, Galois defined and used the group-theoretical properties of normality, simplicity, and solvability, which play a significant role in the theory of groups. Moreover, he solved a problem of fields by translating it into a more tractable problem on groups; in so doing, he probably made the earliest application of a method that has become pervasive in algebra, namely, that of studying a mathematical object by suitably relating it to a mathematical object with a simpler structure. Nor is it an exaggeration to say that Galois theory is a prerequisite for much current research in number theory and algebraic geometry.

The story of Galois's life is a topic of considerable controversy. A gifted mathematician who is killed in a duel at the age of 20 presents unlimited opportunities for the creation of a myth. Unfortunately, this is precisely what several well-known authors have done in their writings on Galois. By means of intentional or unintentional omissions and distortions, legends have been created in which Galois is portrayed as a struggling genius unappreciated not only by the general public, but also by some of the leading mathematicians of his time. The recent article by Rothman [1] offers a lively account of such theories, as well as a careful attempt to unravel them.

Galois's ideas were expressed originally within the context of the theory of equations: To each polynomial equation is assigned a group of permutations of its roots. The progress made toward the axiomatic foundations of algebra in the last part of the nineteenth century had a considerable impact on Galois theory. Dedekind [4] observed that a more natural setting for Galois theory is obtained by regarding the groups associated with polynomial equations as groups of automorphisms of the corresponding splitting fields. Furthermore, he pioneered the systematic use of linear algebra in Galois theory. Since the abstract theory of field extensions was not developed until the first decade of the present century, Dedekind had to restrict his considerations to special types of fields. That his formulation of Galois theory remains meaningful for arbitrary fields was shown subsequently by the works of Weber [1] in 1893, of Steinitz [1] in 1910, and of Artin [3] in 1942. It is to these algebraists, and especially to Artin, that we owe what is now considered to be the definitive exposition of the Galois theory of finite groups of field automorphisms. A further contribution that must be mentioned is the generalization of the principal results of this theory to a special type of infinite groups of field automorphisms, discovered by Krull [1] in 1928.

Prerequisites

We shall assume that the reader possesses a certain familiarity with the rudiments of abstract algebra. More specifically, in addition to the basic properties of integers, sets, and mappings, the reader is expected to know the elementary parts of the theory of groups and the theory of rings, and to possess a reasonable background in linear algebra. Suggested references on these prerequisites are the following.

1. Adamson, I. T. *Elementary Rings and Modules*. New York: Harper & Row, 1972.
2. Godement, R. *Cours d'Algèbre*. Paris: Hermann, 1963. (English translation: *Algebra*. New York: Houghton Mifflin, 1968.)
3. Halmos, P. R. *Naive Set Theory*. New York: Springer-Verlag, 1974.
4. Hoffman, K., and Kunze, R. *Linear Algebra*. Englewood Cliffs, NJ: Prentice-Hall, 1971.
5. Ledermann, W. *Introduction to Group Theory*. Edinburgh: Oliver & Boyd, 1973.
6. Rotman, J. J. *The Theory of Groups, an Introduction*. Boston: Allyn & Bacon, 1973.

This list is not intended as an exhaustive bibliography on the basic concepts of algebra. We have simply selected six easily accessible books that, for our purposes, are particularly suitable as references. The books [1] and [2] seem the most convenient: In the first place, we shall adhere almost completely to the terminology and notation used in these books; further-more, taken together, these cover all the required background on rings, ideals, polynomials, modules, and vector spaces. The few facts on ordering and cardinal numbers occasionally used here are contained in the book [3]; and each of the books [5] and [6] contains all the background on groups needed in our presentation of Galois theory. Finally, the book [4] can be used as an alternative reference on linear algebra.

It should be noted that many books on abstract algebra, in chapters dealing separately with sets, groups, rings, and linear algebra, contain all or more of the prerequisites just described. Some of these are listed in the bibliography at the end of this book. (There is one section of the present book that requires additional prerequisites. This is section 3.12, which is devoted to infinite Galois theory, and in which some facts on topological groups are used. This section, however, is intended for readers interested in modern number theory; such readers would have to be well-versed in the theory of topological groups, and so it would be superfluous to give references on this subject.)

We now proceed to state in precise terms the conventions that will be adopted, and to explain the terminology and notation that will be used. Since there is no total agreement on these matters in the literature, the reader should make sure that we are using the same language.

Three types of algebraic structure are considered in our presentation. The first is defined by one operation, the second by two operations, and the third by one operation and one action. The term *operation* is being used here with the same meaning as "law of composition", "internal law of composition", and "binary operation", all of which are standard in the literature; and the term *action* is being used with the same meaning as "external law of composition", which is also of common usage.

We shall be concerned exclusively with operations that are associative and admit a neutral element. Moreover, for the most part, we shall use the multiplicative and additive notations. In the former case, the neutral element is called the **unit element** and is denoted by 1; and in the latter, it is called the **zero element** and is denoted by 0.

In the case of groups, subgroups, and group-homomorphisms, we shall usually follow [5] and [6]. In particular, the operation of a group will be written multiplicatively; the only exception to this occurs when reference is being made to the additive group of a ring, where the context always makes the intended meaning clear.

On the other hand, it will not be necessary for us to use the concept of ring in its full generality. First, our rings, subrings, and ring-homomor-phisms will be restricted as in [2]: Rings possess a unit element; ring and subring have the same unit element; and ring-homomorphisms send unit element to unit element. Also, the nature of our subject dictates that we restrict our consideration to commutative rings in which the zero and unit elements are distinct. Whenever we speak of rings, subrings, and ring-homo-morphisms, it will be tacitly understood that all these restrictions apply.

Finally, in the case of modules and vector spaces, we shall follow [1], [2], and [4]. As usual, the operation and action of a module are referred to as its **vector addition** and **scalar multiplication**, respectively. In view of the conventions just adopted, it will not be necessary to distinguish between left and right modules. We shall speak of A-modules, A-submodules, and A-linear mappings whenever we wish to indicate that the ring of scalars is A. The general concept of module will play only an ancillary role in this book, since we shall be concerned primarily with vector spaces; if the field of scalars is A, we shall speak of A**-spaces** instead of vector spaces over A. It is hoped that this departure from standard terminology will not cause misunderstandings.

So far, for each of the prerequisites, we have made reference to certain books whose terminology and notation we shall generally follow. We shall now indicate the few instances where deviations occur.

A relation is said to **order** a set when it is reflexive, antisymmetric, and transitive on the elements of the set. By an **ordered set** we shall understand a set provided with a relation that orders it.

Let E be an ordered set. If $x, y \in E$, we write $x \le y$ or $y \ge x$ to express that the pair (x, y) is in the given relation ordering E; and we write $x < y$ or $y > x$ to express that (x, y) is in this relation and $x \ne y$. If $(x_i)_{i \in I}$ is a family of elements of E, to say that $(x_i)_{i \in I}$ is **filtered** means that for all $i, j \in I$, there exists a $k \in I$ for which $x_k \ge x_i$ and $x_k \ge x_j$; and to say that $(x_i)_{i \in I}$ is a **chain** means that for all $i, j \in I$, we have $x_i \le x_j$ or $x_i \ge x_j$. If $S \subseteq E$, then S is said to be **filtered** when the family $(x)_{x \in S}$ is filtered; and similarly, S is said to be a **chain** when $(x)_{x \in S}$ is a chain. If $S \subseteq E$ and $b \in E$, then b is an **upper bound for** S when $b \ge x$ for every $x \in S$.

If E is an ordered set, there can be in E at most one upper bound for E; when it exists, it is said to be the **largest element of** E. A **maximal element of** E is an $x \in E$ such that $x < y$ for no $y \in E$. Note that if the largest element of E exists, it is the only maximal element of E; but when E does not admit a largest element, it may admit more than one maximal element.

The preceding considerations on ordered sets apply, in particular, to sets of sets. Whenever we speak of a set of sets as being ordered by the inclusion relation, it will be understood that the relation in question is $\subseteq$. It is clear, therefore, what is meant when we speak of a **filtered family of sets**, a **filtered set of sets**, a **chain of sets**, the **largest element of a set of sets**, and a **maximal element of a set of sets**.

It should be noted, on the other hand, that every set of sets is also ordered by the opposite inclusion relation $\supseteq$. This, however, will be applied in only two instances: when we speak of the **smallest element of a set of sets** and of a **minimal element of a set of sets**.

To conclude these remarks on ordered sets, we shall state the result called **Zorn's lemma**. By an **inductive set** we shall understand an ordered set in which every nonempty chain admits an upper bound. The result in question asserts the following:

Every nonempty inductive set admits a maximal element.

This is a powerful set-theoretical tool that we shall use to derive important properties of algebraically closed fields and to establish the extendibility of certain mappings. It is not an "intuitive" statement, and does not yield "constructive" proofs. It is known to be equivalent to the "more intuitive" **axiom of choice** in the theory of sets, which asserts that the cartesian product of every nonempty family of nonempty sets is nonempty. The reader interested in a detailed study of these questions may wish to consult the book [3]. We shall simply accept Zorn's lemma as a valid result, and apply it without further comment.

A group consisting of a single element will be called **trivial**. If G is a group and H is a subgroup of G, a **left transversal of H in G** is a subset of G having exactly one element in common with each left coset of H in G; a **right transversal of H in G** is defined similarly, using right cosets.

Let A be a ring. There exists a unique homomorphism from the ring $\mathbf{Z}$ of integers to A; this is the mapping $n \to n1$ from $\mathbf{Z}$ to A. It is customary to denote by the same symbol n the value of this homomorphism at an integer n; this is only a notational convenience, and it should be noted that if m and n are distinct integers, the equality $m = n$ may be valid in A. The image of this homomorphism is called the **image of $\mathbf{Z}$ in A**; it is the smallest element of the set of all subrings of A.

If A is a ring, the **invertible elements of A** are the multiplicatively invertible elements of A. The set of all invertible elements of A is multiplicatively stable, and, provided with the operation defined by restriction of the multiplication of A, is a group. This group is denoted by A^*; its neutral element is 1, the unit element of A. The subgroups of A^* are called the **multiplicative groups in A**. The elements of finite order in A^* are the **roots of unity in A**; and if n is a positive integer, an nth **root of unity in A** is an $\alpha \in A$ for which $\alpha^n = 1$, that is, a root of unity in A with order dividing n.

An ideal in a ring is **null** when it consists of a single element; **prime** when it is a proper ideal and its complement in the ring is multiplicatively stable; and **maximal** when it is a maximal element of the set of all proper ideals.

We shall speak of **domains** instead of integral domains, and of **factorial domains** instead of unique factorization domains. By a **system of representatives of irreducible elements** in a factorial domain we shall understand a set of irreducible elements having exactly one element in common with the set of all associates of each irreducible element.

Polynomials play an essential role in our subject. The letters X, Y, Z —with or without subscripts—will be reserved for the variables in our rings of polynomials. Polynomials in infinitely many variables will be required only occasionally in this book (and in the only important instance, alternatives are indicated); the reader who is not familiar with this more general type of polynomial should read 0.0.5 below, where it is explained how to construct rings of polynomials in infinitely many variables.

An injective group-homomorphism or ring-homomorphism will be called a **monomorphism** or an **embedding**. Given two groups or two rings A and B, to say that A is **embeddable in** B will mean that there exists a monomorphism from A to B. This terminology is particularly convenient when dealing with fields, since it serves as a constant reminder of the fact that every homomorphism from a field to a ring is injective.

A module or vector space consisting of a single element will be called **null**. If A is a field and E is an A-space, the symbol $[E:A]$ will denote the dimension of E over A. Incidentally, the reader in need of a rapid review of the theory of dimension for general vector spaces may wish to learn Steinitz's axiomatic approach; this is given in section 4.1 and requires set-theoretical prerequisites exclusively, so that it can be read without reference to any other section.

If A is a ring and I is a set, the symbol $A^{(I)}$ will be used to denote the **free A-module based on** I. In order to define this module, we recall that if $(P_i)_{i \in I}$ is a family of statements, we say that P_i holds **for almost every** $i \in I$ when the set of all $i \in I$ for which P_i does not hold is finite. This being so, the elements of $A^{(I)}$ are the families $(\lambda_i)_{i \in I}$ of elements of A such that $\lambda_i = 0$ for almost every $i \in I$; and the vector addition and scalar multiplication of $A^{(I)}$ are defined "coordinate-wise":

$$(\lambda_i)_{i \in I} + (\mu_i)_{i \in I} = (\lambda_i + \mu_i)_{i \in I} \quad \text{and} \quad \alpha(\lambda_i)_{i \in I} = (\alpha\lambda_i)_{i \in I}.$$

For each $i \in I$, let ε_i denote the element of $A^{(I)}$ with 1 as its ith coordinate and with 0 as its jth coordinate for every $j \in I - \{i\}$. Then $(\varepsilon_i)_{i \in I}$ is a base of $A^{(I)}$, and so $A^{(I)}$ is indeed a free A-module; we refer to $(\varepsilon_i)_{i \in I}$ as the **standard base of** $A^{(I)}$.

If A is a ring and n is a positive integer, then the free A-module based on $\{1, 2, \ldots, n\}$ is none other than the familiar A-module $A^{(n)}$ of "vectors" with n coordinates in A.

If a ring A is a subring of a ring B, then B can be regarded as an A-module in a natural way: The vector addition is the addition of B, and the scalar multiplication is the action of A on B defined by restriction of the multiplication of B. Whenever a ring is viewed as a module over a subring, it will be understood that the linear structure under consideration is defined in this manner.

If a ring A is a common subring of rings B and C, it is customary to define an **A-homomorphism from** B **to** C as a homomorphism from B to C

such that $\alpha \to \alpha$ for every $\alpha \in A$. It is readily seen that a homomorphism from B to C has this property if and only if it is an A-linear mapping from the A-module B to the A-module C.

Two notions of an independent family of elements arise in the present context. Let a ring A be a subring of a ring B, and let $(\beta_1, \beta_2, \ldots, \beta_n)$ be a finite sequence of elements of B —it is sufficient to consider only finite sequences, since the properties in question are of "finite character". First, it is meaningful to speak of $(\beta_1, \beta_2, \ldots, \beta_n)$ as a linearly independent sequence of elements of the A-module B; we shall express this by saying that $(\beta_1, \beta_2, \ldots, \beta_n)$ is **linearly independent over** A. On the other hand, recall that we have the A-homomorphism $f(X_1, X_2, \ldots, X_n) \to f(\beta_1, \beta_2, \ldots, \beta_n)$ from $A[X_1, X_2, \ldots, X_n]$ to B, and that its kernel is said to be the **ideal of algebraic relations of** $(\beta_1, \beta_2, \ldots, \beta_n)$ **over** A; then $(\beta_1, \beta_2, \ldots, \beta_n)$ is said to be **algebraically independent over** A when this ideal is null. Of course, these two notions are not identical; generally speaking, algebraic independence implies linear independence, but not conversely.

For each of the algebraic objects considered previously, there is a notion of quotient object. In the case of a group, it is defined by a normal subgroup; in that of a ring, by a proper ideal; and in that of a module, by a submodule. The elements of the quotient object are the cosets of the elements in the original object relative to the defining normal subgroup, proper ideal, or submodule. The mapping from the original object to the quotient object that to each element assigns its coset is called the **natural projection**; it is a surjective homomorphism in the case of groups and rings, and a surjective linear mapping in the case of modules.

To close the present discussion on prerequisites, we shall derive some isolated special results that will be used in this book and for which we were unable to give suitable references.

0.0.1.   Given a set A, a ring R, and a bijection u from A to R, there is a unique ring structure on A relative to which u is an isomorphism. The addition and multiplication defining this ring structure are the operations

$$(\alpha, \beta) \to u^{-1}(u(\alpha) + u(\beta)) \quad \text{and} \quad (\alpha, \beta) \to u^{-1}(u(\alpha)u(\beta))$$

on A.

It is clear that these are the only possible choices for the required operations: If A is provided with a ring structure in such a way that u is an isomorphism, then for all $\alpha, \beta \in A$ we have

$$u(\alpha + \beta) = u(\alpha) + u(\beta) \quad \text{and} \quad u(\alpha\beta) = u(\alpha)u(\beta),$$

and hence

$$\alpha + \beta = u^{-1}(u(\alpha) + u(\beta)) \quad \text{and} \quad \alpha\beta = u^{-1}(u(\alpha)u(\beta)).$$

This being said, we can now state and prove the following result on ring-monomorphisms.

Let A and R be rings, and let u be a monomorphism from A to R. Then there exist a ring B and an isomorphism v from B to R such that A is a subring of B and such that v extends u.

Proof. First, choose a set S such that S and $R - \text{Im}(u)$ are equipotent and such that A and S are disjoint. Then choose a bijection t from S to $R - \text{Im}(u)$, and put $B = A \cup S$.

Denote by v the mapping from B to R, such that

$$\alpha \to u(\alpha) \quad \text{for } \alpha \in A \quad \text{and} \quad \alpha \to t(\alpha) \quad \text{for } \alpha \in S.$$

Then v is a bijection extending u, and the desired conclusion follows by providing B with the ring structure relative to which v is an isomorphism. $\square$

0.0.2. The following useful result relates the linear structures defined by a chain of rings.

Let A be a ring, let B be a ring having A as a subring, and let C be a ring having B as a subring.

(i) *If $(\beta_i)_{i \in I}$ and $(\gamma_j)_{j \in J}$ are, respectively, generating systems of the A-module B and of the B-module C, then $(\beta_i \gamma_j)_{(i, j) \in I \times J}$ is a generating system of the A-module C.*

(ii) *If $(\beta_i)_{i \in I}$ and $(\gamma_j)_{j \in J}$ are, respectively, families of elements of B and C that are linearly independent over A and B, then $(\beta_i \gamma_j)_{(i, j) \in I \times J}$ is linearly independent over A.*

(iii) *If $(\beta_i)_{i \in I}$ and $(\gamma_j)_{j \in J}$ are, respectively, bases of the A-module B and of the B-module C, then $(\beta_i \gamma_j)_{(i, j) \in I \times J}$ is a base of the A-module C.*

Proof. It suffices to prove (i) and (ii), since these clearly imply (iii).

First, let $(\beta_i)_{i \in I}$ and $(\gamma_j)_{j \in J}$ be as in (i), and let $\xi \in C$. We can write $\xi = \sum_{j \in J} \mu_j \gamma_j$, where $\mu_j \in B$ for every $j \in J$ and $\mu_j = 0$ for almost every $j \in J$. Then for every $j \in J$, we can write $\mu_j = \sum_{i \in I} \alpha_{ij} \beta_i$, where $\alpha_{ij} \in A$ for every $i \in I$ and $\alpha_{ij} = 0$ for almost every $i \in I$; and furthermore, when $\mu_j = 0$, we can take $\alpha_{ij} = 0$ for every $i \in I$. It then follows that $\alpha_{ij} = 0$ for almost every $(i, j) \in I \times J$ and

$$\xi = \sum_{j \in J} \mu_j \gamma_j = \sum_{(i, j) \in I \times J} \alpha_{ij} \beta_i \gamma_j,$$

which shows that ξ is a linear combination of $(\beta_i \gamma_j)_{(i, j) \in I \times J}$ with coefficients in A.

To conclude, let $(\beta_i)_{i \in I}$ and $(\gamma_j)_{j \in J}$ be as in (ii). Assume that

$$\sum_{(i, j) \in I \times J} \alpha_{ij} \beta_i \gamma_j = 0,$$

where $\alpha_{ij} \in A$ for every $(i, j) \in I \times J$ and $\alpha_{ij} = 0$ for almost every $(i, j) \in I \times J$. Put $\mu_j = \sum_{i \in I} \alpha_{ij} \beta_i$ for every $j \in J$; then $\mu_j \in B$ for every $j \in J$ and

$\mu_j = 0$ for almost every $j \in J$, and

$$\sum_{j \in J} \mu_j \gamma_j = \sum_{(i, j) \in I \times J} \alpha_{ij} \beta_i \gamma_j = 0.$$

The linear independence of $(\gamma_j)_{j \in J}$ over B now implies that

$$\sum_{i \in I} \alpha_{ij} \beta_i = \mu_j = 0$$

for every $j \in J$; and the linear independence of $(\beta_i)_{i \in I}$ over A implies then that $\alpha_{ij} = 0$ for every $(i, j) \in I \times J$. Thus, $(\beta_i \gamma_j)_{(i, j) \in I \times J}$ is linearly independent over A. □

0.0.3. The following proposition, which will be used in the study of algebraic extensions, gives a sufficient condition for a domain to be a field.

Let A be a domain. If there exists a subfield K of A such that A is a finite-dimensional K-space, then A is a field.

Proof. Indeed, let $\alpha \in A$ and $\alpha \neq 0$. If K is a subfield of A with the indicated property, then the mapping $\xi \to \alpha\xi$ from A to A is K-linear. Since A is a domain, this mapping is injective, and the assumed finite dimensionality of A as a K-space implies that it is bijective. In particular, we have $\alpha\beta = 1$ for some $\beta \in A$, and so α is invertible in A. □

0.0.4. As a typical illustration of how Zorn's lemma is used in algebra, we shall now derive a result on the existence of maximal ideals. This will be used to prove that every field admits an algebraic closure.

Every proper ideal in a ring is contained in a maximal ideal.

Proof. Let A be a ring, and let $\mathcal{J}$ be a proper ideal of A. Denote by Ω the set of all proper ideals of A containing $\mathcal{J}$, and order Ω by inclusion. Then the maximal ideals of A containing $\mathcal{J}$ are precisely the maximal elements of Ω; therefore, by Zorn's lemma, it suffices to verify that Ω is nonempty and inductive.

To do this, note first that $\mathcal{J} \in \Omega$, so that Ω is nonempty. Now let Ψ be a nonempty chain in Ω, and put $\mathcal{M} = \cup_{\mathcal{X} \in \Psi} \mathcal{X}$. Since Ψ is filtered, it is clear that $\mathcal{M}$ is an ideal of A; and since $1 \notin \mathcal{X}$ for every $\mathcal{X} \in \Psi$, we have $1 \notin \mathcal{M}$. Therefore $\mathcal{M}$ is a proper ideal of A such that $\mathcal{M} \supseteq \mathcal{X}$ for every $\mathcal{X} \in \Psi$, which implies that $\mathcal{M}$ is an upper bound for Ψ in Ω. We conclude that Ω is inductive. □

Applying this to the null ideal, we see that every ring possesses maximal ideals. Also, since an element in a ring is noninvertible if and only if it generates a proper ideal, it follows that the invertible elements in a ring are the elements belonging to no maximal ideal.

0.0.5. It is sometimes required to know how given algebraic structures of the same type can be combined in order to obtain an algebraic structure suitably related to the given ones. To prove that every field admits an algebraic closure, we shall use the following special case of a general proposition in universal algebra.

Let $(A_i)_{i \in I}$ be a nonempty family of rings such that for all $i, j \in I$, there exists a $k \in I$ such that A_k contains A_i and A_j as subrings. Then $\cup_{i \in I} A_i$ can be uniquely provided with a ring structure in such a way that it contains A_i as a subring for every $i \in I$.

Proof. Let $A = \cup_{i \in I} A_i$. To define the two required operations on A, let $\alpha, \beta \in A$. The hypothesis implies, first, that there exists an $i \in I$ for which $\alpha, \beta \in A_i$; it also implies that if $i, j \in I$ and $\alpha, \beta \in A_i \cap A_j$, then A_i and A_j are subrings of one and the same ring, and so the symbols $\alpha + \beta$ and $\alpha\beta$ have the same meaning in A_i and in A_j.

It is meaningful, therefore, to speak of the addition and of the multiplication on A that to each $(\alpha, \beta) \in A \times A$ assign, respectively, the elements $\alpha + \beta$ and $\alpha\beta$ of A_i whenever $i \in I$ and $\alpha, \beta \in A_i$. Note also that for all $i, j \in I$, the rings A_i and A_j have the same unit element. The details involved in verifying that the addition and the multiplication on A just described satisfy all the required conditions are now seen to be straightforward, and can be omitted.   □

The proposition just proved can be used in order to define rings of polynomials in infinitely many variables.

To see this, consider a ring A and an infinite set I. Denote by Ω the set of all finite subsets of I; and for every $S \in \Omega$, write $A_S = A[X_i]_{i \in S}$. It is clear that if $S, T \in \Omega$, then $S \cup T \in \Omega$ and $A_{S \cup T}$ contains A_S and A_T as subrings. The ring $A[X_i]_{i \in I}$ is defined as the ring obtained when $\cup_{S \in \Omega} A_S$ is provided with the ring structure described in the preceding proof.

The few general properties of rings of polynomials in infinitely many variables that will be used in this book are then seen to be immediate consequences of the corresponding properties of rings of polynomials in finitely many variables.

0.0.6. Recall that if F is a field and n is a positive integer, the symbol $GL_n(F)$ denotes the multiplicative group of nonsingular $n \times n$ matrices with entries in F. In the computations of certain Galois groups, it will be necessary to know the order of $GL_2(F)$ when F is a finite field. As we now show, this order can be determined by an elementary counting argument.

If F is a finite field, then $GL_2(F)$ is a group of order
$(\mathrm{Card}(F)^2 - \mathrm{Card}(F))(\mathrm{Card}(F)^2 - 1)$.

Proof. Let us write $q = \mathrm{Card}(F)$.

The elements of $GL_2(F)$ are the 2×2 matrices whose rows are linearly independent vectors in $F^{(2)}$. Therefore, the first row of such a matrix can be any nonzero vector in $F^{(2)}$; and the second row can be any vector in $F^{(2)}$ that is not a scalar multiple of the first row. Since there are $q^2 - 1$ choices for the first row, and since for each of these there are $q^2 - q$ for the second row, we conclude that the number of elements of $GL_2(F)$ is $(q^2 - 1)(q^2 - q)$, as claimed. $\square$

0.0.7.   Here we shall collect a few facts on the elementary symmetric polynomials that will be used in some of the classical illustrations of Galois theory.

Given a positive integer n and an integer k such that $0 \le k \le n$, let $[n; k]$ denote the set of all subsets of $\{1, 2, \ldots, n\}$ with cardinality k.

If A is a ring and n is a positive integer, the **elementary symmetric polynomials** in $A[X_1, X_2, \ldots, X_n]$ are the polynomials $e_0, e_2, \ldots, e_n$ defined by

$$e_k(X_1, X_2, \ldots, X_n) = \sum_{S \in [n; k]} \prod_{i \in S} X_i \quad \text{for } 0 \le k \le n.$$

Thus, we have

$$e_0(X_1, X_2, \ldots, X_n) = 1,$$

$$e_1(X_1, X_2, \ldots, X_n) = \sum_{i=1}^{n} X_i,$$

$$e_2(X_1, X_2, \ldots, X_n) = \sum_{1 \le i < j \le n} X_i X_j,$$

$$e_3(X_1, X_2, \ldots, X_n) = \sum_{1 \le i < j < k \le n} X_i X_j X_k,$$

$$\vdots \qquad \vdots \qquad \vdots$$

$$e_n(X_1, X_2, \ldots, X_n) = \prod_{i=1}^{n} X_i.$$

We begin with an auxiliary result showing that the elementary symmetric polynomials satisfy simple recurrence relations. These will be used in order to derive two important facts about the elementary symmetric polynomials.

Let A be a ring, let n be an integer such that $n > 1$, and let $\bar{e}_0, \bar{e}_1, \ldots, \bar{e}_{n-1}$ and $e_0, e_1, \ldots, e_n$ denote, respectively, the elementary symmetric polynomials in $A[X_1, X_2, \ldots, X_{n-1}]$ and $A[X_1, X_2, \ldots, X_n]$. Then

$$e_n(X_1, X_2, \ldots, X_n) = X_n \bar{e}_{n-1}(X_1, X_2, \ldots, X_{n-1}),$$

and

$$e_k(X_1, X_2, \ldots, X_n) = \bar{e}_k(X_1, X_2, \ldots, X_{n-1}) + X_n \bar{e}_{k-1}(X_1, X_2, \ldots, X_{n-1})$$

for $1 \le k \le n - 1$.

Proof.   The first equality is evident. Suppose now that $1 \le k \le n - 1$, and let Ω denote the set of all sets of the form $S \cup \{n\}$ with $S \in [n-1; k-1]$. Since

$$[n; k] = [n-1; k] \cup \Omega \quad \text{and} \quad [n-1; k] \cap \Omega = \varnothing,$$

it follows that

$$e_k(X_1, X_2, \ldots, X_n) = \sum_{S \in [n; k]} \prod_{i \in S} X_i$$

$$= \sum_{S \in [n-1; k]} \prod_{i \in S} X_i + \sum_{S \in \Omega} \prod_{i \in S} X_i$$

$$= \bar{e}_k(X_1, X_2, \ldots, X_{n-1}) + X_n \left(\sum_{S \in [n-1; k-1]} \prod_{i \in S} X_i \right)$$

$$= \bar{e}_k(X_1, X_2, \ldots, X_{n-1}) + X_n \bar{e}_{k-1}(X_1, X_2, \ldots, X_{n-1}),$$

as required.   □

Let A be a ring, and let n be a positive integer. If $\alpha_1, \alpha_2, \ldots, \alpha_n \in A$ and $e_0, e_1, \ldots, e_n$ denote the elementary symmetric polynomials in $A[X_1, X_2, \ldots, X_n]$, then

$$\prod_{i=1}^{n} (X - \alpha_i) = \sum_{k=0}^{n} (-1)^{n-k} e_{n-k}(\alpha_1, \alpha_2, \ldots, \alpha_n) X^k.$$

Proof. We shall proceed by induction on n. The conclusion is evident when $n = 1$. Assume that $n > 1$, and that it holds for $n - 1$.

Let $\alpha_1, \alpha_2, \ldots, \alpha_n \in A$, and let $\bar{e}_0, \bar{e}_1, \ldots, \bar{e}_{n-1}, e_0, e_1, \ldots, e_n$ be as in the preceding result. By the latter and by the induction hypothesis, we have

$$\prod_{i=1}^{n} (X - \alpha_i) = \left(\prod_{i=1}^{n-1} (X - \alpha_i) \right) (X - \alpha_n)$$

$$= \left(\sum_{k=0}^{n-1} (-1)^{n-1-k} \bar{e}_{n-1-k}(\alpha_1, \alpha_2, \ldots, \alpha_{n-1}) X^k \right) (X - \alpha_n)$$

$$= \sum_{k=0}^{n-1} (-1)^{n-1-k} \bar{e}_{n-1-k}(\alpha_1, \alpha_2, \ldots, \alpha_{n-1}) X^{k+1}$$

$$- \sum_{k=0}^{n-1} (-1)^{n-1-k} \alpha_n \bar{e}_{n-1-k}(\alpha_1, \alpha_2, \ldots, \alpha_{n-1}) X^k$$

$$= (-1)^n \alpha_n \bar{e}_{n-1}(\alpha_1, \alpha_2, \ldots, \alpha_{n-1}) + X^n$$

$$+ \sum_{k=1}^{n-1} (-1)^{n-k} (\bar{e}_{n-k}(\alpha_1, \alpha_2, \ldots, \alpha_{n-1}) +$$

$$\alpha_n \bar{e}_{n-1-k}(\alpha_1, \alpha_2, \ldots, \alpha_{n-1})) X^k$$

$$= (-1)^n e_n(\alpha_1, \alpha_2, \ldots, \alpha_n)$$

$$+ \sum_{k=1}^{n-1} (-1)^{n-k} e_{n-k}(\alpha_1, \alpha_2, \ldots, \alpha_n) X^k + X^n$$

$$= \sum_{k=0}^{n} (-1)^{n-k} e_{n-k}(\alpha_1, \alpha_2, \ldots, \alpha_n) X^k,$$

which is what we wanted.   □

If A is a ring and n is a positive integer, then the nonconstant elementary symmetric polynomials in $A[X_1, X_2, \ldots, X_n]$ are algebraically independent over A. More precisely, if $e_0, e_1, \ldots, e_n$ denote the elementary symmetric polynomials in $A[X_1, X_2, \ldots, X_n]$, then $(e_1, e_2, \ldots, e_n)$ is algebraically independent over A.

Proof. Again, we shall use induction on n. The conclusion is obvious when $n = 1$. Suppose then that it holds for $n - 1$, where n is an integer such that $n > 1$.

Let $\bar{e}_0, \bar{e}_1, \ldots, \bar{e}_{n-1}, e_0, e_1, \ldots, e_n$ have the same meaning as in the preceding proofs. Assume, for a contradiction, that $(e_1, e_2, \ldots, e_n)$ is algebraically dependent over A, and choose a nonzero polynomial f of the smallest possible total degree in the ideal of algebraic relations of $(e_1, e_2, \ldots, e_n)$ over A. Applying the A-endomorphism of $A[X_1, X_2, \ldots, X_n]$ such that $X_i \to X_i$ for $1 \le i \le n - 1$ and $X_n \to 0$ to the relation

$$f(e_1(X_1, X_2, \ldots, X_n), e_2(X_1, X_2, \ldots, X_n), \ldots, e_n(X_1, X_2, \ldots, X_n)) = 0$$

we get from the first result above

$$f(\bar{e}_1(X_1, X_2, \ldots, X_{n-1}), \bar{e}_2(X_1, X_2, \ldots, X_{n-1}), \ldots,$$
$$\bar{e}_{n-1}(X_1, X_2, \ldots, X_{n-1}), 0) = 0,$$

which means that $f(X_1, X_2, \ldots, X_{n-1}, 0)$ is in the ideal of algebraic relations of $(\bar{e}_1, \bar{e}_2, \ldots, \bar{e}_{n-1})$ over A.

By the induction hypothesis, this implies that

$$f(X_1, X_2, \ldots, X_{n-1}, 0) = 0,$$

and so there exists a $g \in A[X_1, X_2, \ldots, X_n]$ for which

$$f(X_1, X_2, \ldots, X_n) = X_n g(X_1, X_2, \ldots, X_n).$$

Applying to this equality the A-endomorphism of $A[X_1, X_2, \ldots, X_n]$ such that $X_i \to e_i(X_1, X_2, \ldots, X_n)$ for $1 \le i \le n$, we obtain

$$e_n(X_1, X_2, \ldots, X_n) g(e_1(X_1, X_2, \ldots, X_n), e_2(X_1, X_2, \ldots, X_n), \ldots,$$
$$e_n(X_1, X_2, \ldots, X_n)) = 0,$$

whence

$$g(e_1(X_1, X_2, \ldots, X_n), e_2(X_1, X_2, \ldots, X_n), \ldots, e_n(X_1, X_2, \ldots, X_n)) = 0.$$

This means that g is in the ideal of algebraic relations of $(e_1, e_2, \ldots, e_n)$ over A, which is impossible because $g \ne 0$ and $\mathrm{Deg}(g) < \mathrm{Deg}(f)$. $\square$

Let A be a ring, and let n be a positive integer. A polynomial f in $A[X_1, X_2, \ldots, X_n]$ is said to be **symmetric** when

$$f(X_1, X_2, \ldots, X_n) = f(X_{c(1)}, X_{c(2)}, \ldots, X_{c(n)})$$

for every permutation $c \in \mathrm{Sym}(n)$. It is evident that the symmetric polynomials in $A[X_1, X_2, \ldots, X_n]$ make up a subring of $A[X_1, X_2, \ldots, X_n]$ containing the elementary symmetric polynomials. In fact, we have the following more precise statement, known as the **fundamental theorem on symmetric polynomials**.

Let A be a ring, let n be a positive integer, and let $e_0, e_1, \ldots, e_n$ denote the elementary symmetric polynomials in $A[X_1, X_2, \ldots, X_n]$. Then the subring of $A[X_1, X_2, \ldots, X_n]$ consisting of the symmetric polynomials in $A[X_1, X_2, \ldots, X_n]$ is generated over A by $e_1, e_2, \ldots, e_n$; in other words, it is identical with $A[e_1, e_2, \ldots, e_n]$.

We shall not prove this result here, since it will not be used in this book. The interested reader can consult the book [2], where a simple proof is sketched in one of the problems on polynomials.

0.0.8.   The last section of this book is devoted to the study of derivations of algebraic function fields. This will be based on a technical result in whose proof use is made of linear mappings of finite index. Here we shall define these mappings, and derive the properties that will be needed in that proof.

Let F be a field. An F-linear mapping is said to have **finite index** when its kernel and cokernel are finite-dimensional F-spaces; if u is such a mapping, we put

$$[u:F] = [\operatorname{Ker}(u):F] - [\operatorname{Coker}(u):F],$$

and say that $[u:F]$ is the **index of u over F**.

Note that this notion is devoid of interest for linear mappings of finite-dimensional vector spaces. For if F is a field, and if L and M are finite-dimensional F-spaces, then every F-linear mapping u from L to M has finite index; and furthermore, since $L/\operatorname{Ker}(u)$ and $\operatorname{Im}(u)$ are isomorphic F-spaces, we see that

$$[u:F] = [\operatorname{Ker}(u):F] - [\operatorname{Coker}(u):F]$$

$$= [\operatorname{Ker}(u):F] - [M/\operatorname{Im}(u):F]$$

$$= ([L:F] - [L/\operatorname{Ker}(u):F]) - ([M:F] - [\operatorname{Im}(u):F])$$

$$= [L:F] - [M:F].$$

Note also that every isomorphism of vector spaces has finite index 0, because its kernel and cokernel are null.

To conclude, we shall state and prove the following basic result on linear mappings of finite index.

Let F be a field, let L, M, N be F-spaces, and let u and v be, respectively, F-linear mappings from L to M and from M to N. If u and v have finite index, then $v \circ u$ has finite index, and

$$[v \circ u:F] = [u:F] + [v:F].$$

Proof. It will be well to apply some notational simplifications in the argument. For every F-space V let us write $\dim(V)$ instead of $[V:F]$; and for every F-linear mapping w let us write, respectively, $K(w)$, $I(w)$, $C(w)$ instead of $\mathrm{Ker}(w)$, $\mathrm{Im}(w)$, $\mathrm{Coker}(w)$.

Denote by $\bar{u}$ the F-linear mapping from $K(v \circ u)$ to M defined by restriction of u; and by $\bar{v}$ the F-linear mapping from M to $C(v \circ u)$ obtained by composing v with the natural projection from N to $C(v \circ u)$. It is readily verified that

$$K(\bar{u}) = K(u) \quad \text{and} \quad I(\bar{u}) = K(v) \cap I(u);$$

therefore, the F-spaces $K(v \circ u)/K(u)$ and $K(v) \cap I(u)$ are isomorphic, and hence

$$\dim(K(v \circ u)/K(u)) = \dim(K(v) \cap I(u)).$$

Similarly, it is seen that

$$K(\bar{v}) = K(v) + I(u) \quad \text{and} \quad I(\bar{v}) = I(v)/I(v \circ u),$$

whence the F-spaces $M/(K(v) + I(u))$ and $I(v)/I(v \circ u)$ are isomorphic; consequently

$$\dim(M/(K(v) + I(u))) = \dim(I(v)/I(v \circ u)).$$

Now note that since

$$I(v \circ u) \subseteq I(v) \subseteq N \quad \text{and} \quad I(u) \subseteq K(v) + I(u) \subseteq M,$$

we have

$$\dim(C(v \circ u)) = \dim(N/I(v \circ u)) = \dim(N/I(v)) + \dim(I(v)/I(v \circ u))$$
$$= \dim(C(v)) + \dim(M/(K(v) + I(u)))$$

and

$$\dim(C(u)) = \dim(M/I(u))$$
$$= \dim(M/(K(v) + I(u))) + \dim((K(v) + I(u))/I(u)),$$

which implies that

$$\dim(C(v \circ u)) + \dim((K(v) + I(u))/I(u)) = \dim(C(u)) + \dim(C(v)).$$

Next note that

$$\dim(K(v \circ u)) = \dim(K(v \circ u)/K(u)) + \dim(K(u))$$
$$= \dim(K(v) \cap K(u)) + \dim(K(u))$$

and

$$\dim(K(v)) = \dim(K(v)/(K(v) \cap I(u))) + \dim(K(v) \cap I(u)),$$

whence

$$\dim(K(v \circ u)) + \dim(K(v)/(K(v) \cap I(u))) = \dim(K(u)) + \dim(K(v)).$$

Finally, since the two dimensions appearing in the right-hand side of each of the last equalities of the preceding two paragraphs are finite, the same is true of those appearing in the left-hand side. We conclude that $v \circ u$ has finite index; and taking into account that

$$\dim((K(v) + I(u))/I(u)) = \dim(I(u)/(K(v) \cap I(u))),$$

we get

$$\dim(K(v \circ u)) - \dim(C(v \circ u))$$
$$= (\dim(K(u)) - \dim(C(u)) + (\dim(K(v)) - \dim(C(v)),$$

as was to be shown.   □

Notation

NOTATION PERTAINING TO THE PREREQUISITES

In this part we list the symbols that are used for concepts assumed to be known. If a symbol is unfamiliar, its meaning will be clear from the brief accompanying description.

Integers

$m \mid n$	m divides n
$m \nmid n$	m does not divide n
$m \equiv n \,(\mathrm{mod}\, d)$	m is congruent to n modulo d
$n!$	n factorial
$\binom{n}{k}$	binomial coefficient defined by n and k
φ	the Euler function

Sets

$a \in A$	a belongs to A
$A \subseteq B$ or $B \supseteq A$	A is a subset of B

$A \subset B$ or $B \supset A$	A is a proper subset of B
$\{a, b, c, \ldots\}$	set consisting of $a, b, c, \ldots$
(a, b)	pair with a and b as its first and second components
$A \cup B$	union of A and B
$A \cap B$	intersection of A and B
$A - B$	complement of B in A
$A \times B$	cartesian product of A and B
$u(a)$	value of u at a
$u(A)$	direct image of A by u
$\mathrm{Im}(u)$	image of u
$u^{-1}(A)$	inverse image of A by u
$v \circ u$	composition of u and v
u_S	mapping from S to S defined by restriction of u
i_A	identity mapping on A
$i_{A \to B}$	inclusion mapping from A to B
$(a_i)_{i \in I}$	family indexed by I and with a_i as its ith coordinate for every i in I
$\bigcup_{i \in I} A_i$	union of $(A_i)_{i \in I}$

$\displaystyle\bigcap_{i \in I} A_i$	intersection of $(A_i)_{i \in I}$
$\displaystyle\mathop{\times}_{i \in I} A_i$	Cartesian product of $(A_i)_{i \in I}$
$(a_i)_{m \le i \le n}$ or $(a_m, a_{m+1}, \ldots, a_n)$	sequence indexed by the interval $[m, n]$ of the integers
$(a_i)_{i \ge m}$ or $(a_m, a_{m+1}, a_{m+2}, \ldots)$	sequence indexed by the interval $[m, \rightarrow)$ of the integers
$(a_d)_{d \mid n}$	family indexed by the positive divisors of n
$A^{(n)}$	set of n-tuples of elements of A
$\mathscr{P}(A)$	power set of A
$\mathrm{Card}(A)$	cardinality of A
$[n; k]$	set consisting of the subsets of $\{1, 2, \ldots, n\}$ with cardinality k
$a \le b$ or $b \ge a$	a precedes b in an ordering
$a < b$ or $b > a$	a strictly precedes b in an ordering
$\varnothing$	the empty set
$\mathbf{N}$	the set of natural numbers
$\aleph_0$	the cardinality of $\mathbf{N}$

Groups

$\displaystyle\prod_{i \in I} \alpha_i$	product of $(\alpha_i)_{i \in I}$

$G \times H$	external direct product of G and H
$\langle A \rangle$	subgroup generated by A
$\langle \alpha, \beta, \gamma, \ldots \rangle$	subgroup generated by $\{\alpha, \beta, \gamma, \ldots\}$
$H \vee K$	subgroup generated by $H \cup K$
$\bigvee_{i \in I} H_i$	subgroup generated by $\cup_{i \in I} H_i$
αH	left coset of α relative to H
$H\alpha$	right coset of α relative to H
$[G : H]$	index of H in G
G/H	quotient group of G by H
$\mathrm{Ker}(u)$	kernel of u
$\mathrm{Sym}(A)$	symmetric group on A
$O_G(\alpha)$	orbit of α under G
$\mathrm{Sym}(n)$	symmetric group of degree n
$\mathrm{Alt}(n)$	alternating group of degree n
G^+	group consisting of the even permutations in G
$\mathrm{sgn}(c)$	sign of c
$(i_1 i_2)$	transposition defined by i_1 and i_2
$(i_1 i_2 \cdots i_r)$	r-cycle defined by $(i_1, i_2, \ldots, i_r)$

V	Klein's Vierergruppe (four-group)

Rings

$\displaystyle\sum_{i \in I} \alpha_i$	sum of $(\alpha_i)_{i \in I}$
$\displaystyle\sum_{d \mid n} \alpha_d$	sum of $(\alpha_d)_{d \mid n}$
$\displaystyle\prod_{i \in I} \alpha_i$	product of $(\alpha_i)_{i \in I}$
$\displaystyle\prod_{d \mid n} \alpha_d$	product of $(\alpha_d)_{d \mid n}$
A^+	additive group of A
A^*	multiplicative group of invertible elements of A
$A[D]$	subring generated by D over A
$A[\alpha, \beta, \gamma, \ldots]$	subring generated by $\{\alpha, \beta, \gamma, \ldots\}$ over A
AB	subring generated by $A \cup B$
SA	ideal of A generated by S
αA	principal ideal of A generated by α
$\mathcal{J} + \mathcal{K}$	ideal generated by $\mathcal{J} \cup \mathcal{K}$
$\displaystyle\sum_{i \in I} \mathcal{J}_i$	ideal generated by $\displaystyle\bigcup_{i \in I} \mathcal{J}_i$
$\alpha + \mathcal{J}$	coset of α relative to $\mathcal{J}$
$A / \mathcal{J}$	quotient ring of A by $\mathcal{J}$

$\mathrm{Ker}(u)$	kernel of u
$A[X]$	ring of polynomials in the variable X with coefficients in A
$\deg(f)$	degree of f
f'	derivative of f
$f(\rho)$	value of f at ρ
$A[X_i]_{i \in I}$	ring of polynomials with family $(X_i)_{i \in I}$ of variables and coefficients in A
$\mathrm{Deg}(f)$	total degree of f
$\deg_{X_i}(f)$	degree of f with respect to X_i
$\dfrac{\partial f}{\partial X_i}$	partial derivative of f with respect to X_i
$f(\rho_i)_{i \in I}$	value of f at $(\rho_i)_{i \in I}$
uf	polynomial obtained by applying u to the coefficients of f
$[\alpha_{ij}]_{1 \le i \le m, 1 \le j \le n}$	$m \times n$ matrix with entry α_{ij} in the ith row and jth column
$[\alpha_{ij}]_{1 \le i, j \le n}$	$n \times n$ matrix with entry α_{ij} in the ith row and jth column
$\mathrm{Mat}_{m,n}(A)$	set of $m \times n$ matrices with entries in A
$\mathrm{Mat}_n(A)$	set of $n \times n$ matrices with entries in A

$GL_n(A)$	multiplicative group of nonsingular matrices in $\text{Mat}_n(A)$
$\det([\alpha_{ij}]_{1 \le i, j \le n})$	determinant of $[\alpha_{ij}]_{1 \le i, j \le n}$
$\mathbf{Z}$	the domain of integers
$\mathbf{Z}/d$	the ring of integers modulo d

Fields

$P \vee Q$	subfield generated by $P \cup Q$
$\displaystyle\bigvee_{i \in I} P_i$	subfield generated by $\cup_{i \in I} P_i$
$\mathbf{Q}$	the field of rational numbers
$\mathbf{R}$	the field of real numbers
$\mathbf{C}$	the field of complex numbers
$\sqrt{-1}$	the imaginary unit

Modules and Vector Spaces

$\displaystyle\sum_{i \in I} x_i$	sum of $(x_i)_{i \in I}$
AD	A-submodule generated by D
Ax	A-submodule generated by $\{x\}$
$M + N$	submodule generated by $M \cup N$
$\displaystyle\sum_{i \in I} M_i$	submodule generated by $\cup_{i \in I} M_i$
$x + M$	coset of x relative to M

E/M	quotient module of E by M	
$\hat{E}$	dual of E	
$A^{(I)}$	free A-module based on I	
$A^{(n)}$	A-module of vectors with n coordinates in A	
$\mathrm{Ker}(u)$	kernel of u	
$\mathrm{Coker}(u)$	cokernel of u	
$[E:A]$	dimension of E over A	
$[u:A]$	index of u over A	

NOTATION PERTAINING TO THE THEORY
OF FIELD EXTENSIONS AND GALOIS THEORY

The symbols listed here are used for notions whose definitions are given in this book.

$K(X)$	field of rational functions in the variable X and with coefficients in K	5
$K(X_i)_{i \in I}$	field of rational functions with family $(X_i)_{i \in I}$ of variables and with coefficients in K	5
$\mathrm{Char}(A)$	characteristic of A	5
K^p	subfield consisting of the pth powers of the elements of K	10
$K(D)$	subfield generated by D over K	13
$K(\alpha, \beta, \gamma, \ldots)$	subfield generated by $\{\alpha, \beta, \gamma, \ldots\}$ over K	13

Field Extensions
and Galois Theory

Chapter 1

Preliminaries on Fields and Polynomials

1.1. FIELDS OF FRACTIONS

A basic relationship between the field $\mathbf{Q}$ and its subdomain $\mathbf{Z}$ with which the reader is already familiar is that every element of $\mathbf{Q}$ can be expressed as a fraction with numerator and denominator in $\mathbf{Z}$. It is clear that such a connection can be meaningfully formulated in the more general context of an arbitrary field and a subdomain. This leads to the general concept of a field of fractions, which we shall discuss in this section.

Let A be a domain. By a **field of fractions of** A we understand a field K having A as a subdomain and such that every element of K is expressible in the form α/β with $\alpha, \beta \in A$ and $\beta \neq 0$.

It follows that if A is a domain, and if K is a field of fractions of A, then no proper subfield of K contains A, and K is a field of fraction of every intermediate domain between A and K.

In particular, a field is its only field of fractions.

The preceding definition immediately suggests the questions of existence and essential uniqueness of fields of fractions of domains. In the discussion that follows we shall see that these can be settled completely.

If a domain is given as a subdomain of a field, there is no difficulty in showing that it admits a field of fractions. In fact, we have the following result.

1.1.1. Proposition. *Let K be a field, and let A be a subdomain of K. Then the subfield of K generated by A is the only subfield of K that is a field of fractions of A; it consists of all elements of K of the form α/β with $\alpha, \beta \in A$ and $\beta \neq 0$.*

Proof. Let F denote the set of all elements of K of the form α/β with $\alpha, \beta \in A$ and $\beta \neq 0$. It is clear that $A \subseteq F$. Moreover, in view of the equalities

$$(\alpha/\beta) \pm (\gamma/\delta) = (\alpha\delta \pm \beta\gamma)/\beta\delta \quad \text{and} \quad (\alpha/\beta)(\gamma/\delta) = \alpha\gamma/\beta\delta,$$

which hold when $\alpha, \beta, \gamma, \delta \in A$ and $\beta \neq 0 \neq \delta$, we see that F is a subdomain of K. Finally, since $(\alpha/\beta)^{-1} = \beta/\alpha \in F$ whenever $\alpha, \beta \in A$ and $\alpha \neq 0 \neq \beta$, we conclude that F is a subfield of K.

It follows from its definition that F is the only subfield of K that is a field of fractions of A; and since it is contained in every subfield of K containing A, it is the subfield of K generated by A.   □

The preceding proposition shows that if K is a field, and if A is a subdomain of K, we are justified in speaking of the subfield of K generated by A as **the field of fractions of A in K**.

We shall now establish the existence of a field of fractions of an arbitrary domain, without assuming *a priori* that it is contained in a field.

1.1.2. Theorem. *Every domain admits a field of fractions.*

Proof. This is an important theorem, but its proof may appear to be artificial. We shall first make some observations that may help to motivate the argument.

If K is a field, and if $\alpha, \beta, \gamma, \delta \in K$ and $\beta \neq 0 \neq \delta$, then to say that $\alpha/\beta = \gamma/\delta$ means that $\alpha\beta^{-1} = \gamma\delta^{-1}$, which in turn means that $\alpha\delta = \beta\gamma$. If the four elements $\alpha, \beta, \gamma, \delta$ belong to a subdomain A of K, we see that the equality $\alpha/\beta = \gamma/\delta$ in K is equivalent to the equality $\alpha\delta = \beta\gamma$ in A.

If we are given a domain A, then an element of a field of fractions of A is determined by a pair in $A \times (A - \{0\})$. We are led, therefore, to think of every such pair (α, β) as determining the element α/β of that field, and to take into consideration that two such pairs (α, β) and (γ, δ) determine the same element of the latter if and only if $\alpha\delta = \beta\gamma$. Note that this does *not* require A to be given as a subdomain of some field.

We can now proceed with the formal argument. It will be clear to the reader that we shall be merely imitating the familiar method of constructing $\mathbf{Q}$ from $\mathbf{Z}$. For this reason, we shall omit the tedious details required in the verification of some assertions.

Consider a domain A, and let $E = A \times (A - \{0\})$. If $(\alpha, \beta), (\gamma, \delta) \in E$, let us write $(\alpha, \beta) \sim (\gamma, \delta)$ if and only if $\alpha\delta = \beta\gamma$. It is easily verified that this defines an equivalence relation on E. Let F denote the resulting quotient set; and for each $(\alpha, \beta) \in E$, let $\langle \alpha, \beta \rangle$ denote the equivalence class in F determined by (α, β).

Now suppose that $(\alpha, \beta), (\gamma, \delta), (\bar{\alpha}, \bar{\beta}), (\bar{\gamma}, \bar{\delta}) \in E$, and that $(\alpha, \beta) \sim (\gamma, \delta)$ and $(\bar{\alpha}, \bar{\beta}) \sim (\bar{\gamma}, \bar{\delta})$. Then $\alpha\delta = \beta\gamma$ and $\bar{\alpha}\bar{\delta} = \bar{\beta}\bar{\gamma}$, so that

$$(\alpha\bar{\beta} + \bar{\alpha}\beta)(\delta\bar{\delta}) = (\alpha\delta)(\bar{\beta}\bar{\delta}) + (\bar{\alpha}\bar{\delta})(\beta\delta)$$
$$= (\beta\gamma)(\bar{\beta}\bar{\delta}) + (\bar{\beta}\bar{\gamma})(\beta\delta) = (\gamma\bar{\delta} + \bar{\gamma}\delta)(\beta\bar{\beta})$$

and

$$(\alpha\bar{\alpha})(\delta\bar{\delta}) = (\alpha\delta)(\bar{\alpha}\bar{\delta}) = (\beta\gamma)(\bar{\beta}\bar{\gamma}) = (\gamma\bar{\gamma})(\beta\bar{\beta}),$$

which implies that

$$(\alpha\bar{\beta} + \bar{\alpha}\beta, \beta\bar{\beta}) \sim (\gamma\bar{\delta} + \bar{\gamma}\delta, \delta\bar{\delta})$$

and

$$(\alpha\bar{\alpha}, \beta\bar{\beta}) \sim (\gamma\bar{\gamma}, \delta\bar{\delta}).$$

Consequently, there exist two operations in F such that

$$(\langle \alpha, \beta \rangle, \langle \bar{\alpha}, \bar{\beta} \rangle) \rightarrow \langle \alpha\bar{\beta} + \bar{\alpha}\beta, \beta\bar{\beta} \rangle$$

and

$$(\langle \alpha, \beta \rangle, \langle \bar{\alpha}, \bar{\beta} \rangle) \rightarrow \langle \alpha\bar{\alpha}, \beta\bar{\beta} \rangle$$

for all $(\alpha, \beta), (\bar{\alpha}, \bar{\beta}) \in E$. Let us use, respectively, the additive and multiplicative notations for these operations; we then have

$$\langle \alpha, \beta \rangle + \langle \bar{\alpha}, \bar{\beta} \rangle = \langle \alpha\bar{\beta} + \bar{\alpha}\beta, \beta\bar{\beta} \rangle$$

and

$$\langle \alpha, \beta \rangle \langle \bar{\alpha}, \bar{\beta} \rangle = \langle \alpha\bar{\alpha}, \beta\bar{\beta} \rangle$$

whenever $\alpha, \beta, \bar{\alpha}, \bar{\beta} \in A$ and $\beta \neq 0 \neq \bar{\beta}$.

A tedious, but completely elementary, computation now shows that this addition and this multiplication define a field structure on F with respect to which the zero and unit elements are, respectively, $\langle 0, 1 \rangle$ and $\langle 1, 1 \rangle$. For all $\alpha, \beta \in A$ with $\beta \neq 0$, the additive inverse of $\langle \alpha, \beta \rangle$ is $\langle -\alpha, \beta \rangle$; and for all $\alpha, \beta \in A$ with $\alpha \neq 0 \neq \beta$, the multiplicative inverse of $\langle \alpha, \beta \rangle$ is $\langle \beta, \alpha \rangle$.

It is readily seen that if F is provided with this field structure, the mapping $\alpha \rightarrow \langle \alpha, 1 \rangle$ from A to F is a monomorphism. It then follows from 0.0.1 that there exist a field K and an isomorphism u from K to F such that A is a subdomain of K and such that u extends this monomorphism from A to F.

We now claim that K is a field of fractions of A. Indeed, if $\theta \in K$, we can write $u(\theta) = \langle \alpha, \beta \rangle$ with $\alpha, \beta \in A$ and $\beta \neq 0$; then

$$u(\theta) = \langle \alpha, \beta \rangle = \langle \alpha, 1 \rangle \langle 1, \beta \rangle = \langle \alpha, 1 \rangle / \langle \beta, 1 \rangle$$
$$= u(\alpha)/u(\beta) = u(\alpha/\beta),$$

whence $\theta = \alpha/\beta$. $\qquad\square$

The essential uniqueness of fields of fractions will be a consequence of the following fundamental result on the extendibility of monomorphisms.

1.1.3.   Theorem.   *Let A be a domain, and let K be a field of fractions of A. If F is a field, and if u is a monomorphism from A to F, then there exists a mapping from K to F such that $\alpha/\beta \to u(\alpha)/u(\beta)$ whenever $\alpha, \beta \in A$ and $\beta \neq 0$; and this mapping is the only monomorphism from K to F extending u.*

Proof.   First note that if v is a monomorphism from K to F extending u, then

$$v(\alpha/\beta) = v(\alpha)/v(\beta) = u(\alpha)/u(\beta)$$

when $\alpha, \beta \in A$ and $\beta \neq 0$. This shows that there exists at most one monomorphism from K to F extending u, and explains why we are led to consider the possibility of defining a mapping from K to F such that $\alpha/\beta \to u(\alpha)/u(\beta)$ whenever $\alpha, \beta \in A$ and $\beta \neq 0$.

If $\alpha, \beta \in A$ and $\beta \neq 0$, then $u(\beta) \neq 0$, and hence it is meaningful to form the fraction $u(\alpha)/u(\beta)$ in F. Furthermore, if $\alpha, \beta, \gamma, \delta \in A$ and $\beta \neq 0 \neq \delta$, and if $\alpha/\beta = \gamma/\delta$, then $\alpha\delta = \beta\gamma$; therefore

$$u(\alpha)u(\delta) = u(\alpha\delta) = u(\beta\gamma) = u(\beta)u(\gamma),$$

which implies that $u(\alpha)/u(\beta) = u(\gamma)/u(\delta)$.

It follows that there exists a mapping from K to F such that $\alpha/\beta \to u(\alpha)/u(\beta)$ when $\alpha, \beta \in A$ and $\beta \neq 0$. This mapping obviously extends u; and an easy computation shows that it is a homomorphism from K to F. Consequently, it is a monomorphism from K to F extending u.   □

The following two important corollaries are immediate.

1.1.4.   Corollary.   *Let A and B be domains, and let K and L be, respectively, fields of fractions of A and B. Then every isomorphism from A to B is uniquely extendible to an isomorphism from K to L.*

1.1.5.   Corollary.   *Let A be a domain, and let K and L be fields of fractions of A. Then there exists a unique A-isomorphism from K to L.*

The property of unique extendibility of monomorphisms stated in the preceding theorem actually characterizes fields of fractions (problem 1).

The second corollary shows that two fields of fractions K and L of a domain A are related in the strongest possible way. Note that the A-isomorphism from K to L "looks like" an identity mapping; for if $\alpha, \beta \in A$ and $\beta \neq 0$, it sends the fraction α/β in K to the fraction α/β in L.

Sometimes we speak of "the" field of fractions of a domain A. What is meant, of course, is that every two fields of fractions of A are being "identified" with each other by means of the A-isomorphism between them.

Note that if such "identifications" are made, then whenever A and B are domains such that A is a subdomain of B, "the" field of fractions of A is a subfield of "the" field of fractions of B.

Let K be a field, and let I be a set. Then the polynomial ring $K[X_i]_{i \in I}$ is a domain. "Its" field of fractions is denoted by $K(X_i)_{i \in I}$, and its elements are called **rational functions**. When I is nonempty and finite, we apply the same notational changes as in the case of polynomials: If $n = \text{Card}(I)$ and $I = \{i_1, i_2, \ldots, i_n\}$, we write $K(X_{i_1}, X_{i_2}, \ldots, X_{i_n})$ instead of $K(X_i)_{i \in I}$. In particular, we write $K(X)$ for "the" field of fractions of $K[X]$.

PROBLEMS

1. Let K be a field, and let A be a subdomain of K. Suppose that for every field F, every monomorphism from A to F is extendible to a monomorphism from K to F. Prove that K is a field of fractions of A.
2. Let A be a domain, and let K be a field of fractions of A. Prove that if $(\theta_i)_{i \in I}$ is a finite family of elements of K, then there exist a family $(\alpha_i)_{i \in I}$ of elements of A and a $\beta \in A - \{0\}$ such that $\theta_i = \alpha_i / \beta$ for every $i \in I$.

    Give an example that shows that this assertion does not remain valid if the finiteness assumption is omitted.
3. Let A be a domain, let K be a field of fractions of A, and let I be a set. Show that $K(X_i)_{i \in I}$ is a field of fractions of $A[X_i]_{i \in I}$.
4. Let A be a domain, let $\mathscr{P}$ be a prime ideal of A, and let K be a field of fractions of A. Let R denote the subset of K consisting of the fractions α / β with $\alpha \in A$ and $\beta \in A - \mathscr{P}$; and let $\mathscr{M}$ denote the subset of K consisting of the fractions α / β with $\alpha \in \mathscr{P}$ and $\beta \in A - \mathscr{P}$. Verify the following assertions:
    a. R is an intermediate domain between A and K.
    b. $\mathscr{M}$ is the only maximal ideal of R.
    c. $\mathscr{P} = A \cap \mathscr{M}$.
    d. There exists a monomorphism from $A/\mathscr{P}$ to $R/\mathscr{M}$ such that $\alpha + \mathscr{P} \rightarrow \alpha + \mathscr{M}$ for every $\alpha \in A$.
    e. If F is a field of fractions of $A/\mathscr{P}$, then the monomorphism from F to $R/\mathscr{M}$ extending the monomorphism described in assertion d is an isomorphism.
5. Let A be a factorial domain, and let K be a field of fractions of A. Prove that every element of K that is a zero of a monic polynomial in $A[X]$ belongs to A.

1.2. THE CHARACTERISTIC

Let A be a domain. We know that there exists a unique homomorphism $\mathbf{Z}$ to A. The kernel of this homomorphism is a prime ideal of $\mathbf{Z}$; its nonnegative generator is called the **characteristic of** A and is denoted by $\text{Char}(A)$. There are two possibilities for Char(A): It is either 0 or a prime, according as the homomorphism from $\mathbf{Z}$ to A is or is not injective.

Let us recall that, given a domain A and an integer n, it is customary to denote by the same symbol n the element of A assigned to the integer n by the homomorphism from $\mathbf{Z}$ to A.

When A has prime characteristic p, we have to keep in mind that distinct integers may represent a single element of A. In fact, if m and n are integers, the equality $m = n$ holds in A if and only if $m \equiv n \pmod{p}$, since each of these conditions expresses the fact that $m - n$ belongs to the kernel of the homomorphism from $\mathbf{Z}$ to A. In particular, if n is an integer, then the equality $n = 0$ holds in A if and only if $p \mid n$.

This mild notational ambiguity, which should not cause confusion, does not occur when A has characteristic 0. In this case, we can simply regard $\mathbf{Z}$ as a subdomain of A; in other words, $\mathbf{Z}$ can be "identified" with its image in A.

The following examples show that there exist fields of every possible characteristic.

1.2.1. Examples

a. Every domain admitting $\mathbf{Z}$ as a subdomain has characteristic 0.

This is obvious: If A is such a domain, then the homomorphism from $\mathbf{Z}$ to A is the inclusion mapping $i_{\mathbf{Z} \to A}$, which is injective.

b. If p is a prime, then the field $\mathbf{Z}/p$ has characteristic p.

Indeed, the homomorphism from $\mathbf{Z}$ to $\mathbf{Z}/p$ is the natural projection, which has $p\mathbf{Z}$ as its kernel.   □

We shall now derive the basic properties of the characteristic.

1.2.2. Proposition.   Let A be a domain. Then the following conditions are equivalent:
    (a) Char(A) = 0.
    (b) $\mathbf{Z}$ *is embeddable in A.*
    (c) *The image of $\mathbf{Z}$ in A is not a field.*

Proof.   Let u denote the homomorphism from $\mathbf{Z}$ to A. To prove the equivalence of (a) and (b), we need only note that $\mathbf{Z}$ is embeddable in A if and only if u is injective, hence if and only if Char(A) = 0.

Since $\mathbf{Z}/\mathrm{Ker}(u)$ and $\mathrm{Im}(u)$ are isomorphic, to say that $\mathrm{Im}(u)$ is a field means that $\mathbf{Z}/\mathrm{Ker}(u)$ is a field, which in turn means that $\mathrm{Ker}(u)$ is nonnull. This shows that (a) and (c) are equivalent.   □

1.2.3. Corollary.   If K is a field, then Char(K) = 0 *if and only if* $\mathbf{Q}$ *is embeddable in K.*

Proof. Since Q is a field of fractions of Z, we know from 1.1.3 that the embeddability of Z in K is equivalent to that of Q in K. $\square$

1.2.4. Corollary. *Every finite field has prime characteristic.*

Proof. By the preceding corollary, a field of characteristic 0 contains a subfield equipotent to Q, and hence it is not finite. $\square$

1.2.5. Proposition. *Let A be a domain, and let p be a prime. Then the following conditions are equivalent:*
- (a) $\mathrm{Char}(A) = p$.
- (b) *The equality $p = 0$ holds in A.*
- (c) *Z/p is embeddable in A.*
- (d) *The image of Z in A and Z/p are isomorphic.*

Proof. As in the proof of proposition 1.2.2, let u denote the homomorphism from Z to A.

We already know that (a) implies (b). To verify the opposite implication, note that if the equality $p = 0$ holds in A, then $\mathrm{Char}(A) \neq 0$ and $\mathrm{Char}(A)|p$, whence $\mathrm{Char}(A) = p$. Thus, (a) and (b) are equivalent.

To prove that (a) implies (d), suppose that $\mathrm{Char}(A) = p$. Then $\mathrm{Ker}(u) = pZ$, and hence $Z/\mathrm{Ker}(u) = Z/p$. Since $Z/\mathrm{Ker}(u)$ and $\mathrm{Im}(u)$ are isomorphic, it follows that $\mathrm{Im}(u)$ and Z/p are isomorphic.

It is obvious that (d) implies (c). For if Z/p is isomorphic to the subring $\mathrm{Im}(u)$ of A, then it is embeddable in A.

To conclude, we now show that (c) implies (a). Suppose that Z/p is embeddable in A, and choose a monomorphism v from Z/p to A. If w denotes the natural projection from Z to Z/p, it is clear that $u = v \circ w$. Furthermore, since v is injective, we have

$$\mathrm{Ker}(u) = \mathrm{Ker}(v \circ w) = \mathrm{Ker}(w) = pZ,$$

whence $\mathrm{Char}(A) = p$. $\square$

1.2.6. Proposition. *If A and B are domains such that A is embeddable in B, then $\mathrm{Char}(A) = \mathrm{Char}(B)$.*

Proof. Since every ring that is embeddable in A is embeddable in B, this follows at once from 1.2.2 and 1.2.5. $\square$

In regard to 1.2.4, it should be noted that there exist infinite fields of every possible characteristic. For if K is a field and I is a nonempty set, then the field $K(X_i)_{i \in I}$ of rational functions is infinite; and by 1.2.6, it has the same characteristic as K.

Given a domain A and a positive integer n, the mapping $\alpha \to \alpha^n$ from A to A is usually of very limited interest, because it does not have good *additive* properties.

It is known that some beginners in algebra, with complete disregard for the classical binomial theorem, are quite prepared to accept the validity of the equality

$$(\alpha \pm \beta)^n = \alpha^n \pm \beta^n.$$

As a consequence, they reach a number of interesting conclusions.

The "freshman's dream", as the illusion just described is sometimes called, actually contains an element of truth: It will be shown presently that the equality

$$(\alpha \pm \beta)^{p^n} = \alpha^{p^n} \pm \beta^{p^n} \qquad .$$

holds in domains of prime characteristic p. This will exemplify one of the fundamental differences between fields of characteristic 0 and fields of prime characteristic.

Let A be a domain of prime characteristic p. The mapping $\alpha \to \alpha^p$ from A to A is called the **Frobenius mapping of** A.

The following theorem explains why the Frobenius mapping is of interest in the theory of fields.

1.2.7.   Theorem.   *If A is a domain of prime characteristic, then the Frobenius mapping of A is an injective endomorphism.*

Proof.   Let us put $p = \mathrm{Char}(A)$. Since the conditions $\alpha \in A$ and $\alpha^p = 0$ imply that $\alpha = 0$, we need only prove that the Frobenius mapping of A is an endomorphism; and to do this, it suffices to verify that

$$(\alpha + \beta)^p = \alpha^p + \beta^p$$

for all $\alpha, \beta \in A$.

Let $1 \le k \le p - 1$; since

$$p \nmid k!, \quad p \nmid (p-k)!, \quad \text{and} \quad p \mid p!,$$

and since

$$p! = k!(p-k)!\binom{p}{k},$$

we conclude next that $p \mid \binom{p}{k}$, and hence the equality $\binom{p}{k} = 0$ holds in A.

Now it is easy to complete the proof. For if $\alpha, \beta \in A$, then

$$(\alpha + \beta)^p = \sum_{k=0}^{p} \binom{p}{k} \alpha^{p-k} \beta^k$$

$$= \alpha^p + \sum_{k=1}^{p-1} \binom{p}{k} \alpha^{p-k} \beta^k + \beta^p = \alpha^p + \beta^p,$$

which is what was needed.   □

1.2.8. Corollary. *If A is a domain of prime characteristic p, and if n is a nonnegative integer, then the mapping $\alpha \to \alpha^{p^n}$ from A to A is an injective endomorphism.*

Proof. This is obvious: If $n = 0$, the mapping in question is i_A; and if $n > 0$, it is the nth iterate of the Frobenius mapping of A. $\square$

1.2.9. Example. Let A be a domain of prime characteristic p, If $\alpha \in A$ and if n is a nonnegative integer, then the polynomial $X^{p^n} - \alpha$ in $A[X]$ admits at most one zero in A.

For if β and γ are zeros of $X^{p^n} - \alpha$ in A, then

$$\beta^{p^n} = \alpha = \gamma^{p^n},$$

and the preceding corollary shows that $\beta = \gamma$. $\square$

PROBLEMS

1. Let p be a prime. Use the fact that $\mathrm{Char}(\mathbf{Z}/p) = p$ in order to prove that
    $$n^p \equiv n \,(\mathrm{mod}\ p)$$
    for every integer n. (This result is known as **Fermat's little theorem**.)
2. Let p be a prime. Verify that 1 and $p - 1$ are the zeros of $X^2 - 1$ in $\mathbf{Z}/p$, and then prove that
    $$(p - 1)! \equiv -1 (\mathrm{mod}\ p).$$
    (This result is known as **Wilson's theorem**.)
3. Let A be a domain of prime characteristic p. Show that
    $$(\alpha - \beta)^{p-1} = \sum_{k=0}^{p-1} \alpha^{p-1-k}\beta^k$$
    for all $\alpha, \beta \in A$.
4. Let A and B be domains. Show that if A has prime characteristic and if there exists a homomorphism from A to B, then $\mathrm{Char}(A) = \mathrm{Char}(B)$.
5. Give an example of two fields of characteristic 0 neither of which is embeddable in the other.
6. Let A be a domain. Prove the following assertions:
    a. $\mathrm{Char}(A) = 2$ if and only if $-\alpha = \alpha$ for every $\alpha \in A$.
    b. If A^* is embeddable in A^+, then $\mathrm{Char}(A) = 2$.
    c. A^* and A^+ are not isomorphic.
7. Let K be a field such that $\mathrm{Char}(K) \neq 2$, and let u be a mapping from K to K such that $u(1) = 1, u(\alpha + \beta) = u(\alpha) + u(\beta)$ for all $\alpha, \beta \in K$, and $u(\alpha)u(1/\alpha) = 1$ for all $\alpha \in K^*$. Show that u is an endomorphism.

1.3.   PERFECT FIELDS AND PRIME FIELDS

The two special types of field appearing in the title of this section arise naturally in connection with the characteristic.

In general, the Frobenius mapping of a domain of prime characteristic is not an automorphism, because it may fail to be surjective (problem 1). On the other hand, there exist domains of prime characteristic for which the Frobenius mapping reduces to the identity mapping (problem 2).

We say that a field is **perfect** when either it has characteristic 0 or it has prime characteristic and its Frobenius mapping is an automorphism.

If K is a field of prime characteristic p, the symbol K^p will be used to denote the image of the Frobenius mapping of K. Thus, K^p is the subfield of K consisting of the elements of K that admit pth roots in K; and the mapping $\alpha \to \alpha^p$ from K to K^p is an isomorphism.

It follows that a field K of prime characteristic p is perfect if and only if $K^p = K$, hence if and only if every element of K admits a pth root in K.

The same comments made in the proof of 1.2.8 can now be used in order to prove the following proposition.

1.3.1.   Proposition.   If K is a perfect field of prime characteristic p, and if n is a nonnegative integer, then the mapping $\alpha \to \alpha^{p^n}$ from K to K is an automorphism.

As the next proposition shows, there exist perfect fields of every possible prime characteristic.

1.3.2.   Proposition.   Every finite field is perfect.

Proof. We have already seen that every finite field has prime characteristic; and since every injective mapping from a finite set to itself is bijective, the Frobenius mapping of every finite field is an automorphism.   □

It can also be shown that there exist fields that are not perfect of every possible prime characteristic (problem 1).

One of the important properties of field extensions that we shall study in detail is that of separability. It will be seen that perfect fields can be described as the fields that do not admit inseparable extensions.

We say that a field is **prime** when it is its only subfield. Equivalently, a field is prime when it possesses no proper subfields.

It is not difficult to determine the prime fields. In fact, we have the following result.

1.3.3.   Proposition.   A field of characteristic 0 is prime if and only if it is isomorphic to $\mathbf{Q}$; and a field of prime characteristic p is prime if and only if it is isomorphic to $\mathbf{Z}/p$.

Proof. If E is a subfield of $\mathbf{Q}$, then E contains $\mathbf{Z}$, and hence $E = \mathbf{Q}$. Therefore $\mathbf{Q}$ is prime.

Now let p be a prime. If E is a subfield of $\mathbf{Z}/p$, then $1 + p\mathbf{Z} \in E$, and hence $n + p\mathbf{Z} = n(1 + p\mathbf{Z}) \in E$ for every $n \in \mathbf{Z}$; consequently, $E = \mathbf{Z}/p$. It follows that $\mathbf{Z}/p$ is prime.

Finally, let P be a prime field, and apply 1.2.3 and 1.2.5: If P has characteristic 0, then $\mathbf{Q}$ is embeddable in P; and if P has prime characteristic p, then $\mathbf{Z}/p$ is embeddable in P. Since a field is embeddable in P if and only if it is isomorphic to a subfield of P, hence if and only if it is isomorphic to P, we conclude that $\mathbf{Q}$ is isomorphic to P in the first case, and that $\mathbf{Z}/p$ is isomorphic to P in the second case. □

1.3.4. Corollary. *Every prime field is perfect.*

Proof. By the proposition, a prime field either has characteristic 0 or is finite. □

The preceding proposition justifies the common practice of referring to $\mathbf{Q}$ as "the" prime field of characteristic 0, and to $\mathbf{Z}/p$ as "the" prime field of prime characteristic p.

1.3.5. Proposition. *Let P be a prime field.*
   (i) *If A is a ring, then there exists at most one homomorphism from P to A.*
   (ii) *If A and B are rings admitting P as a common subfield, then every homomorphism from A to B is a P-homomorphism.*
   (iii) *The only endomorphism of P is i_P.*

Proof. We need only verify (i). If A is a ring, then every two homomorphisms from P to A agree on $\{1\}$; therefore, they also agree on P, because P is identical with the subfield of P generated by $\{1\}$. □

It is not difficult to give an example of a field that is not prime and admits a unique endomorphism (section 1.4, problem 3). Therefore, the property stated in (iii) of proposition 1.3.5 does not characterize prime fields.

1.3.6. Proposition. *If a ring admits a subfield, then it admits a unique prime subfield. Moreover, for every such ring A, the prime subfield of A is the smallest element of the set of all subfields of A.*

Proof. Let A be a ring that admits a subfield. According to 1.2.3, 1.2.5, and 1.3.3, this subfield admits a prime subfield, and hence A admits a prime subfield.

Let P be a prime subfield of A. If E is a subfield of A, then $E \cap P$ is a subfield of P; therefore $E \cap P = P$, which means that $P \subseteq E$. This shows that P is the smallest element of the set of all subfields of A; as such, P is seen to be the only prime subfield of A. □

If a domain A has prime characteristic, then 1.2.5, 1.3.3, and 1.3.6 show that A possesses a unique prime subfield, namely, the image of $\mathbf{Z}$ in A. On the other hand, it is possible for a domain of characteristic 0 to contain no subfields; this is true, for example, of every proper subdomain of $\mathbf{Q}$.

We now show that the prime subfield of a domain of prime characteristic can be described by means of the Frobenius mapping.

1.3.7. Proposition. *If A is a domain of prime characteristic, then the prime subfield of A consists of the fixed points of the Frobenius mapping of A.*

Proof. Let $p = \mathrm{Char}(A)$, and let P and D denote, respectively, the prime subfield of A and the subdomain of A consisting of the fixes points of the Frobenius mapping of A. As noted in the foregoing remarks, P is the image of $\mathbf{Z}$ in A, and hence $P \subseteq D$.

The elements of D are the zeros in A of the polynomial $X^p - X$ in $A[X]$. Since the latter has degree p, it follows that $\mathrm{Card}(D) \le p$.

In conclusion, we have

$$\mathrm{Card}(P) = p, \quad \mathrm{Card}(D) \le p, \quad \text{and} \quad P \subseteq D;$$

and these conditions obviously imply that $P = D$. $\square$

Let m be a positive integer. In elementary number theory we say that a set D of integers is a **complete residue system modulo** m when every integer is congruent modulo m to exactly one integer in D. Every complete residue system modulo m has m elements; and to say that integers $r_1, r_2, \ldots, r_m$ form a complete residue system modulo m amounts to saying that the cosets

$$r_1 + m\mathbf{Z}, r_2 + m\mathbf{Z}, \ldots, r_m + m\mathbf{Z}$$

are the m elements of the ring $\mathbf{Z}/m$.

If A is a domain of prime characteristic p, and if $r_1, r_2, \ldots, r_p$ are integers making up a complete residue system modulo p, it is evident that $r_1, r_2, \ldots, r_p$ represent the p elements of the prime subfield of A.

1.3.8. Example. Let A be a domain of prime characteristic p, and let $\alpha \in A$. We now show that if β is a zero in A of the polynomial $X^p - X - \alpha$ in $A[X]$, then $\beta + 1, \beta + 2, \ldots, \beta + p$ are the zeros of $X^p - X - \alpha$ in A.

Since the integers $1, 2, \ldots, p$ make up a complete residue system modulo p, they represent the p elements of the prime subfield of A. According to 1.3.7, it is then clear that for $1 \le i \le p$ the equality $i^p = i$ holds in A.

We know that $\beta^p - \beta - \alpha = 0$. If $1 \le i \le p$, we also have

$$(\beta + i)^p - (\beta + i) - \alpha = (\beta^p + i^p) - (\beta + i) - \alpha$$
$$= (\beta^p - \beta - \alpha) + (i^p - i) = 0,$$

which shows that $\beta + i$ is a zero of $X^p - X - \alpha$ in A. This proves our assertion, because $X^p - X - \alpha$ has degree p and admits the p distinct zeros $\beta + 1, \beta + 2, \ldots, \beta + p$ in A. $\square$

PROBLEMS

1. Prove that if K is a field of prime characteristic p, then $X \notin K(X)^p$.
2. Prove that if A is a domain of prime characteristic, then the Frobenius mapping of A reduces to i_A if and only if A is a prime field.
3. Show that a finite field is prime if and only if its cardinality is a prime.

1.4. FIELD EXTENSIONS

The primary purpose of this section is to introduce the basic generalities on field extensions that will be used throughout the book.

Let K be a field. By an **extension field of** K we understand a field having K as a subfield; and by a **proper extension field of** K we understand a field having K as a proper subfield.

We shall occasionally speak, somewhat loosely, of a **field extension**. By this we mean a pair consisting of a field and an extension field, to which we refer, respectively, as the **bottom field** and the **top field**.

Let K be a field, and let L be an extension field of K. If $D \subseteq L$, we denote by $K(D)$ the subfield of L generated by $K \cup D$, and say that $K(D)$ is the **subfield of** L **generated by** D **over** K or the **subfield of** L **obtained by adjoining** D **to** K. If n is a positive integer and $\alpha_1, \alpha_2, \ldots, \alpha_n \in L$, we write $K(\alpha_1, \alpha_2, \ldots, \alpha_n)$ instead of $K(\{\alpha_1, \alpha_2, \ldots, \alpha_n\})$; in particular, if $\alpha \in L$, we write $K(\alpha)$ instead of $K(\{\alpha\})$.

It is clear that if K is a field, if L is an extension field of K, and if $D \subseteq L$, then $K(D)$ is the field of fractions of $K[D]$ in L.

1.4.1. Examples

a. Let us show that the subfield $Q(\sqrt{2})$ of R consists of all real numbers of the form $r + s\sqrt{2}$ with $r, s \in Q$.

Indeed, denote by F the set of all real numbers of the indicated form. It is easily verified that F is a subdomain of R and $Q \cup \{\sqrt{2}\} \subseteq F \subseteq Q(\sqrt{2})$.

Let $\alpha \in F$ and $\alpha \neq 0$, and write $\alpha = r + s\sqrt{2}$ with $r, s \in Q$. The irrationality of $\sqrt{2}$ implies that $r^2 - 2s^2 \neq 0$; and an elementary computation shows that

$$1/\alpha = \left(r/(r^2 - 2s^2) \right) + \left(-s/(r^2 - 2s^2) \right)\sqrt{2},$$

which implies that $1/\alpha \in F$.

It then follows that F is a subfield of R. In view of the inclusions above, this shows that $F = Q(\sqrt{2})$.

b. Proceeding as in example a, it is seen that if K is a subfield of R, then $K(\sqrt{-1})$ is the subfield of C consisting of all complex numbers of the form $r + s\sqrt{-1}$ with $r, s \in K$.

In particular, we have $C = R(\sqrt{-1})$.   □

The following properties of the operation of adjunction, which we state for the sake of reference, are immediate from the definitions.

1.4.2. Proposition. *Let K be a field, and let L be an extension field of K.*

(i) *If $C \subseteq D \subseteq L$, then $K(C) \subseteq K(D)$.*

(ii) *If $C, D \subseteq L$, then $K(C \cup D) = K(C)(D)$.*

(iii) *If $(D_i)_{i \in I}$ is a nonempty family of subsets of L, then*

$$K\left(\bigcup_{i \in I} D_i \right) = \bigvee_{i \in I} K(D_i).$$

1.4.3. Proposition. *Let K and $\overline{K}$ be fields, let L and $\overline{L}$ be, respectively, extension fields of K and $\overline{K}$, and let u be a monomorphism from L to $\overline{L}$ such that $u(K) = \overline{K}$. Then $u(K(D)) = \overline{K}(u(D))$ for every subset D of L.*

1.4.4. Corollary. *Let K be a field, let L and M be extension fields of K, and let u be a K-monomorphism from L to M. Then $u(K(D)) = K(u(D))$ for every subset D of L.*

Let K be a field, and let L be an extension field of K. We say that L is **finitely generated over** K, or that L is a **finitely generated extension** of K, when there exists a finite subset D of L such that $L = K(D)$.

1.4.5. Example. An easy counting argument can be used to show that R is not finitely generated over Q.

Let n be a positive integer, and let $\alpha_1, \alpha_2, \ldots, \alpha_n \in R$. Since the mapping $f(X_1, X_2, \ldots, X_n) \to f(\alpha_1, \alpha_2, \ldots, \alpha_n)$ from $Q[X_1, X_2, \ldots, X_n]$ to $Q[\alpha_1, \alpha_2, \ldots, \alpha_n]$ is surjective, it is clear that Q is equipotent to $Q[\alpha_1, \alpha_2, \ldots, \alpha_n]$, and hence also to $Q(\alpha_1, \alpha_2, \ldots, \alpha_n)$. Since Q and R are not equipotent, it follows that $R \neq Q(\alpha_1, \alpha_2, \ldots, \alpha_n)$.   □

The next two propositions contain the elementary properties of finitely generated extensions.

1.4.6. Proposition. *Let K be a field, let L be an extension field of K, and let M be an extension field of L.*

(i) *If L is finitely generated over K, and if M is finitely generated over L, then M is finitely generated over K.*

(ii) *If M is finitely generated over K, then M is finitely generated over L.*

Proof. To prove (i), suppose that $L = K(C)$ and $M = L(D)$, where C and D are, respectively, finite subsets of L and M. By 1.4.2, we then have

$$M = L(D) = K(C)(D) = K(C \cup D);$$

since $C \cup D$ is a finite subset of M, this shows that M is finitely generated over K.

To prove (ii) we need only observe that if D is a finite subset of M such that $M = K(D)$, then

$$M = K(D) \subseteq L(D) \subseteq M,$$

whence $M = L(D)$.   □

In reality, the complete conclusion in the preceding proposition is that M is finitely generated over K if and only if M is finitely generated over L and L is finitely generated over K.

The part that we have not stated is, "If M is finitely generated over K, then L is finitely generated over K". We are unaware of any proof of this assertion that does not depend on the notion of a transcendence base of a field extension, which we have not yet discussed.

1.4.7. Proposition. *Let K be a field, let N be an extension field of K, and let L and M be intermediate fields between K and N.*

(i) If L is finitely generated over K, then $L \vee M$ is finitely generated over M.

(ii) If L and M are finitely generated over K, then $L \vee M$ is finitely generated over K.

Proof. It is clear that (ii) is a consequence of (i) and 1.4.6.

To prove (i), suppose that D is a finite subset of L such that $L = K(D)$. Then

$$L \vee M = M(L) = M(K(D)) = M(D),$$

which shows that $L \vee M$ is finitely generated over M.   □

We now introduce commonly used terminology for the most elementary type of finitely generated extension.

Let K be a field, and let L be an extension field of K. By a **primitive element of L over K** we understand an $\alpha \in L$ such that $L = K(\alpha)$.

If K is a field, by a **simple extension of K** we understand an extension field of K that admits a primitive element over K.

It is not always easy to decide when a field extension is simple. We shall see later that, under suitable hypotheses, necessary and sufficient conditions for the existence of primitive elements can be given.

1.4.8. Example. It can be shown that $Q(\sqrt{2}, \sqrt{3})$ is a simple extension of Q. In fact, we shall see that $\sqrt{2} + \sqrt{3}$ is a primitive element of $Q(\sqrt{2}, \sqrt{3})$ over Q.

It is clear, first, that $\sqrt{2} + \sqrt{3} \in Q(\sqrt{2}, \sqrt{3})$, and hence $Q(\sqrt{2} + \sqrt{3}) \subseteq Q(\sqrt{2}, \sqrt{3})$. To prove the opposite inclusion, note that

$$(\sqrt{2} + \sqrt{3})(\sqrt{2} - \sqrt{3}) = -1,$$

so that

$$\sqrt{3} - \sqrt{2} = 1/(\sqrt{2} + \sqrt{3}) \in Q(\sqrt{2} + \sqrt{3});$$

therefore

$$\sqrt{3} = ((\sqrt{3} + \sqrt{2}) + (\sqrt{3} - \sqrt{2}))/2 \in \mathbf{Q}(\sqrt{2} + \sqrt{3})$$

and

$$\sqrt{2} = ((\sqrt{3} + \sqrt{2}) - (\sqrt{3} - \sqrt{2}))/2 \in \mathbf{Q}(\sqrt{2} + \sqrt{3}),$$

which implies that $\mathbf{Q}(\sqrt{2}, \sqrt{3}) \subseteq \mathbf{Q}(\sqrt{2} + \sqrt{3})$. $\square$

Finding a primitive element does not necessarily yield a better understanding of the field extension under consideration. For instance, it will be seen that for certain purposes, it is more advantageous to regard $\mathbf{Q}(\sqrt{2}, \sqrt{3})$ as the result of the successive adjunctions of $\sqrt{2}$ and $\sqrt{3}$ to $\mathbf{Q}$ than of the single adjunction of $\sqrt{2} + \sqrt{3}$ to $\mathbf{Q}$.

We saw in 1.4.5 that $\mathbf{R}$ is not a finitely generated extension of $\mathbf{Q}$, and hence is not a simple extension of $\mathbf{Q}$. In Section 2.4, we shall be able to give an example of a finitely generated extension that is not simple.

Let K be a field, and let L be an extension field of K. A base of the K-space L will also be called a **linear base of L over K**; and the dimension $[L:K]$ of this K-space will be called the **linear degree of L over K**. We say that L is **finite over K**, or that L is a **finite extension of K**, when $[L:K]$ is finite; and we say that L is **infinite over K**, or that L is an **infinite extension of K**, when $[L:K]$ is infinite.

The finite extensions of $\mathbf{Q}$ are called **algebraic number fields**. These are the main objects of study in algebraic number theory.

It is evident that every field is finite over itself; in fact, if K is a field, then K is the only extension field of K having linear degree 1 over K. For finite extensions of small linear degree, it is common to apply special terminology: The words **quadratic, cubic, quartic,** and **quintic** are used, respectively, in the cases of linear degree 2, 3, 4, and 5.

If K is a field and L is an extension field of K, and if D is a linear base of L over K, then $L = KD \subseteq K(D) \subseteq L$, and hence $L = K(D)$. It follows, in particular, that every finite extension is finitely generated. It is easy to see, on the other hand, that there exist infinite finitely generated extensions of every field: If K is a field, and if n is a positive integer, then $K(X_1, X_2, \ldots, X_n)$ is finitely generated and infinite over K.

To end this section, we shall now discuss the elementary properties of finite extensions.

1.4.9. Proposition. *If K is a field, and if L is a finite extension of K,* then

$$\mathrm{Card}(L) = \mathrm{Card}(K)^{[L:K]}.$$

Proof. This is obvious: If $n = [L:K]$, then the K-spaces L and $K^{(n)}$ are isomorphic, and hence equipotent. $\square$

It follows, in particular, that if K is a finite field, then every finite extension of K is a finite field.

1.4.10. Proposition. *Let K be a field, let L be an extension field of K, and let M be an extension field of L.*

(i) *If $(\alpha_i)_{i \in I}$ and $(\beta_j)_{j \in J}$ are, respectively, linear bases of L over K and of M over L, then $(\alpha_i \beta_j)_{(i,j) \in I \times J}$ is a linear base of M over K.*

(ii) *$[M:K] = [M:L][L:K]$.*

(iii) *M is finite over K if and only if M is finite over L and L is finite over K.*

Proof. We need only note that (i) is a particular case of 0.0.2; and that (i) and (ii) imply (ii) and (iii), respectively. □

Let us point out some evident consequences of the preceding proposition that will be used frequently in the sequel without further comment. Let K be a field, let L be a finite extension of K, and let E be an intermediate field between K and L. Then $[E:K]$ and $[L:E]$ are divisors of $[L:K]$. Also, $[E:K] = [L:K]$ implies $E = L$, and $[L:E] = [L:K]$ implies $E = K$. Finally, when $[L:K]$ is prime, either $E = K$ or $E = L$.

It is customary to express the last assertion in the proposition by saying that the transitivity property is satisfied by finite extensions. As noted previously, this is also true of finitely generated extensions, but only the partial result 1.4.6 could be proved with the tools available to us at present.

1.4.11. Proposition. *Let K be a field, let N be an extension field of K, and let L and M be intermediate fields between K and N.*

(i) *If either L or M is finite over K, then $LM = L \vee M$.*

(ii) *If L is finite over K, then $L \vee M$ is finite over M, and $[L \vee M : M] \leq [L:K]$.*

(iii) *If L and M are finite over K, then $L \vee M$ is finite over K, and $[L \vee M : K] \leq [L:K][M:K]$.*

Proof. Since $[L \vee M : K] = [L \vee M : M][M:K]$, we see that (iii) follows from (ii).

To prove (i) and (ii), we now assume that L is finite over K. Since every generating system of the K-space L is also a generating system of the M-space LM, every base of the former contains a base of the latter, whence $[LM:M] \leq [L:K]$. Since $[L:K]$ is finite, so is $[LM:M]$. Thus, the domain LM is a finite-dimensional M-space. But this, according to 0.0.3, shows that LM is a field, which means that $LM = L \vee M$. □

It is not difficult to show that the inequalities appearing in the preceding proposition may be strict (section 2.1, problem 6).

PROBLEMS

1. Show that $Q(\sqrt[3]{2})$ consists of all real numbers of the form $r + s\sqrt[3]{2} + t\sqrt[3]{4}$ with $r, s, t \in Q$.
2. Prove that $\sqrt{2} + \sqrt{-1}$ is a primitive element of $Q(\sqrt{2}, \sqrt{-1})$ over Q.
3. Find an irrational real number α such that the only endomorphism of $Q(\alpha)$ is $i_{Q(\alpha)}$.
4. Let K be a field such that $\mathrm{Char}(K) \neq 2$, and let L be a quadratic extension of K. Show that there exists an $\alpha \in L$ such that $L = K(\alpha)$ and $\alpha^2 \in K$.
5. Let K be a field, and let L be an extension field of K such that $[L:K]$ is a prime. Show that L is a simple extension of K, and determine the primitive elements of L over K.
6. Let K be a field, and let L be a finite extension of K. Prove that K is perfect if and only if L is perfect.
7. Let K be a field, let N be an extension field of K, and let L and M be intermediate fields between K and N. Verify the following assertions:
   a. If L is finite over K and $[L \vee M : M] = [L:K]$, then $[L \vee M : L] = [M:K]$.
   b. If M is finite over K and $[L \vee M : K] = [L:K][M:K]$, then $[L \vee M : M] = [L:K]$.
   c. If L and M are finite over K and $[L \vee M : K] = [L:K][M:K]$, then $L \cap M = K$.
   d. If L and M are finite over K, and if $[L:K]$ and $[M:K]$ are relatively prime, then $[L \vee M : K] = [L:K][M:K]$.

1.5. FACTORIZATION OF POLYNOMIALS

One of the prerequisites for the study of algebraic extensions is the factorization theory for polynomials. This theory rests upon the remarkable fact that the polynomials in one variable over a field form a principal ideal domain. As we now show, this follows from the division properties of polynomials.

    1.5.1. Theorem. *If K is a field, then $K[X]$ is a principal ideal domain.*

    Proof. We have to verify that every nonnull ideal $\mathscr{J}$ of $K[X]$ is principal. To do this, let f be a nonzero polynomial belonging to $\mathscr{J}$ and of the smallest possible degree. We claim that $\mathscr{J} = fK[X]$.

    Since $f \in \mathscr{J}$, it is clear that $fK[X] \subseteq \mathscr{J}$. To prove the opposite inclusion, now let $g \in \mathscr{J}$. By the division algorithm in $K[X]$, we can write $g = fu + v$, where $u, v \in K[X]$ and $\deg(v) < \deg(f)$. Since $v = g - fu$ and

$f, g \in \mathcal{J}$, we have $v \in \mathcal{J}$. If $v \neq 0$, then we would have

$$v \neq 0, \quad v \in \mathcal{J}, \quad \text{and} \quad \deg(v) < \deg(f),$$

in contradiction to the choice of f. Therefore $v = 0$, so that $g = fu \in fK[X]$.

□

Since every nonzero polynomial possesses a unique monic associate when the domain of coefficients is a field, the following two corollaries are immediate.

1.5.2. Corollary. *If K is a field, then every nonnull ideal of $K[X]$ admits a unique monic polynomial as a generator.*

1.5.3. Corollary. *If K is a field, then $K[X]$ is a factorial domain in which the monic irreducible polynomials make up a system of representatives of irreducible elements.*

Let K be a field, and let f be a nonconstant polynomial in $K[X]$. It follows from 1.5.3 that the number of monic irreducible factors of f in $K[X]$ is finite. Denoting by n this number, and by $f_1, f_2, \ldots, f_n$ the n distinct monic irreducible factors of f in $K[X]$, we have

$$f = \alpha \prod_{i=1}^{n} f_i^{e_i},$$

where α is the leading coefficient of f and $e_1, e_2, \ldots, e_n$ are positive integers. Moreover, this representation for f is unique up to a permutation of $\{1, 2, \ldots, n\}$. Finally, note that the monic factors of f in $K[X]$ are the polynomials of the form $\prod_{i=1}^{n} f_i^{d_i}$, where $0 \leq d_i \leq e_i$ for $1 \leq i \leq n$.

When considering a field extension, it will sometimes be necessary to know about the connections between the generators of an ideal of polynomials over the bottom field and of the corresponding ideal of polynomials over the top field. We shall now derive two useful results dealing with this equation.

1.5.4. Corollary. *Let K be a field, let L be an extension field of K, and let $\mathcal{J}$ be a nonnull ideal of $K[X]$. Then the monic polynomial in $K[X]$ generating $\mathcal{J}$ and the monic polynomial in $L[X]$ generating $\mathcal{J}L[X]$ coincide. In particular, $\mathcal{J} = K[X]$ if and only if $\mathcal{J}L[X] = L[X]$.*

Proof. This is an obvious consequence of 1.5.2. Indeed, if f denotes the monic polynomial in $K[X]$ generating $\mathcal{J}$, then f is a monic polynomial in $L[X]$ and

$$\mathcal{J}L[X] = fK[X]L[X] = fL[X],$$

which implies our assertion.   □

1.5.5. Corollary. *Let K be a field, let L be an extension field of K, and let $f, g \in K[X]$. Then a monic polynomial in $K[X]$ is a highest common* *factor of f and g in $K[X]$ if and only if it is a highest common factor of f and g in $L[X]$. In particular, f and g are relatively prime in $K[X]$ if and only if they are relatively prime in $L[X]$.*

Proof. Let h be a monic polynomial in $K[X]$, and write $\mathscr{J} = fK[X] + gK[X]$. Applying corollary 1.5.4 to $\mathscr{J}$ and h, and taking into account that $\mathscr{J}L[X] = fL[X] + gL[X]$, we see that $fK[X] + gK[X] = hK[X]$ if and only if $fL[X] + gL[X] = hL[X]$, which is a restatement of the desired conclusion.   □

The final assertion in the conclusion of the last corollary can be loosely stated by saying that for two given polynomials over a field, the property of being relatively prime is of an "absolute" character, in the sense that its validity is not altered by enlarging the field of coefficients.

In regard to theorem 1.5.1, it should also be mentioned that a polynomial ring is not a principal ideal domain if the ring of coefficients is not a field (problem 1).

We shall now proceed to the study of polynomials over factorial domains. Our main objective is to derive the fundamental theorem stating that the polynomials over a factorial domain form a factorial domain. We shall follow a classical argument—essentially due to Gauss—based on a sequence of auxiliary results. Some of these will occasionally be useful, especially in the construction of interesting examples.

Let A be a factorial domain. By a **primitive polynomial in** $A[X]$ we shall understand a nonzero polynomial in $A[X]$ having the property that the unit element of A is a highest common factor in A of the set of all its coefficients.

Given a factorial domain A, it is clear that every polynomial in $A[X]$ with a coefficient in A^* is primitive in $A[X]$; this applies, in particular, to every monic polynomial in $A[X]$. Also, every factor in $A[X]$ of a primitive polynomial in $A[X]$ and every irreducible polynomial in $A[X]$ are primitive in $A[X]$.

1.5.6. Proposition. Let A be a factorial domain.

(i) *If f and g are nonzero polynomials in $A[X]$, if $\alpha \in A$, and if $g = \alpha f$, then α is a highest common factor in A of the set of all coefficients of g if and only if f is primitive in $A[X]$.*

(ii) *Every nonzero polynomial in $A[X]$ is expressible as the product of a nonzero element of A and a primitive polynomial in $A[X]$.*

Proof. Let f, g, and α be as in (i). Denote by n the common degree of f and g, and write

$$f(X) = \sum_{i=0}^{n} \gamma_i X^i \quad \text{and} \quad g(X) = \sum_{i=0}^{n} \delta_i X^i$$

with $\gamma_0, \gamma_1, \ldots, \gamma_n, \delta_0, \delta_1, \ldots, \delta_n \in A$. Then let β be a highest common factor in A of $\{\gamma_0, \gamma_1, \ldots, \gamma_n\}$. Since $\delta_i = \alpha\gamma_i$ for $0 \le i \le n$, it follows that $\alpha\beta$ is a highest common factor in A of $\{\delta_0, \delta_1, \ldots, \delta_n\}$. Consequently, α is a highest common factor in A of $\{\delta_0, \delta_1, \ldots, \delta_n\}$ if and only if α and $\alpha\beta$ are associates in A, hence if and only if $\beta \in A^*$. This proves (i), since the latter condition means that f is primitive in $A[X]$.

To prove (ii), let g be a nonzero polynomial in $A[X]$, and write $g(X) = \sum_{i=0}^n \delta_i X^i$, where $n = \deg(g)$ and $\delta_0, \delta_1, \ldots, \delta_n \in A$. Now let α be a highest common factor in A of $\{\delta_0, \delta_1, \ldots, \delta_n\}$. For $0 \le i \le n$, write $\delta_i = \alpha\gamma_i$ with $\gamma_i \in A$; and then define a polynomial f in $A[X]$ by $f(X) = \sum_{i=0}^n \gamma_i X^i$. It is clear that $g = \alpha f$; and by (i), f is primitive in $A[X]$. $\qquad\square$

1.5.7. Corollary. *Let A be a factorial domain, and let K be a field of fractions of A.*

(i) *Every nonzero polynomial in $K[X]$ is expressible as the product of an element of K^* and a primitive polynomial in $A[X]$.*

(ii) *If $\alpha, \beta \in K^*$, if f and g are primitive polynomials in $A[X]$, and if $\alpha f = \beta g$, then there exists a $\lambda \in A^*$ such that $\lambda\alpha = \beta$ and $f = \lambda g$.*

(iii) *If two primitive polynomials in $A[X]$ are associates in $K[X]$, then they are associates in $A[X]$.*

Proof. To prove (i), let $g \in K[X]$ and $g \ne 0$. Write $g(X) = \sum_{i=0}^n \delta_i X^i$, where $n = \deg(g)$ and $\delta_0, \delta_1, \ldots, \delta_n \in K$. Express $\delta_0, \delta_1, \ldots, \delta_n$ as fractions with numerators and denominators in A, and denote by τ the product of these denominators. It then follows that $\tau g \in A[X]$ and $\tau g \ne 0$; and the proposition shows that $\tau g = \sigma f$, where $\sigma \in A$ and f is a primitive polynomial in $A[X]$. Now we need only take $\alpha = \sigma/\tau$, for then $\alpha \in K^*$ and $g = \alpha f$.

To prove (ii), let α, β, f, g be as described in its hypothesis. Writing $\alpha = \pi/\rho$ and $\beta = \sigma/\tau$ with $\pi, \rho, \sigma, \tau \in A$, it is clear from $\alpha f = \beta g$ that $\tau\pi f = \rho\sigma g$. Since f and g are primitive in $A[X]$, the proposition now shows that $\tau\pi$ and $\rho\sigma$ are highest common factors in A of the set of all coefficients of the same nonzero polynomial in $A[X]$. Therefore $\tau\pi$ and $\rho\sigma$ are associates in A, so that $\rho\sigma = \lambda\tau\pi$ for some $\lambda \in A^*$. But then

$$\lambda\alpha = \lambda(\pi/\rho) = \sigma/\tau = \beta;$$

and since $\alpha f = \beta g = \lambda\alpha g$, we also get $f = \lambda g$.

Finally, note that (iii) is a consequence of (ii). For if f and g are primitive polynomials in $A[X]$ that are associates in $K[X]$, we have $f = \beta g$ for some $\beta \in K^*$. But, by (ii), this implies that $\beta \in A^*$, whence f and g are indeed associates in $A[X]$. $\qquad\square$

1.5.8. Proposition. *Let A be a factorial domain. If f and g are primitive polynomials in $A[X]$, then fg is primitive in $A[X]$.*

Proof. If $\alpha, \beta \in A$, we shall write $\alpha|\beta$ or $\alpha \nmid \beta$ according as α divides or does not divide β in A.

Let us write

$$f(X) = \sum_{i=0}^{m} \gamma_i X^i \quad \text{and} \quad g(X) = \sum_{j=0}^{n} \delta_j X^j,$$

where $m = \deg(f)$, $n = \deg(g)$, and $\gamma_0, \gamma_1, \ldots, \gamma_m, \delta_0, \delta_1, \ldots, \delta_n \in A$. Then $f(X)g(X) = \sum_{k=0}^{m+n} \alpha_k X^k$, where $\alpha_k = \sum_{r=0}^{k} \gamma_r \delta_{k-r}$ for $0 \leq k \leq m+n$.

Assume that fg is not primitive in $A[X]$. Then there exists an irreducible element π in A such that $\pi | \alpha_k$ for $0 \leq k \leq m+n$. By hypothesis, it is not true that $\pi | \gamma_i$ for $0 \leq i \leq m$; and similarly, it is not true that $\pi | \delta_j$ for $0 \leq j \leq n$. Therefore, indices s and t can be chosen so that

$$0 \leq s \leq m, \quad \pi \nmid \gamma_s, \quad \text{and} \quad \pi | \gamma_i \quad \text{for } 0 \leq i < s;$$

and

$$0 \leq t \leq n, \quad \pi \nmid \delta_t, \quad \text{and} \quad \pi | \delta_j \quad \text{for } 0 \leq j < t.$$

We then have

$$\alpha_{s+t} = (\gamma_0 \delta_{s+t} + \cdots + \gamma_{s-1}\delta_{t+1}) + \gamma_s \delta_t + (\gamma_{s+1}\delta_{t-1} + \cdots + \gamma_{s+t}\delta_0)$$

and

$$\pi | \alpha_{s+t}; \quad \pi | \gamma_0, \ldots, \pi | \gamma_{s-1}; \quad \pi | \delta_{t-1}, \ldots, \pi | \delta_0.$$

Consequently $\pi | \gamma_s \delta_t$, and the irreducibility of π in A implies that either $\pi | \gamma_s$ or $\pi | \delta_t$. But this is incompatible with the choice of s and t. $\qquad \square$

1.5.9. Proposition. *Let A be a factorial domain, and let K be a field of fractions of A. Then every nonconstant irreducible polynomial in $A[X]$ is irreducible in $K[X]$.*

Proof. Assume that the conclusion does not hold, and choose a nonconstant irreducible polynomial f in $A[X]$ that is reducible in $K[X]$. Write $f = gh$, where g and h are nonconstant polynomials in $K[X]$; and then use 1.5.7 to obtain primitive polynomials u and v in $A[X]$ and $\alpha, \beta \in K^*$ such that $g = \alpha u$ and $h = \beta v$.

Since f is irreducible in $A[X]$, it is primitive in $A[X]$; and, by 1.5.8, uv is also primitive in $A[X]$. Furthermore, since $f = gh = \alpha\beta uv$ and $\alpha\beta \in K^*$, we see that f and uv are associates in $K[X]$. It then follows from 1.5.7 that f and uv are associates in $A[X]$. Therefore $f = \gamma uv$, where $\gamma \in A^*$. Clearly, γu and v are polynomials in $A[X]$; and they are not constants, because their degrees are equal, respectively, to those of g and h. The equality $f = (\gamma u)v$ implies then that f is reducible in $A[X]$, contrary to assumption. $\qquad \square$

The last two propositions, which constitute the principal part of Gauss's argument, are frequently referred to as **Gauss's lemmas**. Before proceeding to the main theorem, it will be well to illustrate how the results in the foregoing discussion can be used to obtain useful information in certain particular situations.

1.5.10. Examples

Let A be a factorial domain, and let K be a field of fractions of A.

a. If f and g are monic polynomials in $K[X]$ such that $fg \in A[X]$, then $f, g \in A[X]$.

To see this, we first invoke 1.5.7 to obtain $\alpha, \beta \in K$ and primitive polynomials u and v in $A[X]$ so that $f = \alpha u$ and $g = \beta v$. We shall show that $\alpha, \beta \in A$, which will imply the desired conclusion.

Let σ and τ denote, respectively, the leading coefficients of u and v. Since f and g are monic, it is clear that $\alpha\sigma = 1 = \beta\tau$. According to 1.5.8, uv is primitive in $A[X]$; also, since fg is monic and belongs to $A[X]$, it is primitive in $A[X]$; finally, we have $\sigma\tau fg = \sigma\tau\alpha\beta uv = uv$. It now follows from 1.5.7 that $\sigma\tau \in A^*$. But then, since $\sigma, \tau \in A$, we have $\sigma, \tau \in A^*$; therefore $\alpha = 1/\sigma \in A$ and $\beta = 1/\tau \in A$, which is what we wanted.

b. Every element of K that is a zero of a monic polynomial in $A[X]$ belongs to A.

Indeed, let f be a monic polynomial in $A[X]$, and let $\alpha \in K$ and $f(\alpha) = 0$. We then have $f(X) = (X - \alpha)g(X)$ for some $g \in K[X]$. Consequently, $X - \alpha$ and $g(X)$ are monic polynomials in $K[X]$, and their product belongs to $A[X]$. What was shown in example a now implies that $X - \alpha \in A[X]$, whence $\alpha \in A$.

c. Every root of unity in K belongs to A.

This is a particular case of what we just showed in example b. If α is a root of unity in K, then there exists a positive integer n such that α is a zero of the polynomial $X^n - 1$ in $A[X]$. Since this polynomial is monic, we conclude that $\alpha \in A$.

d. We now give a sufficient condition for the irreducibility of a polynomial. This result is known as **Eisenstein's irreducibility criterion**.

Given $\alpha, \beta \in A$, we shall write $\alpha | \beta$ or $\alpha \nmid \beta$ according as α divides or does not divide β in A. The result in question is the following:

Let f be a nonconstant polynomial in $A[X]$; and write $f(X) = \sum_{i=0}^{n} \gamma_i X^i$, where $n = \deg(f)$ and $\gamma_0, \gamma_1, \ldots, \gamma_n \in A$. If there exists an irreducible element π in A such that

$$\pi | \gamma_i \quad \text{for } 0 \le i < n, \qquad \pi \nmid \gamma_n, \quad \text{and} \quad \pi^2 \nmid \gamma_0,$$

then f is irreducible in $K[X]$.

First, we contend that we may take f to be primitive in $A[X]$. To see this, use 1.5.6 to express f in the form $f = \alpha g$, where α is a highest common factor in A of $\{\gamma_0, \gamma_1, \ldots, \gamma_n\}$ and g is a primitive polynomial in $A[X]$. Then

$\deg(g) = n$, and we can write $g(X) = \sum_{i=0}^{n} \delta_i X^i$ with $\delta_0, \delta_1, \ldots, \delta_n \in A$. Clearly, $\gamma_i = \alpha \delta_i$ for $0 \le i \le n$, so that

$$\pi | \alpha \delta_i \quad \text{for } 0 \le i < n, \qquad \pi \nmid \alpha \delta_n, \quad \text{and} \quad \pi^2 \nmid \alpha \delta_0;$$

since $\alpha | \gamma_n$ and $\pi \nmid \gamma_n$, we have $\pi \nmid \alpha$, and from the preceding conditions we deduce that

$$\pi | \delta_i \quad \text{for } 0 \le i < n, \qquad \pi \nmid \delta_n, \quad \text{and} \quad \pi^2 \nmid \delta_0.$$

But this means that g and π satisfy the conditions stated in the hypothesis for f and π. Since the irreducibility in $K[X]$ of f is equivalent to that of g, our claim is proved.

Thus, we now take f to be primitive in $A[X]$. Let us assume that f is reducible in $K[X]$. According to 1.5.9, f is reducible in $A[X]$; and furthermore, since f is primitive in $A[X]$, its only constant factors in $A[X]$ are the elements of A^*. It then follows that $f = gh$, where g and h are nonconstant polynomials in $A[X]$. Write

$$g(X) = \sum_{i=0}^{s} \alpha_i X^i \quad \text{and} \quad h(X) = \sum_{i=0}^{t} \beta_i X^i,$$

where $s = \deg(g)$, $t = \deg(h)$, and $\alpha_0, \alpha_1, \ldots, \alpha_s, \beta_0, \beta_1, \ldots, \beta_t \in A$.

Since $\gamma_0 = \alpha_0 \beta_0$ and $\pi | \gamma_0$, we have either $\pi | \alpha_0$ or $\pi | \beta_0$; and, by the symmetry of the situation, we may assume that $\pi | \alpha_0$. Also, since $\gamma_n = \alpha_s \beta_t$ and $\pi \nmid \gamma_n$, we have $\pi \nmid \alpha_s$. Consequently, an index j can be chosen so that

$$0 < j \le s, \quad \pi \nmid \alpha_j, \quad \text{and} \quad \pi | \alpha_i \quad \text{for } 0 \le i < j.$$

Since

$$\gamma_j = \alpha_0 \beta_j + \alpha_1 \beta_{j-1} + \cdots + \alpha_{j-1} \beta_1 + \alpha_j \beta_0$$

and

$$\pi | \gamma_j, \pi | \alpha_0, \pi | \alpha_1, \ldots, \pi | \alpha_{j-1},$$

it then follows that $\pi | \alpha_j \beta_0$; but $\pi \nmid \alpha_j$, and hence $\pi | \beta_0$. To sum up, we now have

$$\pi | \alpha_0, \quad \pi | \beta_0, \quad \text{and} \quad \gamma_0 = \alpha_0 \beta_0;$$

therefore $\pi^2 | \gamma_0$, which contradicts the hypothesis. We conclude that f is irreducible in $K[X]$. $\square$

Eisenstein's criterion will be used occasionally in examples. Some of its easy applications will be left as exercises (problems 6, 7, and 8).

1.5.11.   Example.   If K is a field, then every root of unity in $K(X)$ belongs to K.

Since $K(X)$ is a field of fractions of the factorial domain $K[X]$, it follows from example 1.5.10c that every root of unity in $K(X)$ belongs to $K[X]$; therefore it is an invertible element of $K[X]$, and hence it belongs to K. $\square$

1.5.12.   Example.   Let K be a field. If f and g are relatively prime polynomials in $K[X]$, then $f(X) - Yg(X)$ is irreducible in $K(Y)[X]$.

Assume, on the contrary, that $f(X) - Yg(X)$ is reducible in $K(Y)[X]$. Since $K(Y)$ is a field of fractions of the factorial domain $K[Y]$, it follows from 1.5.9 that $f(X) - Yg(X)$ is reducible in $K[Y][X]$, and hence in $K[X, Y]$. We then have

$$f(X) - Yg(X) = u(X, Y)v(X, Y),$$

where u and v are nonconstant polynomials in $K[X, Y]$. Therefore

$$\deg_Y(u(X, Y)) + \deg_Y(v(X, Y)) = \deg_Y(u(X, Y)v(X, Y))$$
$$= \deg_Y(f(X) - Yg(X)) = 1,$$

and so we may take

$$\deg_Y(u(X, Y)) = 0 \quad \text{and} \quad \deg_Y(v(X, Y)) = 1.$$

This means that there exist $a, b, c \in K[X]$ such that

$$u(X, Y) = a(X) \quad \text{and} \quad v(X, Y) = b(X) + c(X)Y.$$

But then

$$f(X) - Yg(X) = a(X)(b(X) + c(X)Y)$$
$$= a(X)b(X) + a(X)c(X)Y,$$

which shows that

$$f(X) = a(X)b(X) \quad \text{and} \quad g(X) = a(X)c(X).$$

Consequently, $a(X)$ is a common factor in $K[X]$ of $f(X)$ and $g(X)$. Since the latter are relatively prime, this implies that $a(X)$ is constant, which leads to the contradictory conclusion that $u(X, Y)$ is constant.   □

To complete our discussion, we shall now prove the result for which the preceding background was developed.

1.5.13.   Theorem.   *If A is a factorial domain, and if n is a positive integer, then $A[X_1, X_2, \ldots, X_n]$ is a factorial domain.*

Proof.   An obvious induction shows that we need only concern ourselves with the case in which $n = 1$. We shall divide the proof into two parts, the first dealing with the existence of factorizations into irreducible elements in $A[X]$, and the second with the essential uniqueness of such factorizations.

Existence.   By 1.5.6, every nonzero polynomial in $A[X]$ is the product of an element of A and a primitive polynomial in $A[X]$. Since every nonzero element of A has a factorization into irreducible elements in A, and since the latter are also irreducible elements in $A[X]$, we need only verify that every primitive polynomial in $A[X]$ has a factorization into irreducible elements in $A[X]$.

We shall do this by induction on the degree of the primitive polynomial in $A[X]$ under consideration. The assertion is trivially true when this degree is 0, the polynomial in question being then an element of A^*. Next, assume that n is a positive integer, and that it is true for primitive polynomials in $A[X]$ of degree less than n. Consider now a primitive polynomial f in $A[X]$ of degree n. If f is irreducible in $A[X]$, there is nothing to verify. Assume, therefore, that f is reducible in $A[X]$. Since the only constant factors of f in $A[X]$ are the elements of A^*, we can write $f = gh$, where g and h are nonconstant polynomials in $A[X]$. But then g and h are primitive polynomials in $A[X]$ of degree less than n; and hence, by the induction assumption, they possess factorizations into irreducible elements in $A[X]$. Combining these, we obtain a factorization of f into irreducible elements in $A[X]$.

Essential Uniqueness. First, consider a nonconstant primitive polynomial f in $A[X]$, and assume that

$$f = \prod_{i=1}^{m} g_i \quad \text{and} \quad f = \prod_{i=1}^{n} h_i,$$

where m and n are positive integers, and $g_1, g_2, \ldots, g_m, h_1, h_2, \ldots, h_n$ are irreducible polynomials in $A[X]$. Since f is primitive, every irreducible factor of f in $A[X]$ is nonconstant. Therefore $g_1, g_2, \ldots, g_m, h_1, h_2, \ldots, h_n$ are nonconstant; and hence, according to 1.5.9, they are irreducible in $K[X]$. By unique factorization in the principal ideal domain $K[X]$, it now follows that $m = n$, and, by suitably permuting indices, that g_i and h_i are associates in $K[X]$ for $1 \le i \le n$. Finally, since for $1 \le i \le n$ the polynomials g_i and h_i are primitive in $A[X]$, it follows from 1.5.7 that they are associates in $A[X]$. Thus, the required essential uniqueness has been established in our particular case.

To conclude, we consider next a nonzero polynomial f in $A[X]$. There is nothing to prove if f is constant: Indeed, in this case, f is a nonzero element of A, and its irreducible factors in $A[X]$ are irreducible elements in A. Assume, therefore, that f is nonconstant. Since the nonconstant irreducible factors of f in $A[X]$ are primitive in $A[X]$, it follows from 1.5.8 that the product of finitely many such factors is primitive in $A[X]$. Consequently, in a given factorization of f into irreducible elements in $A[X]$, the product of the constant terms is a nonzero element of A, and that of the nonconstant terms is a nonconstant primitive polynomial in $A[X]$. Now it is clear how to complete the argument: Given two factorizations of f into irreducible elements in $A[X]$, we see from 1.5.7 that the two nonzero elements of A obtained by taking the product of the constant terms in each of the factorizations are associates in A, and that the two nonconstant primitive polynomials in $A[X]$ obtained by taking the product of the nonconstant terms in each of the factorizations are associates in $A[X]$; therefore, the unique factorization property of A takes care of the constant terms, and the particular case considered in the preceding paragraph takes care of the nonconstant terms. □

Using this theorem, it is easy to show that the class of all factorial domains is larger than that of all principal ideal domains (problem 2).

We shall require the theorem almost exclusively in the discussion of certain examples in which 1.5.9 has to be used in order to show that if K is a field and n is a positive integer, then the irreducibility in $K[X_1, X_2,\ldots,X_n, Y]$ of a polynomial implies its irreducibility in $K(X_1, X_2,\ldots,X_n)[Y]$. To deduce this from 1.5.9, we take into account that $K(X_1, X_2,\ldots,X_n)$ is a field of fractions of $K[X_1, X_2,\ldots,X_n]$, and that the latter, by the theorem, is a factorial domain.

PROBLEMS

1. Show that if A is a ring such that $A[X]$ is a principal ideal domain, then A is a field.

2. Prove that if A is a factorial domain, and if n is an integer such that $n > 1$, then $A[X_1, X_2,\ldots,X_n]$ is a factorial domain that is not a principal ideal domain.

3. Let K be a field of prime characteristic p, and let $\alpha \in K$. Show that if the polynomial $X^p - X - \alpha$ is reducible in $K[X]$, then each of its irreducible factors in $K[X]$ has degree 1.

4. Let A be a factorial domain, and let K be a field of fractions of A. Verify the following assertions:
   a. If f is a primitive polynomial in $A[X]$, if $g \in A[X]$, and if f divides g in $K[X]$, then f divides g in $A[X]$.
   b. If a primitive polynomial in $A[X]$ is irreducible in $K[X]$, then it is irreducible in $A[X]$.

5. Let A be a factorial domain. If f is a nonzero polynomial in $A[X]$, by a **content of f in A** we understand a highest common factor in A of the set of all coefficients of f.

   Prove that if f and g are nonzero polynomials in $A[X]$, and if α and β are, respectively, contents of f and g in A, then $\alpha\beta$ is a content of fg in A.

6. Use Eisenstein's criterion to prove that if n is a positive integer, and if m is a square-free integer such that $|m| > 1$, then $X^n - m$ is irreducible in $Q[X]$.

7. There are polynomials in $Z[X]$ whose irreducibility in $Q[X]$ cannot be deduced directly from Eisenstein's criterion. In some cases, however, the irreducibility in $Q[X]$ of such a polynomial $f(X)$ can be established by showing that there exists an integer k such that the criterion

applies to $f(X+k)$. Give examples of quadratic and cubic polynomials in $\mathbf{Z}[X]$ for which this happens.

8. Let n be an odd integer such that $n > 1$, let $\gamma_0, \gamma_1, \ldots, \gamma_n \in \mathbf{Z}$ and $f(X) = \sum_{i=0}^{n} \gamma_i X^i$. Show that if $\gamma_{n-2} = \gamma_n = 1$ and $\gamma_{n-1} = 0$, and if $k \in \mathbf{Z}$, then $f(X+k)$ does not satisfy the hypothesis of Eisenstein's criterion.

9. Suppose that, in the statement of Eisenstein's criterion, the hypothesis on π is changed to "if $1 \le m \le n$, and if there exists an irreducible element π in A such that

$$\pi \mid \gamma_i \quad \text{for } 0 \le i < m, \quad \pi \nmid \gamma_m, \quad \pi \nmid \gamma_n, \quad \text{and} \quad \pi^2 \nmid \gamma_0 ".$$

Then prove that f has an irreducible factor in $K[X]$ of degree greater than or equal to m.

One of the elementary techniques used in integral calculus in order to compute indefinite integrals is based on the partial fraction decomposition of rational functions. This decomposition is valid for an arbitrary field of coefficients; the precise statements are given in the next three problems.

10. Let K be a field, let f_1 and f_2 be relatively prime nonconstant polynomials in $K[X]$, and let $g \in K[X]$ and $\deg(g) < \deg(f_1) + \deg(f_2)$. Show that there exist $u_1, u_2 \in K[X]$ such that

$$\deg(u_1) < \deg(f_1) \quad \text{and} \quad \deg(u_2) < \deg(f_2),$$

and such that the equality

$$g/f_1 f_2 = u_1/f_1 + u_2/f_2$$

holds in $K(X)$.

11. Let K be a field, and let $f, g \in K[X]$. Prove that if f is nonconstant and if e is a positive integer for which $\deg(g) < \deg(f^e)$, then there exist $u_1, u_2, \ldots, u_e \in K[X]$ such that

$$\deg(u_i) < \deg(f) \quad \text{for } 1 \le i \le e,$$

and such that the equality

$$\frac{g}{f^e} = \sum_{i=1}^{e} \frac{u_i}{f^i}$$

holds in $K(X)$.

12. Let K be a field, and let $f, g \in K[X]$ and $\deg(g) < \deg(f)$. Show that if f is nonconstant and $f = \prod_{i=1}^{n} f_i^{e_i}$, where n is a positive integer, and $f_1, f_2, \ldots, f_n$ are n distinct irreducible polynomials in $K[X]$, and $e_1, e_2, \ldots, e_n$ are positive integers, then there exists a family $(u_{ij})_{1 \le i \le n, 1 \le j \le e_i}$ of polynomials in $K[X]$ such that

$$\deg(u_{ij}) < \deg(f_i) \quad \text{for } 1 \le i \le n \quad \text{and} \quad 1 \le j \le e_i,$$

and such that the equality

$$\frac{g}{f} = \sum_{i=1}^{n} \sum_{j=1}^{e_i} \frac{u_{ij}}{f_i^j}$$

holds in $K(X)$.

1.6. SPLITTING OF POLYNOMIALS

Having studied the general factorization properties of polynomials, it is now natural to inquire about the polynomials whose irreducible factors are of the simplest type. This leads to the notion of splitting, which is of fundamental importance in the theory of fields.

Let K be a field. A nonconstant polynomial in $K[X]$ is said to **split in** $K[X]$ when it is the product of a finite family of polynomials in $K[X]$ of degree 1. Equivalently, we could say that a nonconstant polynomial in $K[X]$ splits in $K[X]$ when each of its irreducible factors in $K[X]$ has degree 1.

Thus, to say that a nonconstant polynomial f in $K[X]$ splits in $K[X]$ is equivalent to saying that there exists a positive integer n and elements $\alpha, \alpha_1, \alpha_2, \ldots, \alpha_n$ of K such that

$$f(X) = \alpha \prod_{i=1}^{n} (X - \alpha_i).$$

And it is clear that if such a relation holds, then n and α are, respectively, the degree and leading coefficient of f, and $\alpha_1, \alpha_2, \ldots, \alpha_n$ are the zeros of f in K.

Let K be a field, and let L be an extension field of K. Every nonconstant polynomial in $K[X]$ that splits in $K[X]$ also splits in $L[X]$. In fact, it has the same irreducible factors in $L[X]$ as in $K[X]$, and the same zeros in L as in K: No new irreducible factors or zeros of such a polynomial can be obtained by extending K to L. On the other hand, there may exist nonconstant polynomials in $K[X]$ that admit no zeros in K but split in $L[X]$.

1.6.1. Examples

a. If $\alpha \in \mathbf{C}$ and n is a positive integer, and if we write $\alpha = r\exp(\theta\sqrt{-1})$, where $r = |\alpha|$ and $0 \le \theta < 2\pi$, then the nth complex roots of α are given by

$$\alpha_k = \sqrt[n]{r}\exp\left(\frac{\theta + 2\pi k\sqrt{-1}}{n}\right), \qquad 1 \le k \le n.$$

Consequently, in $\mathbf{C}[X]$ we have

$$X^n - \alpha = \prod_{k=1}^{n} (X - \alpha_k),$$

and so $X^n - \alpha$ splits in $\mathbf{C}[X]$.

b. If $\alpha \in \mathbf{R}^*$, and if n is a positive integer such that $n > 2$, then $X^n - \alpha$ does not split in $\mathbf{R}[X]$. For with the notation of example a, we then have $\{\alpha_1, \alpha_2, \ldots, \alpha_n\} \not\subseteq \mathbf{R}$, because α has at most two nth real roots.

c. If $\alpha \in \mathbf{R}$, then $X^2 - \alpha$ splits in $\mathbf{R}[X]$ if and only if $\alpha \geq 0$.

d. If $\alpha \in \mathbf{Z}$, then $X^2 - \alpha$ splits in $\mathbf{Q}[X]$ if and only if α is the square of an integer.   □

1.6.2. Examples

Let K be a field of prime characteristic p, and let $\alpha \in K$.

a. If n is a nonnegative integer, and if the polynomial $X^{p^n} - \alpha$ in $K[X]$ admits a zero in K, then it splits in K.

Indeed, if β is a zero of $X^{p^n} - \alpha$ in K, then $\alpha = \beta^{p^n}$, and hence

$$X^{p^n} - \alpha = X^{p^n} - \beta^{p^n} = (X - \beta)^{p^n}$$

in $K[X]$.

b. If the polynomial $X^p - X - \alpha$ in $K[X]$ admits a zero in K, then it splits in $K[X]$.

This is easily deduced from 1.3.8: If β is a zero of $X^p - X - \alpha$ in K, then $\beta + 1, \beta + 2, \ldots, \beta + p$ are the zeros of $X^p - X - \alpha$ in K, which implies that

$$X^p - X - \alpha = \prod_{i=1}^{p} (X - \beta - i)$$

in $K[X]$.   □

It will occasionally be useful to have at our disposal necessary and sufficient conditions for two polynomials to be relatively prime. Such conditions will now be obtained as consequences of the following result.

1.6.3.   Proposition.   *Let K be a field, and let W be a set of nonconstant polynomials in $K[X]$. If $WK[X] = K[X]$, then the polynomials in W do not admit a common zero in K. Conversely, if the polynomials in W do not admit a common zero in K, and if there exists a polynomial in W that splits in $K[X]$, then $WK[X] = K[X]$.*

Proof. First, suppose that $WK[X] = K[X]$. Then there exist a positive integer n, polynomials $f_1, f_2, \ldots, f_n$ in W, and polynomials $g_1, g_2, \ldots, g_n$ in $k[X]$ such that

$$\sum_{k=1}^{n} f_k(X) g_k(X) = 1.$$

If $\alpha \in K$ and $f(\alpha) = 0$ for every $f \in W$, then $f_k(\alpha) = 0$ for $1 \le k \le n$, and the substitution $X \to \alpha$ in the preceding equality would result in the equality

$$0 = \sum_{k=1}^{n} f_k(\alpha) g_k(\alpha) = 1,$$

an obvious impossibility. We conclude that no element of K is a common zero of the polynomials in W.

Now suppose that no element of K is a common zero of the polynomials in W, and that there exists a polynomial in W that splits in $K[X]$. Let us write $WK[X] = hK[X]$, where $h \in K[X]$. If $WK[X] \subset K[X]$, then h would be a nonconstant factor in $K[X]$ of every polynomial in W, and hence of a nonconstant polynomial that splits in $K[X]$; since $K[X]$ is factorial, this would imply that h splits in $K[X]$, which is a contradiction because every zero of h in K would then be a common zero of the polynomials in W. It follows that $WK[X] = K[X]$, as required. $\square$

1.6.4. Corollary. *Let K be a field, and let f and g be nonconstant polynomials in $K[X]$. If f and g are relatively prime, then f and g do not admit a common zero in K. Conversely, if f and g do not admit a common zero in K, and if either f or g splits in $K[X]$, then f and g are relatively prime.*

Proof. This is the special case of the proposition in which $W = \{f, g\}$. $\square$

A theme that inevitably occurs throughout our subject is that of obtaining zeros of polynomials. We already know that a polynomial may fail to have zeros in the given field of coefficients. That the zeros can always be obtained by suitably extending the latter is implied by the following remarkable result.

1.6.5. Theorem. *Let K be a field, and let W be a finite set of nonconstant polynomials in $K[X]$. Then there exists an extension field L of K such that every polynomial in W splits in $L[X]$.*

Proof. It is clear, first, that we need only concern ourselves with the particular case of a set consisting of a single polynomial. For if n is a positive integer and $f_1, f_2, \ldots, f_n$ are nonconstant polynomials in $K[X]$, and if L is an extension field of K, unique factorization in $L[X]$ implies that the polynomials $f_1, f_2, \ldots, f_n$ split in $L[X]$ if and only if the polynomial $\prod_{i=1}^{n} f_i$ splits in $L[X]$.

We begin by proving the following lemma, with which the names of Cauchy and Kronecker are often associated.

Lemma. *If F is a field, and if f is an irreducible polynomial in $F[X]$, then there exists an extension field of F in which f admits a zero.*

Proof. Let $\mathcal{M} = fF[X]$. The irreducibility of f in $F[X]$ implies that $\mathcal{M}$ is a maximal ideal of $F[X]$, and hence $F[X]/\mathcal{M}$ is a field.

Composing the inclusion mapping from F to $F[X]$ with the natural projection from $F[X]$ to $F[X]/\mathcal{M}$, we obtain a homomorphism u from F to $F[X]/\mathcal{M}$. Then u is a monomorphism, and 0.0.1 implies that u is extendible to an isomorphism v from an extension field E of F to $F[X]/\mathcal{M}$.

To conclude, let $\alpha = v^{-1}(X + \mathcal{M})$. It is now seen that α is a zero of f in E: Indeed, write $f = \sum_{i=0}^{n}\gamma_i X^i$, where n is a positive integer and $\gamma_0, \gamma_1, \ldots, \gamma_n \in F$; since $f \in \mathcal{M}$, the equality $f + \mathcal{M} = 0$ is valid in $F[X]/\mathcal{M}$, and hence

$$v\big(f(\alpha)\big) = v\left(\sum_{i=0}^{n} \gamma_i \alpha^i \right) = \sum_{i=0}^{n} v(\gamma_i) v(\alpha)^i$$

$$= \sum_{i=0}^{n} u(\gamma_i) v(\alpha)^i = \sum_{i=0}^{n} (\gamma_i + \mathcal{M})(X + \mathcal{M})^i$$

$$= \left(\sum_{i=0}^{n} \gamma_i X^i \right) + \mathcal{M} = f + \mathcal{M} = 0,$$

which implies that $f(\alpha) = 0$. This proves the lemma.

We shall now prove the theorem—for a single polynomial—by induction. It is trivially true for polynomials of degree 1. Next assume that n is a positive integer, and that it is true for polynomials of degree n.

Consider then a field K and a polynomial f in $K[X]$ of degree $n + 1$. First, apply the lemma to the field K and an irreducible factor of f in $K[X]$ to obtain an extension field R of K and a zero ρ of the latter in R. Clearly, ρ is also a zero of f in R, and hence we can write $f(X) = (X - \rho)g(X)$, where $g \in R[X]$ and $\deg(g) = n$. Then apply the induction assumption to the field R and the polynomial g in $R[X]$ to obtain an extension field L of R such that g splits in $L[X]$. Since the equality $f(X) = (X - \rho)g(X)$ remains valid in $L[X]$, it now follows that f splits in $L[X]$. This completes the proof of the theorem.   □

An obvious question arises in connection with the theorem just proved: Does it remain valid for *infinite* sets of nonconstant polynomials?

We shall see at a later time, in our discussion on algebraically closed fields, that this question can be answered in the affirmative. In order to prove this, however, it is necessary to make use of the standard transfinite machinery.

As shown by the following examples, the preceding theorem can be useful in establishing the irreducibility of certain polynomials.

1.6.6. Examples

Let K be a field of prime characteristic p, and let $\alpha \in K$.

a. Let n be a positive integer. We shall prove that $X^{p^n} - \alpha$ is irreducible in $K[X]$ if and only if $\alpha \notin K^p$.

If $\alpha \in K^p$, then $\alpha = \beta^p$ for some $\beta \in K$; therefore

$$X^{p^n} - \alpha = X^{p^n} - \beta^p = \left(X^{p^{n-1}} - \beta \right)^p$$

in $K[X]$, so that $X^{p^n} - \alpha$ is reducible in $K[X]$.

Now suppose that $X^{p^n} - \alpha$ is reducible in $K[X]$. Let f be an irreducible factor of $X^{p^n} - \alpha$ in $K[X]$, and put $m = \deg(f)$. Next, apply 1.6.5 to obtain an extension field L of K in which $X^{p^n} - \alpha$ has a zero β. As shown in 1.6.2, in $L[X]$ we have

$$X^{p^n} - \alpha = X^{p^n} - \beta^{p^n} = (X - \beta)^{p^n};$$

since $L[X]$ is factorial, this implies that $f(X) = (X - \beta)^m$. It follows, in particular, that $\beta^{p^n}, \beta^m \in K$.

Since $1 \le m < p^n$, the highest common factor of m and p^n is of the form p^k, where $0 \le k < n$. Writing $p^k = im + jp^n$, where i and j are integers, we see that

$$\beta^{p^k} = \beta^{im + jp^n} = (\beta^m)^i (\beta^{p^n})^j \in K.$$

But then $\beta^{p^{n-1}} = (\beta^{p^k})^{p^{n-k-1}} \in K$; since $\alpha = \beta^{p^n} = (\beta^{p^{n-1}})^p$, this shows that $\alpha \in K^p$.

b. Let us now show that $X^p - X - \alpha$ either is irreducible in $K[X]$ or splits in $K[X]$.

According to 1.6.2, it suffices to verify that if $X^p - X - \alpha$ is reducible in $K[X]$, then it has a zero in K. Thus, assume that

$$X^p - X - \alpha = f(X)g(X),$$

where f and g are nonconstant polynomials in $K[X]$. Applying 1.6.5, we get an extension field L of K in which $X^p - X - \alpha$ possesses a zero β. Then, exactly as in 1.6.2, we see that

$$X^p - X - \alpha = \prod_{i=1}^{p} (X - \beta - i)$$

in $L[X]$.

Let $m = \deg(f)$ Then $1 < m < p$, and

$$f(X)g(X) = \prod_{i=1}^{p} (X - \beta - i)$$

in $L[X]$; unique factorization in $L[X]$ implies that

$$f(X) = \prod_{j=1}^{m} (X - \beta - i_j),$$

where $1 \le i_1 < i_2 < \cdots < i_m \le p$. Since the coefficient of X^{m-1} in the product appearing in the last equality is $-\sum_{j=1}^{m}(\beta + i_j)$, we have

$$m\beta + \sum_{j=1}^{m} i_j = \sum_{j=1}^{m} (\beta + i_j) \in K,$$

whence $m\beta \in K$. Since $p \nmid m$, it now follows that $\beta \in K$, and so β is a zero of $X^p - X - \alpha$ in K. $\square$

PROBLEMS

1. Let p be a prime. Give an example of a polynomial f in $\mathbf{Z}[X]$ of degree $p - 1$ having the property that if α is a zero of f in $\mathbf{C}$, then f splits in $\mathbf{Q}(\alpha)[X]$.
2. Let K be a field of prime characteristic p, and let $\alpha, \beta \in K$. Show that $X^p - \beta^{p-1}X - \alpha$ either is irreducible in $K[X]$ or splits in $K[X]$.
3. Let p be a prime, and let $n \in \mathbf{Z}$ and $p \nmid n$. Prove that $X^p - X - n$ is irreducible in $\mathbf{Q}[X]$.
4. Let p be a prime, let K be a field, and let $\alpha \in K$. Prove that $X^p - \alpha$ is irreducible in $K[X]$ if and only if α does not admit a pth root in K.

1.7. SEPARABLE POLYNOMIALS

Let K be a field, and let q be a positive integer. Recall that the subdomain of $K[X]$ generated by $K \cup \{X^q\}$ coincides with the image of the injective K-endomorphism $f(X) \rightarrow f(X^q)$ of $K[X]$, and that it is denoted by $K[X^q]$. If $f \in K[X]$ and $f = \sum_{i=0}^{n} \gamma_i X^i$, where n is a nonnegative integer and $\gamma_0, \gamma_1, \ldots, \gamma_n \in K$, to say that $f \in K[X^q]$ simply means that $\gamma_i = 0$ whenever $0 \le i \le n$ and $q \nmid i$. Finally, note that if f is a nonzero polynomial in $K[X^q]$, and if g is the nonzero polynomial in $K[X]$ for which $f(X) = g(X^q)$, then $\deg(f) = q\deg(g)$, and the irreducibility in $K[X]$ of f implies that of g.

It will be useful to know the conditions under which the derivative of a polynomial vanishes. The complete answer is provided by the following result.

1.7.1. Proposition. Let K be a field, and let $f \in K[X]$.

(i) If K has characteristic 0, then $f' = 0$ if and only if f is constant.

(ii) If K has prime characteristic p, then $f' = 0$ if and only if $f \in K[X^p]$.

Proof. Write $f(X) = \sum_{i=0}^{n} \gamma_i X^i$, where n is a nonnegative integer and $\gamma_0, \gamma_1, \ldots, \gamma_n \in K$. Then $f'(X) = \sum_{i=0}^{n-1}(i+1)\gamma_{i+1}X^i$; therefore, to say that $f' = 0$ means that $i\gamma_i = 0$ for $1 \le i \le n$.

If K has characteristic 0, then for $1 \le i \le n$ we have $i\gamma_i = 0$ if and only if $\gamma_i = 0$. Consequently, $f' = 0$ if and only if $\gamma_i = 0$ for $1 \le i \le n$, and hence if and only if f is constant.

To conclude, suppose that K has prime characteristic p. For $1 \le i \le n$, we have $i\gamma_i = 0$ if and only if $p \mid i$ or $\gamma_i = 0$. It then follows that $f' = 0$ if and only if $\gamma_i = 0$ whenever $1 \le i \le n$ and $p \nmid i$, and hence if and only if $f \in K[X^p]$. $\qquad\square$

The preceding proposition brings out a fundamental difference between fields of characteristic 0 and fields of prime characteristic. This difference underlies the necessity of introducing the concept of separability in the study of algebraic extensions.

Let K be a field. By a **separable polynomial in** $K[X]$ we shall understand a nonconstant polynomial f in $K[X]$ such that f and f' are relatively prime.

We have chosen this definition because, when $f \in K[X]$, the condition "f and f' are relatively prime" can be tested within $K[X]$. Note also that it allows us to speak of a separable polynomial in the same "absolute" sense as of relatively prime polynomials: If $f \in K[X]$ and L is an extension field of K, then f is separable in $K[X]$ if and only if it is separable in $L[X]$.

There seems to be no complete agreement on the definition of a separable polynomial. However, the various definitions given in the literature agree for irreducible polynomials, which is actually the only essential case. Indeed, several authors define separability for irreducible polynomials exclusively.

The relevance of separable polynomials for the theory of fields lies in the fact that they are the polynomials without multiple zeros (a property that is frequently used to define separability). We state this in precise terms as follows.

1.7.2. Proposition. Let K be a field, and let f be a nonconstant polynomial in $K[X]$. If f is separable, then every zero of f in K is simple. Conversely, if every zero of f in K is simple, and if f splits in $K[X]$, then f is separable.

Proof. Since the multiple zeros of f in K are the common zeros of f and f' in K, our assertions are then immediate consequences of 1.6.4. $\qquad\square$

In a certain sense, the preceding result justifies the use of the term "separable polynomial". For, let K be a field, and let f be an irreducible polynomial in $K[X]$. It will be seen in 2.1.9 that if L is an extension field of K, and if one zero α of f in L is selected, then the zeros of f in L correspond to the K-embeddings of $K(\alpha)$ in L. Therefore, when f possesses no multiple zeros in L, the largest number of such K-embeddings is obtained.

1.7.3. Proposition. *Let K be a field, and let f be a separable polynomial in $K[X]$. Then every nonconstant factor of f in $K[X]$ is separable.*

Proof. Suppose that $f = gh$, where $g, h \in K[X]$. To show that g and g' are relatively prime, we need only note that f and f' are relatively prime, and that, in view of the relations

$$f = gh \quad \text{and} \quad f' = g'h + gh',$$

every common factor in $K[X]$ of g and g' is also a common factor of f and f'. $\square$

1.7.4. Examples

a. Let K be a field, and let n be an integer such that $n > 1$. If $f \in K[X]$, and if there exists a nonconstant polynomial g in $K[X]$ such that $f = g^n$, then f is not separable.

This is clear, since the relations $f = g^n$ and $f' = ng^{n-1}g'$ show that g is a common factor in $K[X]$ of f and f'.

b. Let K be a field of characteristic 0. If $\alpha \in K^*$ and n is a positive integer, then the polynomial $X^n - \alpha$ in $K[X]$ is separable.

Since $(X^n - \alpha)' = nX^{n-1}$, we see that X is the only monic irreducible factor in $K[X]$ of $(X^n - \alpha)'$; since X is not a factor in $K[X]$ of $X^n - \alpha$, our assertion follows.

c. Let K be a field of characteristic 0. If n is a positive integer, then the polynomial $\sum_{i=0}^{n} X^i$ in $K[X]$ is separable.

Since $(\sum_{i=0}^{n} X^i)(X - 1) = X^{n+1} - 1$, our assertion is a consequence of 1.7.3 and the preceding example.

d. Let K be a field of prime characteristic p. If $\alpha \in K^*$ and n is a positive integer, then the polynomial $X^n - \alpha$ in $K[X]$ is separable if and only if $p \nmid n$.

We have $(X^n - \alpha)' = nX^{n-1}$. If $p \nmid n$, we can argue as in example b to show that $X^n - \alpha$ is separable; and if $p \mid n$, then $(X^n - \alpha)' = 0$, whence $X^n - \alpha$ and $(X^n - \alpha)'$ are not relatively prime. $\square$

We now establish a basic criterion for the separability of an irreducible polynomial.

1.7.5. Proposition. *Let K be a field, and let f be a nonconstant polynomial in $K[X]$. If f is separable, then $f' \neq 0$. Conversely, if $f' \neq 0$ and f is irreducible in $K[X]$, then f is separable.*

Proof. If $f' = 0$, then f is a common factor of f and f' in $K[X]$. Since f is noninvertible in $K[X]$, it follows that f and f' are not relatively prime, and hence f is not separable.

Now suppose that f is irreducible in $K[X]$. Then either f is a factor of f' in $K[X]$ or f and f' are relatively prime. In view of the inequality $\deg(f) > \deg(f')$, the first alternative implies that $f' = 0$; and the second implies that f is separable. □

Thus, the vanishing of the derivative, which was discussed earlier in the section, has entered into our considerations on separability in a natural way. In characteristic 0, the derivative of a nonconstant polynomial does not vanish, and hence irreducibility implies separability. As shown by the next proposition, this property characterizes perfect fields.

1.7.6. Proposition. *Let K be a field. Then K is perfect if and only if every irreducible polynomial in $K[X]$ is separable.*

Proof. First, suppose that K is not perfect. Then K has prime characteristic p, and $K^p \subset K$. Choosing an $\alpha \in K - K^p$, we obtain the polynomial $X^p - \alpha$, whose irreducibility in $K[X]$ was established in 1.6.6; and since $(X^p - \alpha)' = 0$, the preceding proposition shows that it is not separable.

Next, suppose that there exists an irreducible polynomial f in $K[X]$ that is not separable. By the preceding proposition, we have $f' = 0$. It then follows from 1.7.1 that K has prime characteristic p and $f \in K[X^p]$. Write $f = \sum_{i=0}^{m} \gamma_i X^{ip}$, where m is a positive integer and $\gamma_0, \gamma_1, \ldots, \gamma_m \in K$. If K were perfect, then there would exist $\beta_0, \beta_1, \ldots, \beta_m \in K$ such that $\beta_i^p = \gamma_i$ for $0 \leq i \leq m$; but then

$$f(X) = \sum_{i=0}^{m} \beta_i^p X^{ip} = \left(\sum_{i=0}^{m} \beta_i X^i \right)^p$$

would hold in $K[X]$, in contradiction to the assumed irreducibility of f in $K[X]$. Therefore, K is not perfect. □

To close this section, we shall now consider some auxiliary notions for fields of prime characteristic that will be useful in the sequel.

Let K be a field of prime characteristic p, and let f be an irreducible polynomial in $K[X]$. If n is a nonnegative integer and $f \in K[X^{p^n}]$, then $p^n | \deg(f)$. Therefore, the set of all nonnegative integers n for which

$f \in K[X^{p^n}]$ is nonempty and finite; its largest element will be called the **exponent of inseparability of** f. Note that f is separable if and only if its exponent of inseparability is 0: Indeed, the latter assertion means that $f \notin K[X^p]$, which is equivalent to $f' \neq 0$.

Next, let e denote the exponent of inseparability of f, and let g be the polynomial in $K[X]$ for which $f(X) = g(X^{p^e})$. Since $f \notin K[X^{p^{e+1}}]$, we clearly have $g \notin K[X^p]$, whence $g' \neq 0$; and since g is irreducible in $K[X]$, we conclude that g is separable. The positive integers $\deg(g)$ and p^e will be called, respectively, the **reduced degree of** f and the **degree of inseparability of** f; and g will be called the **separable polynomial associated with** f.

With the help of the notions just defined, we can easily derive a basic result on the multiplicities of the zeros of an irreducible polynomial.

1.7.7. Proposition. *Let K be a field of prime characteristic, and let f be an irreducible polynomial in $K[X]$. If L is an extension field of K, then the degree of inseparability of f is the multiplicity of every zero of f in L.*

Proof. Let $p = \mathrm{Char}(K)$, and let e and g denote, respectively, the exponent of inseparability of f and the separable polynomial associated with f.

Let α be a zero of f in L. Since $f(X) = g(X^{p^e})$, we have $g(\alpha^{p^e}) = f(\alpha) = 0$, so that α^{p^e} is a zero of g in L. But every zero of g in L is simple, and hence in $L[X]$ we have

$$g(X) = (X - \alpha^{p^e})v(X),$$

where $v \in L[X]$ and $v(\alpha^{p^e}) \neq 0$. Now define a polynomial u in $L[X]$ by $u(X) = v(X^{p^e})$. Then $u(\alpha) = v(\alpha^{p^e}) \neq 0$, and the equalities

$$f(X) = g(X^{p^e}) = (X^{p^e} - \alpha^{p^e})v(X^{p^e}) = (X - \alpha)^{p^e}u(X)$$

hold in $L[X]$. Consequently, α is a zero of multiplicity p^e of f in L. $\square$

1.7.8. Corollary. *Let K be a field of prime characteristic, and let f be an irreducible polynomial in $K[X]$. If L is an extension field of K such that f splits in $L[X]$, then the reduced degree of f is the cardinality of the set of all zeros of f in L.*

Proof. Since the degree of f is the product of the reduced degree of f and the degree of inseparability of f, this follows at once from the proposition. $\square$

We conclude by noting that if K is a field and f is an irreducible polynomial in $K[X]$, and if L is an extension field of K, then the zeros of f in L have the same multiplicity. This is clear from 1.7.7 in the case of prime characteristic; and it follows from 1.7.2 and 1.7.6 in the case of characteristic 0.

PROBLEMS

1. Let K be a field of prime characteristic p, let I be a set, and let $f \in K[X_i]_{i \in I}$. Show that $\partial f / \partial X_i = 0$ for every $i \in I$ if and only if $f \in K[X_i^p]_{i \in I}$.
2. Let K be a field of characteristic 0, and let f be a nonconstant polynomial in $K[X]$ that splits in $K[X]$. Let d be a highest common factor of f and f' in $K[X]$, and write $f = dg$ with $g \in K[X]$. Prove that g is separable, and that f and g have the same zeros in K.
3. Let K be a field of prime characteristic p, and let f be a monic separable polynomial in $K[X]$ that splits in $K[X]$. Show that if the zeros of f in K form a subgroup of K^+, then there exist a positive integer m and $\gamma_0, \gamma_1, \ldots, \gamma_{m-1} \in K$ such that $\gamma_0 \neq 0$ and $f(X) = \sum_{i=0}^{m-1} \gamma_i X^{p^i} + X^{p^m}$.
4. Let k be a field of prime characteristic p, let f be a nonconstant monic polynomial in $K[X]$ that splits in $K[X]$, and let e be a nonnegative integer. Show that the zeros of f in K have the common multiplicity p^e and form a subgroup of K^+ if and only if there exist a positive integer m and $\gamma_0, \gamma_1, \ldots, \gamma_{m-1} \in K$ such that $\gamma_0 \neq 0$ and

$$f(X) = X^{p^e}\left(\sum_{i=0}^{m-1} \gamma_i X^{p^i} + X^{p^m} \right).$$

NOTES

The contents of this chapter are basic and can be found in practically every textbook on abstract algebra.

Sections 1.1–1.4. The material presented in sections 1.1–1.4 appeared, with the terminology and the notation that have become standard, in the classic work of Steinitz [1]: field of fractions (§3); prime field and characteristic (§4); field extension and adjunction (§2); simple extension and primitive element (§5); finite extension and linear degree (§7); and perfect field (§11).

The concepts developed in section 1.4 furnish all the tools required to establish the impossibility of doubling the cube and of trisecting the angle. In fact, as shown in Hadlock [1: sec. 1.1–1.3], only the most elementary considerations on quadratic extensions are needed for this purpose.

Section 1.5. The factorization theory of polynomials with integer coefficients goes back to Gauss [2: art. 42]. No significant changes are required for the generalization to polynomials with coefficients in a factorial domain.

Section 1.6. The main result of this section—and of the chapter—is theorem 1.6.5, which is essential in the study of field extensions. It was first observed by Cauchy [2; 3] that **C** can be defined as the quotient of **R**[X] by the ideal generated by $X^2 + 1$; this construction was extended later by Kronecker [3] in order to derive the result that appears as a lemma in the proof of 1.6.5.

Section 1.7. Separable polynomials were defined and studied by Steinitz [1: §§10, 11, 13]. The term "separable" was introduced later as a substitute for Steinitz's original term, "of the first kind".

Inseparable irreducible polynomials can occur only when the fields under consideration have prime characteristic. For this reason, there is no mention of separability in the early literature on fields, when fields were restricted to subfields of **C**.

Chapter 2

Algebraic Extensions

2.1. ALGEBRAIC EXTENSIONS

The field extensions with which we shall be concerned throughout most of this book are algebraic extensions. These are the field extensions defined by the condition that every element of the top field should be a root of a nontrivial polynomial equation with coefficients in the bottom field. In this section, algebraic elements and extensions will be defined, and their basic properties will be discussed in detail. These properties constitute the foundation of our subject, and are essential for the understanding of every subsequent section.

Let K be a field, and let L be an extension field of K. We recall that for each $\alpha \in L$, the mapping $f(X) \to f(\alpha)$ from $K[X]$ to L is a K-homomorphism having as its image the subdomain $K[\alpha]$ of L; and that its kernel, which is the ideal of $K[X]$ consisting of all polynomials in $K[X]$ admitting α as a zero, is said to be the ideal of algebraic relations of α over K.

We say that an element of L is **algebraic over** K or **transcendental over** K according as its ideal of algebraic relations over K is nonnull or null. Equivalently, we could say that an element of L is algebraic over K or transcendental over K according as it is or is not a zero of a nonzero polynomial in $K[X]$.

41

2.1.1. Examples

a. If $\alpha, \beta \in \mathbf{Q}$ and $\alpha > 0$, then α^β is algebraic over $\mathbf{Q}$.

Write $\beta = m/n$, where $m, n \in \mathbf{Z}$ and $n > 0$. Then $(\alpha^\beta)^n = \alpha^m$, which means that α^β is a zero of the polynomial $X^n - \alpha^m$ in $\mathbf{Q}[X]$.

b. If K is a field, then every root of unity in K is algebraic over the prime subfield of K.

Let P denote the prime subfield of K. If α is a root of unity in K, then $\alpha^n = 1$ for some positive integer n, whence α is a zero of the polynomial $X^n - 1$ in $P[X]$.

c. If K is a finite field, then every element of K is algebraic over the prime subfield of K.

If we let $q = \text{Card}(K)$, then K^* is a group of order $q - 1$. Hence $\alpha^{q-1} = 1$ for every $\alpha \in K^*$, so that every element of K^* is a root of unity in K. By example b, our assertion now follows at once.

d. If K is a field, then the element X of the field $K(X)$ of rational functions is transcendental over K.

This is clear, for if $f \in K[X]$ and X is a zero of f, then $f = f(X) = 0$.
□

Concerning example d, we would like to make two comments. First, the transcendence of X over K does not tell the whole story, which is that every element of $K(X) - K$ is transcendental over K (problem 7). Also, the reader may feel that this is a rather artificial way of illustrating the notion of transcendental element. But in truth, we cannot do much better: As stated in the following proposition (in whose statement implicit use is being made of 1.1.4), which we state for future reference, the notions of indeterminate and transcendental element are essentially identical.

2.1.2. Proposition. Let K be a field, and let L be an extension field of K. If $\xi \in L$ and ξ is transcendental over K, then the K-homomorphism $f(X) \to f(\xi)$ from $K[X]$ to $K[\xi]$ is a K-isomorphism; and the K-isomorphism from $K(X)$ to $K(\xi)$ extending it is the only K-monomorphism from $K(X)$ to $K(\xi)$ such that $X \to \xi$.

We shall now proceed to derive the properties of algebraic elements. The detailed study of transcendental extensions will be taken up in the last chapter of the book. However, the result on transcendental elements just stated will be used occasionally in our discussion on algebraic extensions.

2.1.3. Proposition. *Let K be a field, let L be an extension field of K, and let $\alpha \in L$. Then the following conditions are equivalent:*

(a) *α is algebraic over K.*

(b) *The ideal of algebraic relations of α over K is a maximal ideal of $K[X]$.*

(c) *$K[\alpha] = K(\alpha)$.*

Proof. Let $\mathscr{J}$ denote the ideal of algebraic relations of α over K. Since $\mathscr{J}$ and $K[\alpha]$ are, respectively, the kernel and image of the homomorphism $f(X) \to f(\alpha)$ from $K[X]$ to L, we see that the rings $K[X]/\mathscr{J}$ and $K[\alpha]$ are isomorphic. Since $K[\alpha]$ is a domain, so is $K[X]/\mathscr{J}$, and hence $\mathscr{J}$ is a prime ideal of $K[X]$.

The equivalence of (a) and (b) is now an immediate consequence of the fact that in $K[X]$ the maximal ideals are the nonnull prime ideals. Finally, since $K[\alpha] = K(\alpha)$ if and only if $K[\alpha]$ is a field, hence if and only if $K[X]/\mathscr{J}$ is a field, we see that (b) and (c) are equivalent. □

Let K be a field, and let L be an extension field of K. Let $\alpha \in L$ and suppose that α is algebraic over K. Since the ideal of algebraic relations of α over K is nonnull, it admits a unique monic polynomial in $K[X]$ as a generator. This polynomial is called the **minimal polynomial of α over K**. It can also be described as the monic polynomial in $K[X]$ that admits α as a zero and divides in $K[X]$ every polynomial in $K[X]$ admitting α as a zero.

The next three propositions contain the main facts on the minimal polynomials of algebraic elements.

2.1.4. Proposition. *Let K be a field, let L be an extension field of K, and let M be an extension field of L. If $\alpha \in M$ and α is algebraic over K, then α is algebraic over L, and the minimal polynomial of α over L divides in $L[X]$ the minimal polynomial of α over K.*

Proof. This is clear, because the minimal polynomial of α over K is a polynomial in $L[X]$ having α as a zero. □

2.1.5. Proposition. *Let K be a field, let L be an extension field of K, and let $\alpha \in L$. If α is algebraic over K, then the minimal polynomial of α over K is the only monic irreducible polynomial in $K[X]$ admitting α as a zero.*

Proof. Let f be the minimal polynomial of α over K. Then, according to 2.1.3, $fK[X]$ is a maximal ideal of $K[X]$, and hence f is irreducible in $K[X]$.

To conclude, let g be a monic irreducible polynomial in $K[X]$ having α as a zero. Then f divides g in $K[X]$. Since f and g are monic and irreducible in $K[X]$, this implies that $f = g$. □

2.1.6. Proposition. *Let K be a field, and let L be an extension field of K. Let $\alpha \in L$, suppose that α is algebraic over K, and denote by n the degree of the minimal polynomial of α over K.*

(i) *If g is a nonzero polynomial in $K[X]$ admitting α as a zero, then $\deg(g) \geq n$.*

(ii) *The minimal polynomial of α over K is the only monic polynomial in $K[X]$ of degree n admitting α as a zero.*

(iii) *The sequence $(\alpha^i)_{0 \leq i \leq n-1}$ is a linear base of $K(\alpha)$ over K.*

(iv) *The mapping $g(X) \to g(\alpha)$ from the set of all polynomials in $K[X]$ of degree less than n to $K(\alpha)$ is bijective.*

(v) *$[K(\alpha) : K] = n$.*

Proof. Let f denote the minimal polynomial of α over K.

To verify (i), suppose that $g \in K[X]$, $g \neq 0$, and $g(\alpha) = 0$. It then follows from $g(\alpha) = 0$ that f divides g in $K[X]$; and since $g \neq 0$, this implies that $\deg(g) \geq \deg(f) = n$.

To prove (ii), suppose that g is a monic polynomial in $K[X]$ such that $\deg(g) = n$ and $g(\alpha) = 0$. Then f divides g in $K[X]$, f and g are monic, and $\deg(f) = n = \deg(g)$; and these conditions imply that $f = g$.

It is evident that (iv) is a restatement of (iii), and that (v) follows from (iii). We shall now complete the proof by verifying (iii).

It is seen, first, that $(\alpha^i)_{0 \leq i \leq n-1}$ is linearly independent over K. Indeed, let $(\gamma_i)_{0 \leq i \leq n-1}$ be a sequence of elements of K such that $\sum_{i=0}^{n-1} \gamma_i \alpha^i = 0$. If g is the polynomial in $K[X]$ defined by $g(X) = \sum_{i=0}^{n-1} \gamma_i X^i$, then $g(\alpha) = 0$ and $\deg(g) \leq n - 1 < \deg(f)$. By (i), this implies that $g = 0$, and hence $\gamma_i = 0$ for $0 \leq i \leq n - 1$, as was to be shown.

Finally, let us show that every $\beta \in K(\alpha)$ is a linear combination of $(\alpha^i)_{0 \leq i \leq n-1}$ with coefficients in K. According to 2.1.3, we have $K[\alpha] = K(\alpha)$, so that $\beta \in K[\alpha]$. We can then write $B = g(\alpha)$, where $g \in K[X]$. Next choose $u, v \in K[X]$ so that

$$g = uf + v \quad \text{and} \quad \deg(v) < \deg(f).$$

Since $f(\alpha) = 0$, we have

$$\beta = g(\alpha) = u(\alpha)f(\alpha) + v(\alpha) = v(\alpha);$$

furthermore, since $\deg(v) < n$, we can write $v = \sum_{i=0}^{n-1} \gamma_i X^i$ with $\gamma_0, \gamma_1, \ldots, \gamma_{n-1} \in K$. But then

$$\beta = v(\alpha) = \sum_{i=0}^{n-1} \gamma_i \alpha^i,$$

which is what we wanted. $\square$

2.1.7. Examples

In order to illustrate the ideas involved in the preceding discussion on minimal polynomials, we now look at some special situations.

a. If K is a field, and if $\alpha \in K$, then α is algebraic over K, and its minimal polynomial over K is $X - \alpha$.

This is obvious, since $X - \alpha$ divides in $K[X]$ every polynomial in $K[X]$ having α as a zero.

b. If $\alpha, \beta \in \mathbf{R}$, then $\alpha + \beta\sqrt{-1}$ and $\alpha - \beta\sqrt{-1}$ are the zeros in $\mathbf{C}$ of the polynomial $X^2 - 2\alpha X + (\alpha^2 + \beta^2)$ in $\mathbf{R}[X]$, and hence are algebraic over $\mathbf{R}$; and if $\beta \neq 0$, then this polynomial is the minimal polynomial over $\mathbf{R}$ of $\alpha + \beta\sqrt{-1}$ and $\alpha - \beta\sqrt{-1}$.

That $\alpha + \beta\sqrt{-1}$ and $\alpha - \beta\sqrt{-1}$ are the complex zeros of the polynomial $X^2 - (\alpha + \beta)X + (\alpha^2 + \beta^2)$ in $\mathbf{R}[X]$ can be verified by direct computation; and since $\alpha + \beta\sqrt{-1}$, $\alpha - \beta\sqrt{-1} \notin \mathbf{R}$ when $\beta \neq 0$, the assertion concerning the minimal polynomial follows from 2.1.6.

c. Let p be a prime, and let α be a pth root of unity in $\mathbf{C}$ such that $\alpha \neq 1$. We shall show that the minimal polynomial of α over $\mathbf{Q}$ is $\sum_{i=0}^{p-1} X^i$. (This is a particular case of a result on cyclotomic polynomials that will be proved in section 3.5.)

The conditions

$$(\alpha - 1)\left(\sum_{i=0}^{p-1} \alpha^i \right) = \alpha^p - 1 = 0 \quad \text{and} \quad \alpha \neq 1$$

imply that $\sum_{i=0}^{p-1} \alpha^i = 0$, which means that α is a zero of $\sum_{i=0}^{p-1} X^i$ in $\mathbf{C}$. According to 2.1.5, our assertion will be established if we prove that $\sum_{i=0}^{p-1} X^i$ is irreducible in $\mathbf{Q}[X]$. We shall do this with the help of Eisenstein's irreducibility criterion.

Let $f = \sum_{i=0}^{p-1} X^i$. Since

$$Xf(X+1) = ((X+1) - 1)\left(\sum_{i=0}^{p-1} (X+1)^i \right)$$

$$= (X+1)^p - 1 = \sum_{i=1}^{p} \binom{p}{i} X^i,$$

it follows that

$$f(X+1) = \sum_{i=1}^{p} \binom{p}{i} X^{i-1}.$$

We have already noted, in the proof of 1.2.7, that $p \mid \binom{p}{i}$ for $1 \le i \le p-1$; also, since $\binom{p}{1} = p$ and $\binom{p}{p} = 1$, we have $p^2 \nmid \binom{p}{1}$ and $p \nmid \binom{p}{p}$. By 1.5.10, we now conclude that $f(X+1)$ is irreducible in $\mathbf{Q}[X]$, which proves the irreducibility of $f(X)$ in $\mathbf{Q}[X]$.

d. Let α and β be complex numbers satisfying the relations

$$\alpha^3 + 6\alpha^2 + 1 = 0$$

and

$$\beta = \alpha^7 + 6\alpha^6 + \alpha^4 + \alpha^3 + 7\alpha^2 + 2.$$

We shall show that $\beta \ne 0$, and determine $1/\beta$ in $\mathbf{Q}[\alpha]$.

First note that α is algebraic over $\mathbf{Q}$, because it is a zero of $X^3 + 6X^2 + 1$. We now claim that $X^3 + 6X^2 + 1$ is the minimal polynomial of α over $\mathbf{Q}$. By 2.1.5, to show this it suffices to establish the irreducibility of this polynomial in $\mathbf{Q}[X]$, which presents no difficulty: If it is reducible in $\mathbf{Q}[X]$, then it possesses a zero in $\mathbf{Q}$, and such a zero can be written in the form m/n, where m and n are relatively prime integers; since

$$(m/n)^3 + 6(m/n)^2 + 1 = 0,$$

we have

$$m^3 + 6m^2 n + n^3 = 0,$$

from which it follows that $m \mid n$ and $n \mid m$. But this implies that either $m/n = 1$ or $m/n = -1$, which is impossible because neither 1 nor -1 is a zero of $X^3 + 6X^2 + 1$.

The division algorithm in $\mathbf{Q}[X]$ now yields the equality

$$X^7 + 6X^6 + X^4 + X^3 + 7X^2 + 2 = (X^3 + 6X^2 + 1)(X^4 + 1) + (X^2 + 1).$$

Taking into consideration the given relations satisfied by α and β, the substitution $X \to \alpha$ in this equality results in the equality $\beta = \alpha^2 + 1$.

According to 2.1.6, $(1, \alpha, \alpha^2)$ is a linear base of $\mathbf{Q}(\alpha)$ over $\mathbf{Q}$. Consequently, $\beta = \alpha^2 + 1 \ne 0$. Then $1/\beta \in \mathbf{Q}(\alpha)$, and we can write

$$1/\beta = \rho\alpha^2 + \sigma\alpha + \tau,$$

where $\rho, \sigma, \tau \in \mathbf{Q}$. Thus

$$(\alpha^2 + 1)(\rho\alpha^2 + \sigma\alpha + \tau) = \beta(1/\beta) = 1,$$

whence

$$\rho\alpha^4 + \sigma\alpha^3 + (\rho + \tau)\alpha^2 + \sigma\alpha + \tau = 1.$$

Moreover, from $\alpha^3 + 6\alpha^2 + 1 = 0$ we obtain

$$\alpha^3 = -6\alpha^2 - 1 \quad \text{and} \quad \alpha^4 = 36\alpha^2 - \alpha + 6.$$

Substituting these expressions for α^3 and α^4 in the preceding equality, we now get

$$(37\rho - 6\sigma + \tau)\alpha^2 + (-\rho + \sigma)\alpha + (6\rho - \sigma + \tau) = 1.$$

It follows that

$$37\rho - 6\sigma + \tau = 0$$

$$-\rho + \sigma = 0$$

$$6\rho - \sigma + \tau = 1.$$

The solution of this system is found to be

$$\rho = -1/26, \qquad \sigma = -1/26, \qquad \tau = 31/26;$$

we conclude that $1/\beta = -(\alpha^2 + \alpha - 31)/26$.

   e. Let K be a field of prime characteristic p, let L be an extension field of K, and let $\alpha \in L$. If α is algebraic over K, and if e denotes the exponent of inseparability of the minimal polynomial of α over K, then α^{p^e} is algebraic over K, and its minimal polynomial over K is the separable polynomial associated with that of α over K.

   This is a consequence of 2.1.5: If f denotes the minimal polynomial of α over K, and if g denotes the separable polynomial associated with f, then g is a monic irreducible polynomial in $K[X]$, and

$$g(\alpha^{p^e}) = f(\alpha) = 0.$$

   f. Let K be a field of prime characteristic p, and let f be a monic irreducible polynomial in $K[X]$. Then $f(X^p)$ is reducible in $K[X]$ if and only if $f \in K^p[X]$, in which case $f(X^p)$ is the pth power of a monic irreducible polynomial in $K[X]$.

   We start by forming a simple extension $K(\alpha)$ of K by adjunction of a zero α of f in an extension field of K. Let $n = \deg(f)$, and write $f(X) = \sum_{i=0}^{n-1} \gamma_i X^i + X^n$ with $\gamma_0, \gamma_1, \ldots, \gamma_{n-1} \in K$. According to 2.1.5, f is the minimal polynomial of α over K.

   Next we invoke 1.6.5 in order to obtain an extension field of $K(\alpha)$ in which there exist elements $\beta, \delta_0, \delta_1, \ldots, \delta_{n-1}$ such that

$$\beta^p = \alpha, \quad \delta_0^p = \gamma_0, \quad \delta_1^p = \gamma_1, \ldots, \delta_{n-1}^p = \gamma_{n-1}.$$

Then let $L = K(\delta_0, \delta_1, \ldots, \delta_{n-1})$, and let g be the monic polynomial in $L[X]$ defined by $g(X) = \sum_{i=0}^{n-1} \delta_i X^i + X^n$. Thus $\deg(g) = n$, and the equalities

$$g(X)^p = \left(\sum_{i=0}^{n-1} \delta_i X^i + X^n \right)^p = \sum_{i=0}^{n-1} \delta_i^p X^{pi} + X^{pn}$$

$$= \sum_{i=0}^{n-1} \gamma_i X^{pi} + X^{pn} = f(X^p)$$

hold in $L[X]$.

We now claim that g is irreducible in $L[X]$. To see this, note first that

$$g(\beta)^p = f(\beta^p) = f(\alpha) = 0,$$

whence $g(\beta) = 0$. It follows, in particular, that β is algebraic over L. Let h denote the minimal polynomial of β over L. By 2.1.6, we then have $\deg(g) \geq \deg(h)$. Put $m = \deg(h)$, and write $h(X) = \sum_{i=0}^{m-1} \lambda_i X^i + X^m$ with $\lambda_0, \lambda_1, \ldots, \lambda_{m-1} \in L$. Then

$$\sum_{i=0}^{m-1} \lambda_i^p \alpha^i + \alpha^m = \sum_{i=0}^{m-1} \lambda_i^p \beta^{pi} + \beta^{pm}$$

$$= \left(\sum_{i=0}^{m-1} \lambda_i \beta^i + \beta^m \right)^p = h(\beta)^p = 0.$$

Since it is clear that $L^p \subseteq K$, we see that $\sum_{i=0}^{m-1} \lambda_i^p X^i + X^m$ is a polynomial in $K[X]$ of degree m having α as a zero; and this, by 2.1.6, implies that

$$\deg(h) = m \geq \deg(f) = n = \deg(g).$$

We conclude that $\deg(g) = \deg(h)$. A third application of 2.1.6 then shows that $g = h$, which proves our claim.

The desired conclusion can now be readily derived from the foregoing argument.

First, suppose that $f \in K^p[X]$. Then $\delta_i^p = \gamma_i \in K^p$ for $0 \leq i \leq n - 1$, and hence $\delta_i \in K$ for $0 \leq i \leq n - 1$. Therefore $L = K$, so that the equality $f(X^p) = g(X)^p$ holds in $K[X]$ and g is irreducible in $K[X]$; in particular, $f(X^p)$ is reducible in $K[X]$.

Finally, suppose that $f(X^p)$ is reducible in $K[X]$. If d is a monic irreducible factor of $f(X^p)$ in $K[X]$, then $d(X) \neq f(X^p)$, and in $L[X]$ we have an equality of the form $d(X) = g(X)^r$ with $1 \leq r < p$. Since r and p are relatively prime, we can choose $s, t \in \mathbf{Z}$ so that $rs + pt = 1$. Consequently, in the field $L(X)$ of rational functions, we have the equalities

$$g(X) = g(X)^{rs} g(X)^{pt} = d(X)^s f(X^p)^t.$$

But $d(X)^s, f(X^p)^t \in K(X)$, so that $g(X) \in L[X] \cap K(X)$. Since $L[X] \cap K(X) = K[X]$, this means that $g(X) \in K[X]$, so that $\delta_i \in K$ for $0 \leq i \leq n - 1$. Therefore $\gamma_i = \delta_i^p \in K^p$ for $0 \leq i \leq n - 1$, which implies that $f \in K^p[X]$. $\qquad\qquad\square$

The material developed thus far enables us, at this time, to initiate the discussion of an interesting theme, namely, the extendibility of field monomorphisms. Here we shall confine ourselves to the most elementary case, that is, the extendibility of a field monomorphism to a simple algebraic extension. Later we shall have to consider more general situations. To deal with these, the usual procedure will consist in combining the particular results discussed here with a suitable application of Zorn's lemma.

2.1.8. Theorem. *Let K and F be fields, and let u be a monomorphism from K to F. Let f be an irreducible polynomial in $K[X]$, and let L be an extension field of K. If α and β are, respectively, zeros of f in L and of uf in F, then u is uniquely extendible to a monomorphism from $K(\alpha)$ to F such that $\alpha \to \beta$.*

Proof. It is clear that there exists at most one monomorphism from $K(\alpha)$ to F extending u and such that $\alpha \to \beta$, for any two such agree on $K \cup \{\alpha\}$, and hence on $K(\alpha)$.

We now must establish the existence of a monomorphism with the required properties. To this end, note first that we may assume f to be monic. According to 2.1.5, this assumption and the hypothesis imply that α is algebraic over K and has f as its minimal polynomial over K. Denoting by b the K-homomorphism $g(X) \to g(\alpha)$ from $K[X]$ to $K[\alpha]$, we then have $\mathrm{Ker}(b) = fK[X]$.

Next let c denote the homomorphism $g(X) \to (ug)(\beta)$ from $K[X]$ to F. It is then seen that $\mathrm{Ker}(b) \subseteq \mathrm{Ker}(c)$: indeed, if $g \in \mathrm{Ker}(b)$, we have $g \in fK[X]$, and so $ug \in (uf)F[X]$; since β is a zero of uf, this implies that it is also a zero of ug, whence $c(g) = (ug)(\beta) = 0$, which shows that $g \in \mathrm{Ker}(c)$.

Since b is surjective and $\mathrm{Ker}(b) \subseteq \mathrm{Ker}(c)$, it now follows that there exists a homomorphism v from $K[\alpha]$ to F such that $v \circ b = c$. We then have

$$v(g(\alpha)) = v(b(g)) = c(g) = (ug)(\beta)$$

for every $g \in K[X]$. By specializing this condition to the case of constant polynomials in $K[X]$, we see that $v(\gamma) = u(\gamma)$ for every $\gamma \in K$, which means that v extends u; and by specializing it to the case where $g(X) = X$, we get $v(\alpha) = \beta$.

Finally, we know from 2.1.3 that $K[\alpha] = K(\alpha)$. Consequently, v is a monomorphism from $K(\alpha)$ to F extending u and such that $v(\alpha) = \beta$. $\square$

It will now be convenient, for future reference, to state the following immediate consequence of the theorem just proved.

2.1.9. Corollary. *Let K be a field, and let L and M be extension fields of K. If f is an irreducible polynomial in $K[X]$, and if α and β are, respectively, zeros of f in L and in M, then there exists a unique K-isomorphism from $K(\alpha)$ to $K(\beta)$ such that $\alpha \to \beta$.*

It is not difficult to show that the irreducibility of the polynomial under consideration is essential for the validity of the preceding theorem and its corollary (problem 5).

In regard to the preceding corollary, we would like to make a comment on terminology. Given a field K and an extension field L of K, it is customary to say that two elements of L are **conjugate over** K when they are algebraic over K and have the same minimal polynomial over K. Clearly, an equivalent condition is that the two elements of L should be zeros of an irreducible polynomial in $K[X]$. By the corollary, we can also say that two elements α and β of L are conjugate over K if and only if they are algebraic over K and correspond to each other by a K-isomorphism between $K(\alpha)$ and $K(\beta)$.

In the foregoing discussion we have been concerned with properties of algebraic elements. We shall now shift our attention to properties of sets of algebraic elements.

Let K be a field, and let L be an extension field of K. We shall denote by $A(L/K)$ the set of all elements of L that are algebraic over K, and say that $A(L/K)$ is the **algebraic closure** of K in L. It will be shown presently that $A(L/K)$ is an intermediate field between K and L.

2.1.10. Proposition. *Let K be a field, and let L be an extension field of K. If D is a finite subset of $A(L/K)$, then $K(D)$ is finite over K.*

Proof. This will be proved by induction on the cardinality of the finite subset of $A(L/K)$ under consideration. It obviously is true when this cardinality is 0. We shall next assume that n is a nonnegative integer and that it is true for every subset of $A(L/K)$ of cardinality n.

Consider now a subset D of $A(L/K)$ such that $\text{Card}(D) = n + 1$. Then choose an $\alpha \in D$, and put $C = D - \{\alpha\}$. Since $C \subseteq A(L/K)$ and $\text{Card}(C) = n$, it follows from the induction assumption that $K(C)$ is finite over K. Moreover, since α is algebraic over K, it is algebraic over $K(C)$, and 2.1.6 shows that $K(C)(\alpha)$ is finite over $K(C)$. Therefore $K(C)(\alpha)$ is finite over K; and since $K(C)(\alpha) = K(D)$, we conclude that $K(D)$ is finite over K.   □

2.1.11. Corollary. *Let K be a field, and let L be an extension field of K. If $D \subseteq A(L/K)$, then $K[D] = K(D)$.*

Proof. To prove this, it suffices to show that the subdomain $K[D]$ of L is a field. Thus, suppose that $\alpha \in K[D]$ and $\alpha \neq 0$. Then choose a finite subset C of D such that $\alpha \in K[C]$. By the proposition, $K(C)$ is finite over K; and since $K \subseteq K[C] \subseteq K(C)$, we see that the K-space $K[C]$ is finite-dimensional. It then follows from 0.0.3 that $K[C]$ is a field, and hence $1/\alpha \in K[C]$. But $K[C] \subseteq K[D]$, and we conclude that $1/\alpha \in K[D]$.   □

2.1.12. Corollary. *Let M be a field, and let K and L be subfields of M. If $L \subseteq A(M/K)$, then $KL = K \vee L$.*

Proof. Indeed, we know that $KL = K[L]$ and $K \vee L = K(L)$; and the preceding corollary shows that $K[L] = K(L)$.   □

The first of the preceding corollaries is a generalization of the implication (a) $\Rightarrow$ (c) in 2.1.3 to the case of arbitrary sets of algebraic elements. It is natural to ask if the opposite implication (c) $\Rightarrow$ (a) can also be generalized; or, equivalently, if the converse of our corollary holds.

There is, to be sure, a partial converse. It is known that if K is a field, if L is an extension field of K, and if D is a *finite* subset of L such that $K[D] = K(D)$, then $D \subseteq A(L/K)$.

The innocent appearance of this result is misleading, and should not induce the reader to conclude that it is easy to prove. We shall deduce it from a fundamental result on places in chapter 4.

Let K be a field, and let L be an extension field of K. We say that L is **algebraic over** K, or that L is an **algebraic extension of** K, when every element of L is algebraic over K; that L is **transcendental over** K, or that L is a **transcendental extension of** K, when there exists an element of L that is transcendental over K; and that K is **algebraically closed in** L when every element of L that is algebraic over K belongs to K. Equivalently, we could say that L is algebraic over K when $A(L/K) = L$; that L is transcendental over K when $A(L/K) \subset L$; and that K is algebraically closed in K when $A(L/K) = K$.

It is evident that if K is a field, then K is the only extension field of K that is algebraic over K and in which K is algebraically closed; and furthermore, if I is a nonempty set, then the field $K(X_i)_{i \in I}$ of rational functions is transcendental over K.

The class of all field extensions has been partitioned into two subclasses, one consisting of the algebraic extensions and the other of the transcendental extensions. In the present chapter and the next, we shall deal with algebraic extensions almost exclusively. Transcendental extensions will be considered in the last chapter.

We begin our discussion on algebraic extensions with the following useful characterization of finite extensions.

2.1.13.   Proposition.   *Let K be a field, and let L be an extension field of K. Then L is finite over K if and only if L is algebraic and finitely generated over K.*

Proof.   First suppose that L is finite over K. To verify that L is finitely generated over K, let D be a linear base of L over K. Then D is finite; and since $L = KD \subseteq K(D) \subseteq L$, we have $L = K(D)$. To show that L is algebraic over K, let $\alpha \in L$. If $n = [L : K]$, then the sequence $(\alpha^i)_{0 \le i \le n}$ is linearly dependent over K; therefore, there exists a sequence $(\gamma_i)_{0 \le i \le n}$ of elements of K, not all equal to 0, such that $\sum_{i=0}^n \gamma_i \alpha^i = 0$. But this means that α is a zero of the nonzero polynomial $\sum_{i=0}^n \gamma_i X^i$ in $K[X]$, which implies that α is algebraic over K.

Next suppose, conversely, that L is algebraic and finitely generated over K. Then $L = K(D)$, where D is a finite subset of L. And since every element of D is algebraic over K, it follows from 2.1.10 that L is finite over K. $\square$

As shown by the next result, algebraic extensions enjoy the transitivity property.

2.1.14. Proposition. *Let K be a field, let L be an extension field of K, and let M be an extension field of L. Then M is algebraic over K if and only if M is algebraic over L and L is algebraic over K.*

Proof. By 2.1.4, the direct implication is clear. To prove the opposite, assume that M is algebraic over L and that L is algebraic over K. We have to show that every element of M is algebraic over K.

Let $\alpha \in M$, and let f denote the minimal polynomial of α over L. Write $f(X) = \sum_{i=0}^{n-1} \gamma_i X^i + X^n$, where $n = \deg(f)$ and $\gamma_0, \gamma_1, \ldots, \gamma_{n-1} \in L$; and then let $D = \{\gamma_0, \gamma_1, \ldots, \gamma_{n-1}\}$. Clearly, $f \in K(D)[X]$ and $f(\alpha) = 0$, whence α is algebraic over $K(D)$; and this, by 2.1.10, implies that $K(D)(\alpha)$ is finite over $K(D)$. Moreover, since D is a finite subset of L, it also follows from 2.1.10 that $K(D)$ is finite over K. Consequently $K(D)(\alpha)$ is finite over K; and since $\alpha \in K(D)(\alpha)$, the preceding proposition shows that α is algebraic over K.   □

We can now state and prove the main result on the set of all algebraic elements in a field extension.

2.1.15. Proposition. *Let K be a field, and let L be an extension field of K. Then $A(L/K)$ is the largest element of the set of all intermediate fields between K and L that are algebraic over K. Moreover, $A(L/K)$ is algebraically closed in L.*

Proof. Let us write $A = A(L/K)$; we already know that $K \subseteq A \subseteq L$. To prove the first assertion, we only need to verify that A is a subfield of L. Thus, let $\alpha, \beta \in A$. Applying first 2.1.10 and then 2.1.13, we see that $K(\alpha, \beta)$ is algebraic over K, and hence $K(\alpha, \beta) \subseteq A$. Since $\alpha - \beta, \alpha\beta \in K(\alpha, \beta)$, we now get $\alpha - \beta, \alpha\beta \in A$; and if $\alpha \neq 0$, then $1/\alpha \in K(\alpha, \beta)$, whence $1/\alpha \in A$. This shows that A is indeed a subfield of L.

In order to prove the second assertion, suppose that $\alpha \in L$ and α is algebraic over A. As in the preceding paragraph, it then follows from 2.1.10 and 2.1.13 that $A(\alpha)$ is algebraic over A. Since A is algebraic over K, the preceding proposition now shows that $A(\alpha)$ is algebraic over K. Therefore $A(\alpha) \subseteq A$, which amounts to saying that $\alpha \in A$.   □

The "closure properties" of the class of all algebraic extensions are immediate consequences of the preceding proposition.

2.1.16. Corollary. *Let K be a field, and let L be an extension field of K. If $D \subseteq A(L/K)$, then $K(D)$ is algebraic over K.*

Proof. Indeed, since $K \cup D \subseteq A(L/K)$, we see from the proposition that $K(D) \subseteq A(L/K)$.   □

2.1.17. Corollary. *Let K be a field, and let M be an extension field of K. If* $(L_i)_{i \in I}$ *is a nonempty family of intermediate fields between K and M that are algebraic over K, then* $\vee_{i \in I} L_i$ *is algebraic over K.*

Proof. This follows at once from the preceding corollary: since $L_i \subseteq A(M/K)$ for every $i \in I$, we have $\cup_{i \in I} L_i \subseteq A(M/K)$; and $\vee_{i \in I} L_i = K(\cup_{i \in I} L_i)$. □

2.1.18. Corollary. *Let M be a field, and let K and L be subfields of M. If* $L \subseteq A(M/K)$, *then* $K \vee L$ *is algebraic over K.*

Proof. Since $K \vee L = K(L)$, this is simply a particular case of the first corollary. □

The next result is of occasional interest. It can be used, for instance, in order to prove the existence of algebraic closures (section 2.2, problem 3).

2.1.19. Proposition. *Let K be a field, and let L be an algebraic extension of K.*
  (i) *If K is infinite, then L is denumerable.*
  (ii) *If K is infinite, then K and L are equipotent.*
  (iii) $\mathrm{Card}(L) \leq \aleph_0 \mathrm{Card}(K)$.

Proof. We shall follow a classical counting argument due to Cantor. This is one of the few proofs for which familiarity with elementary properties of cardinal numbers is essential.

For every positive integer n, let P_n denote the set of all monic irreducible polynomials in $K[X]$ of degree n, and let A_n denote the set of all zeros in L of the polynomials in P_n. Since every polynomial $\sum_{i=0}^{n-1} \gamma_i X^i + X^n$ in P_n is uniquely determined by the sequence $(\gamma_i)_{0 \leq i \leq n-1}$ of its coefficients in K, it follows that $\mathrm{Card}(P_n) \leq \mathrm{Card}(K)^n$.

If n is a positive integer, a polynomial in P_n has at most n zeros in L, and hence

$$\mathrm{Card}(A_n) \leq n\,\mathrm{Card}(P_n) \leq n\,\mathrm{Card}(K)^n;$$

consequently, A_n is finite when K is finite, and $\mathrm{Card}(A_n) \leq \mathrm{Card}(K)$ when K is infinite. Taking into account that $L = \cup_{n=1}^{\infty} A_n$, we now conclude that L is denumerable when K is finite, and that K and L are equipotent when K is infinite. This proves (i) and (ii); and it is clear that these imply (iii). □

Following the traditional terminology of number theory, a complex number will be said to be an **algebraic number** or a **transcendental number** according as it is algebraic or transcendental over the rational field. By virtue of 2.1.15, the algebraic numbers form an intermediate field between **Q** and **C**, which is usually denoted by **A**; thus, $\mathbf{A} = A(\mathbf{C}/\mathbf{Q})$.

It is seen from 2.1.15 and 2.1.19 that $\mathrm{Card}(\mathbf{A}) = \aleph_0$; and this implies that $\mathrm{Card}(\mathbf{C} - \mathbf{A}) = 2^{\aleph_0}$. It follows, in particular, that the set $\mathbf{C} - \mathbf{A}$ consisting of the transcendental numbers contains "many more" elements than **A**. In this way, Cantor established the existence of transcendental numbers without actually exhibiting a single one.

To prove that a specific complex number is transcendental, however, is quite a different matter. This problem has attracted the attention of some of the greatest mathematicians; and indeed, very substantial efforts have been made in order to establish the transcendence of certain numbers. We shall merely mention here that in 1844 Liouville gave a method for constructing large classes of transcendental numbers—observe that the theory of cardinal numbers had not been introduced at that time; Cantor's proof of the denumerability of A was given at the end of the nineteenth century. Also, the transcendence of the number e was established by Hermite in 1873, and that of the number π by Lindemann in 1882. The solution of a problem of this type usually requires the use of ingenious "analytic" methods.

The contributions mentioned in the preceding paragraph belong to the nineteenth century, and represent pioneering efforts in the study of transcendental numbers. The theory of transcendental numbers continues to be a fascinating—and most difficult—chapter in number theory.

2.1.20. Example. **A** is an infinite extension of **Q**; in fact, we have $[\mathbf{A}:\mathbf{Q}] = \aleph_0$.

Let n be a positive integer. By Eisenstein's irreducibility criterion, the polynomial $X^n - 2$ is irreducible in $\mathbf{Q}[X]$. Since this polynomial admits $\sqrt[n]{2}$ as a zero, it is the minimal polynomial of $\sqrt[n]{2}$ over **Q**. Consequently, $[\mathbf{Q}(\sqrt[n]{2}):\mathbf{Q}] = n$ and $\mathbf{Q}(\sqrt[n]{2}) \subseteq \mathbf{A}$.

It then follows that

$$[\mathbf{A}:\mathbf{Q}] \geq [\mathbf{Q}(\sqrt[n]{2}):\mathbf{Q}] = n$$

for every positive integer n. Therefore $[\mathbf{A}:\mathbf{Q}] \geq \aleph_0$; and since

$$[\mathbf{A}:\mathbf{Q}] \leq \mathrm{Card}(\mathbf{A}) = \aleph_0,$$

we conclude that $[\mathbf{A}:\mathbf{Q}] = \aleph_0$.   □

Thus, **A** is an infinite algebraic extension of **Q**. It will follow from 2.2.11 and 2.2.14 that every finite field admits an infinite algebraic extension. Consequently, there exist infinite algebraic extensions in every possible characteristic.

To close this section, we now state and prove a useful result on the endomorphisms of an algebraic extension.

2.1.21. Proposition. *Let K be a field, and let L be an algebraic extension of K. Then every K-endomorphism of L is a K-automorphism.*

Proof. Let s be a K-endomorphism of L. We have to show that s is surjective. To this end, let $\alpha \in L$, and denote by D the set of all zeros in L of the minimal polynomial of α over K. Therefore D is finite, $\alpha \in D$, and $s(D) \subseteq D$. Since s is injective, it follows that $s(D) = D$, and hence $\alpha \in s(D)$. Therefore $\alpha \in s(L)$, as required. $\qquad\square$

2.1.22. Corollary. *Let K be a field, and let L be an algebraic extension of K. If M and N are extension fields of L, and if u is a K-monomorphism from M to N such that $u(L) \subseteq L$, then $u(L) = L$.*

The preceding proposition admits a partial converse (problem 10). Its converse, strictly speaking, is false, as the following example shows.

2.1.23. Example. $\mathbf{R}$ is transcendental over $\mathbf{Q}$; and $i_{\mathbf{R}}$ is the only $\mathbf{Q}$-endomorphism of $\mathbf{R}$.

Since $\mathbf{Q}$ and $\mathbf{R}$ are not equipotent, we see from 2.1.20 that $\mathbf{R}$ is not algebraic over $\mathbf{Q}$.

Let s be a $\mathbf{Q}$-endomorphism of $\mathbf{R}$. It is easily seen that s is strictly increasing: indeed, if $\alpha, \beta \in \mathbf{R}$ and $\alpha < \beta$, then $\beta - \alpha > 0$ and

$$s(\beta) - s(\alpha) = s(\beta - \alpha) = s\left(\left(\sqrt{\beta - \alpha}\right)^2\right) = s\left(\sqrt{\beta - \alpha}\right)^2 > 0,$$

whence $s(\alpha) < s(\beta)$.

Now assume that $s \neq i_{\mathbf{R}}$. Choose first an $\alpha \in \mathbf{R}$ so that $s(\alpha) \neq \alpha$, and then a $\rho \in \mathbf{Q}$ lying strictly between α and $s(\alpha)$. Applying what was proved in the preceding paragraph, we have: If $\alpha < s(\alpha)$, then $\alpha < \rho < s(\alpha)$, whence $s(\alpha) < s(\rho) = \rho < s(\alpha)$; and if $s(\alpha) < \alpha$, then $s(\alpha) < \rho < \alpha$, whence $s(\alpha) < \rho = s(\rho) < s(\alpha)$. In either case, we reach the contradictory conclusion that $s(\alpha) < s(\alpha)$. $\qquad\square$

PROBLEMS

1. Let K be a field, and let L be an extension field of K. Show that if $\alpha \in L$ and α is algebraic over K, and if the minimal polynomial of α over K has odd degree, then $K(\alpha) = K(\alpha^2)$.

2. Let K be a field, let L be an extension field of K, and let m and n be positive integers such that $m|n$. Show that if $\alpha \in K$ and $X^n - \alpha$ is irreducible in $K[X]$, and if β is an nth root of α in L, then β^m is algebraic over K, and $X^{n/m} - \alpha$ is its minimal polynomial over K.

3. Prove that if m and n are relatively prime square-free integers such that $|m| > 1$ and $|n| > 1$, then $(1, \sqrt{m}, \sqrt{n}, \sqrt{mn})$ is a linear base of $\mathbf{Q}(\sqrt{m}, \sqrt{n})$ over $\mathbf{Q}$.

4. Let K be a field, and let L be an extension field of K. Let n be a positive integer, and let f be a polynomial in $K[X]$ of degree n. Show that if D denotes the set of all zeros of f in L, then $[K(D):K] \leq n!$.

5. Give an example of a polynomial in $\mathbf{Q}[X]$ of degree 4 possessing two zeros α and β in $\mathbf{R}$ such that $\mathbf{Q}(\alpha)$ and $\mathbf{Q}(\beta)$ are not $\mathbf{Q}$-isomorphic.

6. Let L and M be the subfields of $\mathbf{C}$ defined by $L = \mathbf{Q}(\sqrt[3]{2}\exp(2\pi\sqrt{-1}/3))$ and $M = \mathbf{Q}(\sqrt[3]{2})$. Verify the following assertions:
   a. $[L \vee M : M] < [L : \mathbf{Q}]$.
   b. $[L \vee M : \mathbf{Q}] < [L : \mathbf{Q}][M : \mathbf{Q}]$.
   c. $L \cap M = \mathbf{Q}$.

7. Let K be a field, and let $K(\xi)$ be a simple transcendental extension of K. Prove that K is algebraically closed in $K(\xi)$.

8. Prove that if K is a field of prime characteristic p, then K is algebraic over K^p.

9. Let K be a field, and let L be an extension field of K. Show that L is algebraic over K if and only if every intermediate domain between K and L is a field.

10. Let K be a field, and let L be an extension field of K. Suppose that for every intermediate field E between K and L, every K-endomorphism of E is a K-automorphism. Show that L is algebraic over K.

11. Prove that every algebraic extension of a perfect field is a perfect field.

12. Let K be a field, let L be an extension field of K, and suppose that there exists a positive integer n such that $[K(D):K] \le n$ for every finite subset D of L. Show that L is finite over K.

13. Let K be a field, let L be a finite extension of K, and let f be an irreducible polynomial in $K[X]$. Prove that if $[L:K]$ and $\deg(f)$ are relatively prime, then f is irreducible in $L[X]$.

14. Let K be a field of prime characteristic p, and let $\alpha \in K$. Prove that if $X^p - X - \alpha$ is irreducible in $K[X]$, and if β is a zero of $X^p - X - \alpha$ in an extension field of K, then $X^p - X - \alpha\beta^{p-1}$ is irreducible in $K(\beta)[X]$.

2.2. ALGEBRAICALLY CLOSED FIELDS

The fields that we shall study in this section are those in which all polynomial equations are solvable. Because of their remarkable properties, these fields will play an important role throughout our presentation.

A field K is called **algebraically closed** when every nonconstant polynomial in $K[X]$ admits a zero in K.

The rational and real fields are examples of fields that are not algebraically closed. On the other hand, as we shall now show, the complex field is algebraically closed.

2.2.1. Example. Until the beginning of the present century, when the axiomatic foundations of the theory of fields were developed, the complex field played a central role in algebra, and this was largely because it is algebraically closed. The result stating this fact was called the "fundamental theorem of algebra"; and although the complex field no longer occupies the same central position in algebra, this terminology has been retained.

Here we shall show that the "fundamental theorem of algebra" is an easy consequence of Liouville's theorem in complex analysis stating that a bounded entire function is constant. Other proofs, more "algebraic" in character, will be discussed later.

Let f be a nonconstant polynomial in $\mathbb{C}[X]$. Put $n = \deg(f)$, and write $f = \sum_{i=0}^{n} \gamma_i X^i$ with $\gamma_0, \gamma_1, \ldots, \gamma_n \in \mathbb{C}$ and $\gamma_n \neq 0$.

For $\alpha \in \mathbb{C}^*$, we have

$$f(\alpha) = \sum_{i=0}^{n} \gamma_i \alpha^i = \alpha^n \sum_{i=0}^{n} \gamma_i \alpha^{i-n}.$$

Since

$$\lim_{\alpha \to \infty} \sum_{i=0}^{n} \gamma_i \alpha^{i-n} = \gamma_n \neq 0,$$

it is clear that

$$\lim_{\alpha \to \infty} |f(\alpha)| = \lim_{\alpha \to \infty} \left| \alpha^n \sum_{i=0}^{n} \gamma_i \alpha^{i-n} \right| = +\infty.$$

It then follows that there exists a positive real number R such that $|f(\alpha)| \geq 1$ whenever $\alpha \in \mathbb{C}$ and $|\alpha| > R$. Let D denote the closed disk with center 0 and radius R, and put $M = \inf_{\alpha \in D} |f(\alpha)|$.

Assume now that f admits no zeros in $\mathbb{C}$. We then have $|f(\alpha)| > 0$ for every $\alpha \in \mathbb{C}$; since D is a compact subset of $\mathbb{C}$, this implies that $M > 0$. To conclude, let $g(\alpha) = 1/f(\alpha)$ for $\alpha \in \mathbb{C}$. Then g is an entire function. Moreover, since $|g(\alpha)| = 1/|f(\alpha)| \leq 1$ for every $\alpha \in \mathbb{C} - D$ and $|g(\alpha)| = 1/|f(\alpha)| \leq 1/M$ for every $\alpha \in D$, it is seen that g is bounded by $\sup(1, 1/M)$. By Liouville's theorem, it then follows that g is a constant function, which leads to the absurd conclusion that f is a constant polynomial. □

As the next proposition shows, algebraically closed fields can be said to be the fields having the simplest factorization theory for polynomials.

2.2.2. Proposition. *If K is a field, then the following conditions are equivalent:*

(a) *K is algebraically closed.*

(b) *Every irreducible polynomial in $K[X]$ has degree 1.*

(c) *Every nonconstant polynomial in $K[X]$ splits in $K[X]$.*

Proof. It is evident that (b) implies (c) and that (c) implies (a). To verify that (a) implies (b), we need only recall that if an irreducible polynomial in $K[X]$ admits a zero in K, then it has degree 1. □

We now show that algebraically closed fields can also be described as the fields having no proper algebraic extensions.

2.2.3.   Proposition.   *If a field is algebraically closed, then it is its only algebraic extension. Conversely, if a field is its only finite extension, then it is algebraically closed.*

Proof.   First, suppose that K is an algebraically closed field, and let L be an algebraic extension of K. To show that $K = L$, let $\alpha \in L$. Then the minimal polynomial of α over K splits in $K[X]$, and hence it admits the same zeros in K as in L. Since α is a zero in L of this polynomial, it follows that $\alpha \in K$.

Next consider a field K that is its only finite extension, and let f be a nonconstant polynomial in $K[X]$. To show that f admits a zero in K, choose an extension field L of K in which f has a zero α. Then α is algebraic over K, so that $K(\alpha)$ is a finite extension of K. By our assumption, we now have $K(\alpha) = K$, which means that $\alpha \in K$.   □

The preceding proposition shows, in particular, that a field is algebraically closed if and only if it is algebraically closed in each of its extension fields.

Generally speaking, we cannot deduce that a field is algebraically closed from its being algebraically closed in one of its extension fields. It will now be seen, however, that there is a particular case in which this deduction is valid.

2.2.4.   Proposition.   *Let K be a field. If there exists an algebraically closed extension field of K in which K is algebraically closed, then K is algebraically closed.*

Proof.   Indeed, let f be a nonconstant polynomial in $K[X]$, and choose an extension field L of K having the properties stated in the proposition. Since L is algebraically closed, there exists a zero α of f in L. Then α is algebraic over K; and since K is algebraically closed in L, it now follows that $\alpha \in K$. This shows that f admits a zero in K.   □

One of the many interesting properties of algebraically closed fields is that they are "large" with respect to embeddability of algebraic extensions. The precise meaning of this assertion is contained in the following important theorem.

2.2.5.   Theorem (Steinitz).   *Let K be a field, let L be an algebraic extension of K, and let M be an algebraically closed field. Then every monomorphism from K to M is extendible to a monomorphism from L to M.*

Proof.   This will be a typical application of Zorn's lemma. Let u be a monomorphism from K to M.

Let us denote by Ω the set of all pairs (A, v) consisting of an intermediate field A between K and L and a monomorphism v from A to M extending u. Since $(K, u) \in \Omega$, we see that Ω is nonempty.

We now order Ω as follows: If $(A, v), (B, w) \in \Omega$, then $(A, v) \le (B, w)$ when $A \subseteq B$ and w extends v. It is readily checked that the ordered set Ω is inductive. Thus, by Zorn's lemma, there exists a maximal element (T, c) in Ω.

The proof will be complete if we show that $T = L$, for then c will be a monomorphism from L to M extending u. Since $T \subseteq L$, we need to verify only that $L \subseteq T$. Thus, let $\alpha \in L$. Since α is algebraic over K, it is algebraic over T. Let f denote the minimal polynomial of α over T. Clearly, cf is a nonconstant polynomial in $M[X]$; since M is algebraically closed, there exists a zero β of cf in M. It now follows from 2.1.8 that c is extendible to a monomorphism d from $T(\alpha)$ to M such that $d(\alpha) = \beta$. We then have $(T(\alpha), d) \in \Omega$ and $(T, c) \le (T(\alpha), d)$. This, in view of the maximality of (T, c) in Ω, implies the equality $(T, c) = (T(\alpha), d)$. We conclude, in particular, that $T = T(\alpha)$, and so $\alpha \in T$. $\square$

2.2.6. Corollary. *Let K be a field, let L be an algebraic extension of K, and let M be an algebraically closed extension field of K. Then L is K-embeddable in M.*

It is not difficult to show that the "largeness" expressed by the last theorem and its corollary is characteristic of algebraically closed fields (problem 1).

It should also be noted that the uniqueness of the extending monomorphism cannot be asserted. For example, the identity mapping i_C and complex conjugation are distinct **R**-automorphisms of **C**, and **C** is an algebraically closed algebraic extension of **R**.

To conclude this preliminary discussion on algebraically closed fields, we now state and prove the following elementary result.

2.2.7. Proposition. *Every algebraically closed field is infinite and perfect.*

Proof. If K is a finite field, then it is meaningful to speak of the polynomial $1 + \prod_{\alpha \in K}(X - \alpha)$ in $K[X]$; and it is clear that this polynomial is nonconstant and admits no zeros in K. Consequently, K is not algebraically closed.

Next, consider an algebraically closed field K of prime characteristic p. If $\alpha \in K$, the nonconstant polynomial $X^p - \alpha$ in $K[X]$ admits a zero in K; but this means that α admits a pth root in K, and proves that $\alpha \in K^p$. We conclude that K is perfect. $\square$

We have already discussed the algebraic closure of a field in an extension field; this is a "relative" notion, in the sense that it depends on the bottom and top fields of the field extension under consideration. We shall now define and study the fundamental concept of an algebraic closure of a field; this will be seen to be an "absolute" notion, depending solely on the given field.

Let K be a field. By an **algebraic closure of** K we shall understand an algebraically closed algebraic extension of K.

2.2.8. Examples

a. **C** is an algebraic closure of **R**.

Indeed, we have already seen that **C** is an algebraically closed finite extension of **R**.

b. **A** is an algebraic closure of **Q**.

Since $\mathbf{A} = A(\mathbf{C}/\mathbf{Q})$ and **C** is algebraically closed, this will be an immediate consequence of the next proposition.   □

2.2.9. Proposition. *Let K be a field, and let L be an algebraically closed extension field of K. Then $A(L/K)$ is the only subfield of L that is an algebraic closure of K.*

Proof. By 2.1.15 and 2.2.4, it is seen that $A(L/K)$ is an algebraic closure of K.

Now let F be a subfield of L that is an algebraic closure of K. To prove that $F = A(L/K)$, we need only apply 2.1.15 and 2.2.3: since F is algebraic over K, we first obtain the inclusion $F \subseteq A(L/K)$, which implies that $A(L/K)$ is algebraic over F; and since F is algebraically closed, we then conclude that $F = A(L/K)$.   □

2.2.10. Proposition. *Every algebraic closure of a denumerable field has cardinality $\aleph_0$.*

Proof. Let L be an algebraic closure of a denumerable field K. By 2.2.7 and 2.1.19 we then have

$$\operatorname{Card}(L) \geq \aleph_0 \quad \text{and} \quad \operatorname{Card}(L) \leq \aleph_0 \operatorname{Card}(K) = \aleph_0,$$

which implies that $\operatorname{Card}(L) = \aleph_0$.   □

2.2.11. Corollary. *If K is a finite field, then every algebraic closure of K is an infinite extension of K.*

Proof. Indeed, let L be an algebraic closure of K. If L were a finite extension of K, we would have

$$\operatorname{Card}(L) = \operatorname{Card}(K)^{[L:K]} < \aleph_0;$$

and this, by the proposition, is impossible.   □

2.2.12. Proposition. *Let K be a field, and let L be an algebraic extension of K. If every monic irreducible polynomial in $K[X]$ splits in $L[X]$, then L is an algebraic closure of K.*

Proof. We have to show that L is algebraically closed. By 2.2.3, it suffices to verify that L is its only algebraic extension.

Thus, suppose that M is an algebraic extension of L. If $\alpha \in M$, then α is algebraic over K, and the hypothesis implies that its minimal polynomial over K splits in $L[X]$; therefore, this polynomial has the same zeros in L as in M, and so $\alpha \in L$. It now follows that $M \subseteq L$, whence $L = M$. □

In the foregoing discussion, we have dealt exclusively with elementary properties of algebraic closures. We have not yet addressed ourselves to the fundamental questions concerning the existence and essential uniqueness of algebraic closures. These will now be settled by the following two striking results.

2.2.13.   Theorem (Steinitz). *Let K and $\bar{K}$ be fields, and let L and $\bar{L}$ be, respectively, algebraic closures of K and $\bar{K}$. Then every isomorphism from K to $\bar{K}$ is extendible to an isomorphism from L to $\bar{L}$.*

Proof. This can be readily derived from the results proved previously.

Indeed, let u be an isomorphism from K to $\bar{K}$. According to 2.2.5, there exists a monomorphism v from L to $\bar{L}$ extending u. We have to show that v is surjective.

To do this, note first that $v(L)$ is algebraically closed, because it is isomorphic to L. Also, since v extends u, it is clear that $v(L)$ is an intermediate field between $\bar{K}$ and $\bar{L}$, and therefore $\bar{L}$ is algebraic over $v(L)$. It then follows from 2.2.3 that $v(L) = \bar{L}$, which is what was needed. □

2.2.14.   Theorem (Steinitz). *Every field admits an algebraic closure. Moreover, every two algebraic closures of a field K are K-isomorphic.*

Proof. Since the second assertion is a consequence of the preceding theorem, we need only concern ourselves with the first one. Furthermore, we see from 2.2.9 that in order to prove the latter, it suffices to show that every field admits an algebraically closed extension field.

The argument that follows is due to Artin. It is based on two lemmas.

Lemma 1. *If F is a field, then there exists an extension field of F in which every nonconstant polynomial in $F[X]$ admits a zero.*

Proof. This is one of the few places in our presentation where polynomials in infinitely many variables play an essential role.

Let I denote the set of all nonconstant polynomials in $F[X]$, and let $A = F[Y_f]_{f \in I}$. Thus, we are considering the domain of polynomials with coefficients in F and variables indexed by the nonconstant polynomials in $F[X]$, the symbol Y_f denoting the variable associated with each such index f.

Let $\mathscr{J}$ be the ideal of A defined by

$$\mathscr{J} = \sum_{f \in I} f(Y_f) A.$$

We claim that $\mathscr{J} \subset A$. For otherwise $\mathscr{J} = A$, and hence there exist a positive integer n, elements $g_1, g_2, \ldots, g_n$ of A, and n distinct elements $f_1, f_2, \ldots, f_n$ of I such that $\sum_{i=1}^{n} f_i(Y_{f_i}) g_i = 1$. This polynomial equality in A, however, is readily seen to yield a contradiction by evaluating $Y_{f_1}, Y_{f_2}, \ldots, Y_{f_n}$ at suitably chosen zeros of $f_1, f_2, \ldots, f_n$. More precisely, use 1.6.5 to obtain an extension field T of F in which there are elements $\rho_1, \rho_2, \ldots, \rho_n$ so that

$$f_1(\rho_1) = f_2(\rho_2) = \cdots = f_n(\rho_n) = 0;$$

if for $1 \leq i \leq n$ we now substitute ρ_i for Y_{f_i} in the above polynomial equality, we get the equality $0 = 1$ in T, which is impossible.

Thus, $\mathscr{J}$ is a proper ideal of A. According to 0.0.4, there exists a maximal ideal $\mathscr{M}$ of A such that $\mathscr{J} \subseteq \mathscr{M}$. The remainder of the argument consists in imitating the final part of the proof in the lemma in 1.6.5, and will be merely sketched.

First, note that $A/\mathscr{M}$ is a field and that a monomorphism u from F to $A/\mathscr{M}$ is defined by composing the inclusion mapping from F to A with the natural projection from A to $A/\mathscr{M}$. Next, use 0.0.1 to extend u to an isomorphism v from an extension field E of F to $A/\mathscr{M}$. Finally, for each $f \in I$, let $\alpha_f = v^{-1}(Y_f + \mathscr{M})$; since $f(Y_f) \in \mathscr{J}$ and $\mathscr{J} \subseteq \mathscr{M}$, the equality $f(Y_f) + \mathscr{M} = 0$ is valid in $A/\mathscr{M}$; this implies that

$$v\big(f(\alpha_f)\big) = f(Y_f) + \mathscr{M} = 0,$$

and hence $f(\alpha_f) = 0$, which shows that α_f is a zero of f in E. This completes the proof of the lemma.

Lemma 2. *Let $(F_i)_{i \geq 0}$ be a sequence of fields satisfying the following condition for every $i \geq 0$: F_{i+1} is an extension field of F_i in which every nonconstant polynomial in $F_i[X]$ admits a zero. If $F = \cup_{i=0}^{\infty} F_i$, then F can be provided with a field structure in such a way that it is an extension field of F_i for every $i \geq 0$; furthermore, F is algebraically closed.*

Proof. The first assertion is an immediate consequence of 0.0.5. To show that F is algebraically closed, consider a nonconstant polynomial f in $F[X]$.

Let $n = \deg(f)$, and write $f = \sum_{i=0}^{n} \gamma_i X^i$ with $\gamma_0, \gamma_1, \ldots, \gamma_n \in F$. For $0 \leq i \leq n$, choose a nonnegative integer r_i so that $\gamma_i \in F_{r_i}$. If we put $r = \sup(r_0, r_1, \ldots, r_n)$, then we have $\gamma_0, \gamma_1, \ldots, \gamma_n \in F_r$, which implies that $f \in F_r[X]$. Consequently, f possesses a zero in F_{r+1}, and hence in F. The lemma is proved.

Now we can easily complete the proof. Given a field K, define inductively a sequence $(L_i)_{i \geq 0}$ of fields as follows: Put $L_0 = K$; and if $i \geq 0$

and a field L_i has been defined, use the first lemma to obtain an extension field L_{i+1} of L_i in which every nonconstant polynomial in $L_i[X]$ admits a zero. By the second lemma, such a sequence defines an algebraically closed field L that is an extension field of L_i for every $i \geq 0$. Since $L_0 = K$, we see, in particular, that L is an extension field of K. $\square$

We have chosen to present Artin's proof of Steinitz's theorem because of its simplicity. The first lemma is a natural extension of the lemma in 1.6.5. In the proofs of these two lemmas, it is essential to obtain a suitable maximal ideal in a polynomial domain. This is accomplished quite simply if a single polynomial is under consideration, but a more subtle trick is required in order to deal simultaneously with all nonconstant polynomials.

The proof given originally by Steinitz is by transfinite induction; it can be found, for instance, in the excellent textbook of van der Waerden. There are other well-known proofs, two of which are considered in the problems.

An essential feature of Artin's proof is the use of polynomials in infinitely many variables, which provides more elbowroom to obtain the required maximal ideal. This is unnecessary in the two proofs discussed in the problems. In saying this, it is not our intention to indicate a preference for the latter: In fact, one of them (problem 2) requires some cardinal arithmetic, which many find disagreeable; and the other (problem 3) may appear to be of a somewhat artificial nature.

We would like to mention that in reality, the second lemma in Artin's proof is superfluous. Indeed, the extension field E of K constructed in the proof of the first lemma is an algebraic closure of the given field K! For it is clear that E is algebraic over K, and that every nonconstant polynomial in $K[X]$ possesses a zero in E; and this, as will be shown in 3.6.4, implies that E is an algebraic closure of K.

To conclude, we note that the validity of theorem 1.6.5 for infinite sets of nonconstant polynomials is an immediate consequence of Steinitz's theorem 2.2.14.

PROBLEMS

1. Let K be a field, and let L be an algebraic extension of K. Show that if every finite extension of K is K-embeddable in L, then L is an algebraic closure of K.

2. Prove that every field admits an algebraic closure by verifying the assertions in the following sketch.

    Let K be a field. Choose a set D containing K and such that $\operatorname{Card}(D) > \aleph_0 \operatorname{Card}(K)$. Let Ω denote the set of all algebraic extensions of K having subsets of D as underlying sets. Order Ω as follows: If $E, F \in \Omega$, then $E \leq F$ when E is a subfield of F. Then Ω is a nonempty inductive ordered set, and every maximal element of Ω is an algebraic closure of K.

3. Prove that every field admits an algebraic closure by verifying the assertions in the following sketch.

Let K be field. Denote by $\overline{K}$ the image of the injective mapping $\alpha \rightarrow (X - \alpha, 0)$ from K to $K[X] \times \mathbf{N}$. This mapping defines, by restriction, a bijection from K to $\overline{K}$. As in 0.0.1, provide $\overline{K}$ with the field structure with respect to which this bijection is an isomorphism, and indicate by $f \rightarrow \bar{f}$ the isomorphism from $K[X]$ to $\overline{K}[Y]$ defined by the latter. Now denote by Ω the set of all extension fields L of $\overline{K}$ having subsets of $K[X] \times \mathbf{N}$ as underlying sets and having the following property: If $(f, n) \in L$, then (f, n) is a zero of $\bar{f}$. Next, order Ω as follows: If $E, F \in \Omega$, then $E \leq F$ when E is a subfield of F. Then Ω is a nonempty inductive ordered set, and every maximal element of Ω is an algebraic closure of $\overline{K}$.

4. Show that every irreducible polynomial in $\mathbf{R}[X]$ has degree 1 or 2.

5. Prove that the identity mapping on $\mathbf{C}$ and complex conjugation are the only $\mathbf{R}$-endomorphisms of $\mathbf{C}$.

2.3. NORMAL EXTENSIONS

The class of algebraic extensions that we wish to analyze first is that defined by a condition ensuring good behavior in regard to the splitting of polynomials and to the extendibility of monomorphisms.

Let K be a field, and let L be an extension field of K. We say that L is **normal over** K, or that L is a **normal extension of** K, when L is algebraic over K and the minimal polynomial over K of every element of L splits in $L[X]$. Equivalently, we could say that L is normal over K when L is algebraic over K and every irreducible polynomial in $K[X]$ admitting a zero in L splits in $L[X]$.

It is seen from these definitions that if K is a field, then K and every algebraic closure of K are normal extensions of K.

2.3.1. Examples

a. Let K be a field, and let L be an extension field of K such that $[K(\alpha): K] \leq 2$ for every $\alpha \in L$. Then L is normal over K.

This follows from the fact that if a polynomial in $K[X]$ of degree 1 or 2 possesses a zero in L, then it splits in $L[X]$.

b. Let m and n be positive integers such that m is not the nth power of an integer. Then $\mathbf{Q}(\sqrt[n]{m})$ is normal over $\mathbf{Q}$ if and only if n is even and m is the $(n/2)$th power of an integer.

First suppose that n is even and that m is the $(n/2)$th power of an integer. If $k = n/2$, then $\sqrt[k]{m}$ is a positive integer, and hence $(\sqrt[n]{m})^2 = \sqrt[k]{m} \in \mathbf{Q}$. Since $\sqrt[n]{m} \notin \mathbf{Q}$, this shows that $[\mathbf{Q}(\sqrt[n]{m}):\mathbf{Q}] = 2$, and the normality of $\mathbf{Q}(\sqrt[n]{m})$ over $\mathbf{Q}$ follows from example a.

Next suppose that $\mathbf{Q}(\sqrt[n]{m})$ is normal over $\mathbf{Q}$, and let f denote the minimal polynomial of $\sqrt[n]{m}$ over $\mathbf{Q}$. Then $f(X)$ divides $X^n - m$ in $\mathbf{Q}[X]$, and so every real zero of f is a zero of $X^n - m$. Also, note that when n is odd, the only real zero of $X^n - m$ is $\sqrt[n]{m}$; and that when n is even, the only real zeros of $X^n - m$ are $\sqrt[n]{m}$ and $-\sqrt[n]{m}$. Since f splits in $\mathbf{Q}(\sqrt[n]{m})[X]$ and $\mathbf{Q}(\sqrt[n]{m}) \subset \mathbf{R}$, it then follows that $f(X) = X - \sqrt[n]{m}$ when n is odd, and that

$$f(X) = \left(X - \sqrt[n]{m}\right)\left(X + \sqrt[n]{m}\right) = X^2 - \sqrt[n]{m^2}$$

when n is even. The former implies that $\sqrt[n]{m} \in \mathbf{Q}$, and the latter that $\sqrt[n]{m^2} \in \mathbf{Q}$. Since $\sqrt[n]{m} \notin \mathbf{Q}$, we now conclude that n is even and $\sqrt[n]{m^2} \in \mathbf{Q}$, which implies that m is the $(n/2)$th power of an integer.

c. In order to give an example, in the case of prime characteristic, of a finite extension that is not normal, consider a finite field K and a positive integer n. First form a simple transcendental extension $K(\xi)$ of K, and then choose an nth root σ of ξ in an extension field of $K(\xi)$. The relation $\sigma^n = \xi$ implies that $K(\sigma)$ is a finite extension of $K(\xi)$. We shall now show that if $\mathrm{Char}(K) \nmid n$ and $n > \mathrm{Card}(K)$, then $K(\sigma)$ is not normal over $K(\xi)$.

According to 2.1.2, we can speak of the K-isomorphism from $K[X]$ to $K[\xi]$ such that $X \to \xi$. Its existence implies that $K[\xi]$ is a factorial domain and that ξ is an irreducible element of $K[\xi]$. In this domain, ξ does not divide 1 and divides $-\xi$, and ξ^2 does not divide $-\xi$. Since $K(\xi)$ is a field of fractions of $K[\xi]$, it then follows from Eisenstein's irreducibility criterion that $X^n - \xi$ is irreducible in $K(\xi)[X]$, and hence it is the minimal polynomial of σ over $K(\xi)$. In order to prove that $K(\sigma)$ is not normal over $K(\xi)$, it will suffice to verify that $X^n - \xi$ does not split in $K(\sigma)[X]$.

Assume, on the contrary, that $X^n - \xi$ splits in $K(\sigma)[X]$. We then have $X^n - \xi = \prod_{i=1}^{n}(X - \tau_i)$, where $\tau_1, \tau_2, \ldots, \tau_n \in K(\sigma)$. Since $\mathrm{Char}(K) \nmid n$, we know from 1.7.4 that $X^n - \xi$ is a separable polynomial, which by 1.7.2 implies that every zero of $X^n - \xi$ in $K(\sigma)$ is simple. We conclude that $\tau_1, \tau_2, \ldots, \tau_n$ are n distinct elements of $K(\sigma)$; and since $\tau_i^n = \xi = \sigma^n$ for $1 \le i \le n$, it then follows that $\tau_1/\sigma, \tau_2/\sigma, \ldots, \tau_n/\sigma$ are n distinct nth roots of unity in $K(\sigma)$.

It is seen, on the other hand, that σ is transcendental over K, for otherwise the relation $\xi = \sigma^n$ would imply that ξ is algebraic over K. By 1.5.11 and 2.1.2, it follows that every root of unity in $K(\sigma)$ belongs to K. In particular, the n distinct elements $\tau_1/\sigma, \tau_2/\sigma, \ldots, \tau_n/\sigma$ of $K(\sigma)$ belong to K. This, however, is incompatible with the given inequality $n > \mathrm{Card}(K)$. We conclude that $X^n - \xi$ does not split in $K(\sigma)[X]$, which is what we wanted. $\square$

Consider the four-term chain $\mathbf{Q} \subset \mathbf{Q}(\sqrt{2}) \subset \mathbf{Q}(\sqrt[4]{2}) \subset \mathbf{A}$. It is clear from the foregoing discussion that $\mathbf{Q}(\sqrt{2})$ is normal over $\mathbf{Q}$, that $\mathbf{Q}(\sqrt[4]{2})$ is normal over $\mathbf{Q}(\sqrt{2})$ but not over $\mathbf{Q}$, and that $\mathbf{A}$ is normal over each of the other three fields.

This simple example shows that the following easy result represents the best approximation to an analog for normal extension of the transitivity property stated in 1.4.10 and 2.1.14 for finite and algebraic extensions, respectively.

2.3.2. Proposition. *Let K be a field, let L be an extension field of K, and let M be an extension field of L. If M is normal over K, then M is normal over L.*

Proof. It is clear that M is algebraic over L. Moreover, for each $\alpha \in M$, the minimal polynomial of α over L is a factor in $L[X]$ of the minimal polynomial of α over K; and since the latter splits in $M[X]$, so does the former. $\square$

An important property of normal extensions is that they are "stable" relative to embeddings. This is stated in precise terms in the next proposition and its corollary.

2.2.3. Proposition. *Let K be a field, and let L be a normal extension of K. If E is an intermediate field between K and L, if M is an extension field of L, and if u is a K-monomorphism from E to M, then $u(E) \subseteq L$.*

Proof. If $\alpha \in E$, the minimal polynomial of α over K splits in $L[X]$, and hence admits the same zeros in L as in M. Since $u(\alpha)$ is a zero in M of this polynomial, it then follows that $u(\alpha) \in L$. $\square$

2.3.4. Corollary. *Let K be a field, and let L be a normal extension of K. If M and N are extension fields of L, and if u is a K-monomorphism from M to N, then $u(L) = L$.*

Proof. By the proposition, we have $u(L) \subseteq L$. The desired conclusion then follows at once from 2.1.22. $\square$

The following proposition shows that normality has strong implications concerning the extendibility of monomorphisms.

2.3.5. Proposition. *Let K be a field, and let L be a normal extension of K. If E is an intermediate field between K and L, then every K-monomorphism from E to L is extendible to a K-automorphism of L.*

Proof. Let u be a K-monomorphism from E to L. Choose an algebraic closure M of L; then M is also an algebraic closure of E and $u(E)$, and Steinitz's theorem 2.2.13 implies that u is extendible to a K-automorphism s of M. It now follows from 2.3.4 that $s(L) = L$, whence it is meaningful to speak of the K-automorphism s_L of L; and it is seen that s_L extends u. $\square$

2.3.6. Corollary. *Let K be a field, and let L be a normal extension of K. If $\alpha \in L$, then the zeros in L of the minimal polynomial of α over K are the values at α of the K-automorphisms of L.*

Proof. We already know that the image of α by every K-automorphism of L is a zero of the minimal polynomial of α over K.

Now let β be a zero in L of this polynomial. By 2.1.9, there exists a K-isomorphism u from $K(\alpha)$ to $K(\beta)$ such that $u(\alpha) = \beta$; and by the proposition, u is extendible to a K-automorphism s of L. Therefore $\beta = u(\alpha) = s(\alpha)$. $\square$

The validity of the preceding two propositions and their corollaries depends in an essential way on the assumed normality (problem 4).

The next result can be considered as a partial converse of 2.3.4. It states that the normality of a field extension is a consequence of its "stability" relative to suitably restricted automorphisms.

2.3.7. Proposition. *Let K be a field, and let L be an extension field of K. If there exists an extension field M of L that is normal over K and such that $s(L) \subseteq L$ for every K-automorphism s of M, then L is normal over K.*

Proof. It is clear that L is algebraic over K. We have to show that the minimal polynomial over K of each $\alpha \in L$ splits in $L[X]$. Since this polynomial splits in $M[X]$, we need only verify that each of its zeros in M belongs to L.

Thus, let β be such a zero. By the preceding proposition, there exists a K-automorphism s of M such that $\beta = s(\alpha)$. Therefore $\beta \in s(L)$; and since $s(L) \subseteq L$, it follows that $\beta \in L$. $\square$

We now derive three corollaries of the proposition just proved. The first and second state general "closure properties" of the class of normal extensions. The third states that in order to establish the normality of an algebraic extension, it suffices to verify the condition in the definition for every element of a generating set.

2.3.8. Corollary. *Let K be a field, and let N be an extension field of K. If L and M are intermediate fields between K and N such that L is normal over K, then $L \vee M$ is normal over M.*

Proof. Let F be an algebraic closure of $L \vee M$. We have already seen that $L \vee M$ is algebraic over M. Consequently, F is an algebraic closure of M, and hence is normal over M.

If s is an M-automorphism of F, then $s(M) = M$; and since s is also a K-automorphism of F, we know from 2.3.4 that $s(L) = L$. Therefore

$$s(L \vee M) = s(L) \vee s(M) = L \vee M$$

for every M-automorphism s of F. The normality of $L \vee M$ over M now follows from the proposition. $\square$

2.3.9. Corollary. *Let K be a field, and let M be an extension field of K. If $(L_i)_{i \in i}$ is a nonempty family of intermediate fields between K and M that are normal over K, then $\cap_{i \in I} L_i$ and $\vee_{i \in I} L_i$ are normal over K.*

Proof. We already know that $\cap_{i \in I} L_i$ and $\vee_{i \in I} L_i$ are algebraic over K and $K \subseteq \cap_{i \in I} L_i \subseteq \vee_{i \in I} L_i$. Let F be an algebraic closure of $\vee_{i \in I} L_i$. Then F is also an algebraic closure of K, and therefore is normal over K.

It follows from 2.3.4 that if $i \in I$, then $s(L_i) = L_i$ for every K-automorphism s of F. Consequently

$$s\left(\bigcap_{i \in I} L_i \right) = \bigcap_{i \in I} s(L_i) = \bigcap_{i \in I} L_i$$

and

$$s\left(\bigvee_{i \in I} L_i \right) = \bigvee_{i \in I} s(L_i) = \bigvee_{i \in I} L_i$$

for every K-automorphism s of F. By the proposition, this shows that $\cap_{i \in I} L_i$ and $\vee_{i \in I} L_i$ are normal over K. $\square$

2.3.10. Corollary. *Let K be a field, and let L be an algebraic extension of K. If there exists a subset D of L such that $L = K(D)$ and such that the minimal polynomial over K of every element of D splits in $L[X]$, then L is normal over K.*

Proof. We proceed as in the proofs of the preceding corollaries. Choose an algebraic closure M of L. Then M is an algebraic closure of L, and hence is normal over K.

If $\alpha \in D$ and s is a K-automorphism of M, then the minimal polynomial of α over K splits in $L[X]$ and admits $s(\alpha)$ as a zero in M, whence $s(\alpha) \in L$. We conclude that $s(D) \subseteq L$ for every K-automorphism s of M.

It now follows that

$$s(L) = s(K(D)) = K(s(D)) \subseteq K(L) = L$$

for every K-automorphism s of M; and this, by the proposition, proves that L is normal over K. $\square$

Let K be a field, and let L be an extension field of K. By a **normal closure of L over K** we understand an extension field M of L having the property that M is the only intermediate field between L and M that is normal over K.

This definition clearly implies that if K is a field, and if L is a normal extension of K, then L is the only normal closure of L over K.

We shall now derive the basic properties of normal closures. These will play an important technical role in later considerations.

2.3.11. Proposition. *Let K be a field, let L be an extension field of K, and let M be an extension field of L that is normal over K. Then there exists a unique subfield of M that is a normal closure of L over K; and this subfield of M is the smallest element of the set of all intermediate fields between L and M that are normal over K.*

Proof. Let Ω denote the set of all intermediate fields between L and M that are normal over K. Since $M \in \Omega$, we see that Ω is nonempty.

If we let $E = \cap_{F \in \Omega} F$, then 2.3.9 shows that E is normal over K, and hence $E \in \Omega$. Consequently, E is the smallest element of Ω; as such, it is seen to be the only subfield of M that is a normal closure of L over K. $\qquad \square$

If K is a field, if L is an extension field of K, and if M is an extension field of L that is normal over K, the subfield of M that is a normal closure of L over K will be said to be **the normal closure of L over K in M.**

2.3.12. Proposition. *Let K be a field, and let L be a normal extension of K. If $D \subseteq L$, and if T denotes the set consisting of the elements of L that are zeros of the minimal polynomials over K of the elements of D, then $K(T)$ is the normal closure of $K(D)$ over K in L.*

Proof. The normality of $K(T)$ over K follows directly from 2.3.10: Indeed, the minimal polynomial over K of every element of T coincides with that of some element of D, and hence splits in $K(T)[X]$.

Now let F be an intermediate field between $K(D)$ and L that is normal over K. Then the minimal polynomial over K of every element of D splits in $F[X]$. Therefore $T \subseteq F$, and so $K(T) \subseteq F$.

Since $K(D) \subseteq K(T)$, it now follows that $K(T)$ is the smallest element of the set of all intermediate fields between $K(D)$ and L that are normal over K. And this, by the preceding proposition, implies the required conclusion. $\qquad \square$

2.3.13. Corollary. *Let K be a field, and let L be a finite extension of K. Then every normal closure of L over K is finite over K.*

Proof. Let M be a normal closure of L over K. Choose a finite subset D of L so that $L = K(D)$, and denote by T the set consisting of the elements of M that are zeros of the minimal polynomials over K of the elements of D. Then T is finite, and the proposition shows that $M = K(T)$. Therefore M is finite over K. $\qquad \square$

2.3.14. Corollary. *Let K be a field, and let L be a normal extension of K. Then L is the union of a nonempty filtered family of intermediate fields between K and L that are normal and finite over K.*

Proof. Let Ω denote the set of all finite subsets of L. For each $D \in \Omega$, denote by F_D the normal closure of $K(D)$ over K in L; since $K(D)$ is finite over K, the preceding corollary shows that F_D is normal and finite over K.

Since $K(D) \subseteq F_D \subseteq L$ for every $D \in \Omega$, and since $L = \cup_{D \in \Omega} K(D)$, it follows that $L = \cup_{D \in \Omega} F_D$. It remains, therefore, to show that $(F_D)_{D \in \Omega}$ is a filtered family. Clearly, Ω is filtered, and so it suffices to verify that the conditions $C, D \in \Omega$ and $C \subseteq D$ imply $F_C \subseteq F_D$, which presents no difficulty: indeed, these conditions imply that $K(C) \subseteq K(D) \subseteq F_D$, so that F_D is an intermediate field between $K(C)$ and L that is normal over K; and this, by virtue of 2.3.11, implies that $F_C \subseteq F_D$. $\square$

To end our discussion on normal closures, we shall establish their existence and essential uniqueness.

2.3.15. Proposition. *Let K be a field, and let L be an algebraic extension of K. Then L admits a normal closure over K; and every two normal closures of L over K are L-isomorphic.*

Proof. In view of 2.3.11, the first assertion follows from the fact that L admits extension fields that are normal over K—for example, every algebraic closure of L.

To prove the second assertion, suppose that M and N are normal closures of L over K. Let A and B be, respectively, algebraic closures of M and N. Then A and B are algebraic closures of L; according to Steinitz's theorem 2.2.14, there exists an L-isomorphism u from A to B. Thus, $u(M)$ is a subfield of B that is a normal closure of L over K. But this, by 2.3.11, implies that $u(M) = N$. $\square$

We shall now conclude this section by studying the notion of a splitting field of a set of polynomials, which is of basic importance in our subject. It will be seen presently that the concepts of a normal extension and of a splitting field can be regarded as essentially identical.

Let K be a field, and let W be a set of nonconstant polynomials in $K[X]$. By a **splitting field of W over K** we shall understand an extension field L of K such that every polynomial in W splits in $L[X]$ and that is generated over K by the set of all zeros in L of the polynomials in W.

If f is a nonconstant polynomial in $K[X]$, we shall of course speak of a **splitting field of f over K** instead of a splitting field of $\{f\}$ over K.

Let K be a field, and let L be an extension field of K. It is clear that if L is a splitting field over K of a set W of nonconstant polynomials in $K[X]$, then it is also a splitting field of W over every intermediate field between K and L. Also, two immediate consequences of unique factorization in $L[X]$ should be noted. First, if n is a positive integer, and if $f_1, f_2, \ldots, f_n$ are nonconstant polynomials in $K[X]$, then L is a splitting field of $\{f_1, f_2, \ldots, f_n\}$ over K if and only if it is a splitting field of $\Pi_{i=1}^{n} f_i$ over K. Second, if W is a set of nonconstant polynomials in $K[X]$, and if V is the set of all monic irreducible factors in $K[X]$ of the polynomials in W, then L is a splitting field of W over K if and only if it is a splitting field of V over K.

2.3.16. Examples

a. Since the subfield $Q(\sqrt[3]{2})$ of $\mathbf{C}$ is contained in $\mathbf{R}$, and since $\sqrt[3]{2}$ is the only real zero of $X^3 - 2$, it is clear that $Q(\sqrt[3]{2})$ is not a splitting field of $X^3 - 2$ over $\mathbf{Q}$.

Now let $\omega = (-1 + \sqrt{-3})/2$. Then $\omega \neq 1$ and $\omega^3 = 1$, so that $\sqrt[3]{2}, \sqrt[3]{2}\,\omega$, $\sqrt[3]{2}\,\omega^2$ are the three distinct complex zeros of $X^3 - 2$. Consequently, the subfield $Q(\sqrt[3]{2}, \omega)$ of $\mathbf{C}$ is a splitting field of $X^3 - 2$ over $\mathbf{Q}$.

b. The complex zeros of the polynomial $(X^2 + 1)(X^2 - 2X + 2)$ in $Q[X]$ are easily found to be $\sqrt{-1}, -\sqrt{-1}, 1 + \sqrt{3}, 1 - \sqrt{3}$. Consequently, the subfield $Q(\sqrt{3}, \sqrt{-1})$ of $\mathbf{C}$ is a splitting field of this polynomial over $\mathbf{Q}$.

c. Let α be a zero of the polynomial $X^3 + X^2 + 1$ in $\mathbf{Z}/2[X]$ in an extension field of $\mathbf{Z}/2$. We shall show that $\mathbf{Z}/2(\alpha)$ is a splitting field of this polynomial over $\mathbf{Z}/2$.

To do this, note that since $\mathrm{Char}(\mathbf{Z}/2) = 2$, the relation $\alpha^3 + \alpha^2 + 1 = 0$ implies that

$$\alpha^6 + \alpha^4 + 1 = (\alpha^3 + \alpha^2 + 1)^2 = 0$$

and

$$\alpha^{12} + \alpha^8 + 1 = (\alpha^3 + \alpha^2 + 1)^4 = 0,$$

so that α^2 and α^4 are also zeros of $X^3 + X^2 + 1$ in $\mathbf{Z}/2(\alpha)$. The desired conclusion will be proved by verifying that $\alpha, \alpha^2, \alpha^4$ are three distinct elements of $\mathbf{Z}/2(\alpha)$.

First, we have $\alpha \neq 0$. If $\alpha = \alpha^2$, then $\alpha = 1$, so that $\alpha^3 + \alpha^2 + 1 = 1$; if $\alpha = \alpha^4$, then $\alpha^3 = 1$, whence $\alpha^3 + \alpha^2 + 1 = \alpha^2$; and if $\alpha^2 = \alpha^4$, then $\alpha^2 = 1$, which implies that $\alpha^3 + \alpha^2 + 1 = \alpha^3$. In each case, the result is incompatible with the equality $\alpha^3 + \alpha^2 + 1 = 0$. □

The relationship between the notion of a normal extension and that of a splitting field is explained in the next two results.

2.3.17. Proposition. *Let K be a field, and let L be an extension field of K. If L is normal over K, then L is a splitting field over K of the set of all minimal polynomials over K of the elements of L. Conversely, if L is a splitting field over K of a set of nonconstant polynomials in K[X], then L is normal over K.*

Proof. If L is normal over K, then L consists of all zeros in L of the minimal polynomials over K of its elements, which implies the first assertion.

Next, suppose that L is a splitting field over K of a set W of nonconstant polynomials in $K[X]$. Let D denote the set of all zeros in L of the polynomials in W, so that $L = K(D)$. Since every element of D is algebraic over K, it is seen that L is algebraic over K. Furthermore, each $\alpha \in D$ is a zero of a polynomial in W; therefore, the minimal polynomial of α over K is a factor in $K[X]$ of a polynomial that splits in $L[X]$, and hence it also splits in $L[X]$. The normality of L over K now follows from 2.3.10.   □

2.3.18. Proposition. *Let K be a field. Then an extension field of K is normal and finite over K if and only if it is a splitting field over K of a nonconstant polynomial in K[X].*

Proof. Let L be an extension field of K.

First, assume that L is normal and finite over K. Let $L = K(\alpha_1, \alpha_2, \ldots, \alpha_n)$, where n is a positive integer and $\alpha_1, \alpha_2, \ldots, \alpha_n \in L$. For $1 \leq i \leq n$, let f_i denote the minimal polynomial of α_i over K. Then $f_1, f_2, \ldots, f_n$ split in $L[X]$, and we conclude that L is a splitting field of $\{f_1, f_2, \ldots, f_n\}$ over K. Therefore L is a splitting field of $\prod_{i=1}^{n} f_i$ over K.

Conversely, assume now that L is a splitting field over K of a nonconstant polynomial in $K[X]$. It is clear that L is finite over K, because L is generated over K by the finite set of all zeros in L of this polynomial; and the preceding proposition shows that L is normal over K.   □

As we now show, an algebraic closure can be described as a splitting field of a special set of polynomials.

2.3.19. Proposition. *Let K be a field. Then an extension field of K is an algebraic closure of K if and only if it is a splitting field over K of the set of all monic irreducible polynomials in K[X].*

Proof. Let L be an extension field of K, and denote by W the set of all monic irreducible polynomials in $K[X]$.

If L is an algebraic closure of K, then L is the set of all zeros in L of the polynomials in W. Since every polynomial in W splits in $L[X]$, it follows that L is a splitting field of W over K.

To conclude, suppose that L is a splitting field of W over K. By 2.3.17, L is normal over K. Thus, L is algebraic over K and every polynomial in W splits in $L[X]$. According to 2.2.12, this implies that L is an algebraic closure of K. □

Our two final results, which will be derived from Steinitz's theorems 2.2.13 and 2.2.14, deal with the important questions of existence and essential uniqueness for splitting fields.

2.3.20. Proposition. *Let K and $\overline{K}$ be fields, and let u be an isomorphism from K to $\overline{K}$. Let W be a set of nonconstant polynomials in $K[X]$, and denote by $\overline{W}$ the image of W by the isomorphism from $K[X]$ to $\overline{K}[X]$ defined by u. If L and $\overline{L}$ are, respectively, splitting fields of W over K and of $\overline{W}$ over $\overline{K}$, then u is extendible to an isomorphism from L to $\overline{L}$.*

Proof. First choose, respectively, algebraic closures M and $\overline{M}$ of L and $\overline{L}$. Then M and $\overline{M}$ are, respectively, algebraic closures of K and $\overline{K}$, and therefore we can choose an isomorphism v from M to $\overline{M}$ extending u. To reach the desired conclusion, it will suffice to prove that $v(L) = \overline{L}$.

To do this, note that a polynomial in W has the same zeros in L as in M, and similarly, that a polynomial in $\overline{W}$ has the same zeros in $\overline{L}$ as in $\overline{M}$. Denoting by D the set of all zeros in L of the polynomials in W, and taking into account that v extends u, we now have that $v(D)$ is the set of all zeros in $\overline{L}$ of the polynomials in $\overline{W}$. Consequently, $L = K(D)$ and $\overline{L} = \overline{K}(v(D))$; therefore

$$v(L) = v(K)(v(D)) = u(K)(v(D)) = \overline{K}(v(D)) = \overline{L},$$

as required, □

2.3.21. Proposition. *Let K be a field, and let W be a set of nonconstant polynomials in $K[X]$. Then W admits a splitting field over K; and every two splitting fields of W over K are K-isomorphic.*

Proof. The second assertion is a particular case of the preceding proposition. To prove the first, choose an algebraic closure L of K, and denote by D the set of all zeros in L of the polynomials in W. Then these polynomials split in $K(D)[X]$, which implies that $K(D)$ is a splitting field of W over K. □

PROBLEMS

1. Let K be a field, and let L be an extension field of K. Show that if there exists a subset D of L such that $L = K(D)$ and such that $[K(\alpha): K] \le 2$ for every $\alpha \in D$, then L is normal over K.
2. Show that if K is a field of prime characteristic p, then K is normal over K^p.

3. Let K be a field of prime characteristic p, let $\alpha \in K$, and let β be a zero of $X^p - X - \alpha$ in an extension field of K. Prove that $K(\beta)$ is normal over K.

4. Use the four-term chain $\mathbf{Q} \subset \mathbf{Q}(\sqrt{2}) \subset \mathbf{Q}(\sqrt[4]{2}) \subset \mathbf{A}$ in order to show that 2.3.3, 2.3.4, 2.3.5, and 2.3.6 do not remain valid if the normality assumptions are omitted.

5. Let K be a field, let L be a normal extension of K, and let f be an irreducible polynomial in $K[X]$. Show that if g and h are monic irreducible factors of f in $L[X]$, then there exists a K-automorphism s of L such that $sg = h$.

6. Let K be a field, and let L be an algebraic extension of K. Prove that L is normal over K if and only if the irreducible factors in $L[X]$ of every irreducible polynomial in $K[X]$ have the same degree.

7. Let K be a field, let L be a cubic extension of K that is not normal over K, and let M be a normal closure of L over K. Show that $[M:K] = 6$.

8. Let K be a field, let L be an extension field of K, and let M be an extension field of L that is normal over K. Show that if Ω denotes the set of all K-monomorphisms from L to M, then $\vee_{u \in \Omega} u(L)$ is the normal closure of L over K in M.

9. Prove that if α is a complex cube root of 2, then the subfield $\mathbf{Q}(\alpha)$ of $\mathbf{C}$ is not a splitting field of $X^3 - 2$ over $\mathbf{Q}$.

10. Give an example of two distinct monic irreducible quadratic polynomials in $\mathbf{Q}[X]$ that admit a common splitting field over $\mathbf{Q}$.

11. Let n be an integer such that $n > 1$. Verify the following assertions:
    a. The zeros of $\sum_{i=0}^{n-1} X^i$ in $\mathbf{C}$ are the nth roots of unity in $\mathbf{C}$ different from 1.
    b. If n is prime and α is a zero of $\sum_{i=0}^{n-1} X^i$ in $\mathbf{C}$, then $\mathbf{Q}(\alpha)$ is a splitting field of $\sum_{i=0}^{n-1} X^i$ over $\mathbf{Q}$.

12. Let K be a field, and let n be a positive integer. Prove that if L is a splitting field over K of a polynomial in $K[X]$ of degree n, then $[L:K] \mid n!$.

2.4. PURELY INSEPARABLE EXTENSIONS

The field extensions that we shall discuss here are normal extensions of a very special type. We shall find it convenient to have their properties at our disposal in our study of separability for algebraic extensions, which will be taken up in the next section.

Let K be a field of prime characteristic p, and let L be an extension field of K. If $\alpha \in L$, we say that α is **purely inseparable over K** when $\alpha^{p^n} \in K$ for some nonnegative integer n.

It is evident that every element of L that is purely inseparable over K is algebraic over K: In effect, we can say that an element of L is purely inseparable over K when it is a zero of a polynomial in $K[X]$ of the form $X^{p^n} - \gamma$, where n is a nonnegative integer.

As we now show, it is easy to describe the minimal polynomial of a purely inseparable element.

2.4.1. Proposition. *Let K be a field of prime characteristic p, and let L be an extension field of K. If $\alpha \in L$ and α is purely inseparable over K, and if e is the smallest element of the set of all nonnegative integers n for which $\alpha^{p^n} \in K$, then $X^{p^e} - \alpha^{p^e}$ is the minimal polynomial of α over K.*

Proof. To prove this, it suffices to verify that $X^{p^e} - \alpha^{p^e}$ is irreducible in $K[X]$. This is clear if $e = 0$; and it follows at once from 1.6.6 if $e > 0$: Indeed, in this case we have $\alpha^{p^e} \notin K^p$, for otherwise $\alpha^{p^e} = \beta^p$ for some $\beta \in K$, which would imply that $\alpha^{p^{e-1}} = \beta \in K$, contrary to the definition of e.   □

Next we give a characterization of purely inseparable elements in terms of the splitting of their minimal polynomials.

2.4.2. Proposition. *Let K be a field of prime characteristic, and let L be an extension field of K. If an element of L is algebraic over K, then it is purely inseparable over K if and only if its minimal polynomial splits in $L[X]$ and admits a unique zero in L.*

Proof. Write $p = \mathrm{Char}(K)$. If $\alpha \in L$, then the equality

$$X^{p^n} - \alpha^{p^n} = (X - \alpha)^{p^n}$$

holds in $L[X]$ for every nonnegative integer n. Thus, the direct implication in the conclusion follows from the preceding proposition, and the opposite from 1.7.7.   □

It will now be seen that purely inseparable elements behave in the most restricted manner in regard to embeddability. In some sense, this justifies the adoption of the terminology introduced in this section.

2.4.3. Proposition. *Let K be a field of prime characteristic, let L be an extension field of K, and let M be an extension field of L. If $\alpha \in L$ and α is purely inseparable over K, and if u is a K-monomorphism from L to M, then $u(\alpha) = \alpha$.*

Proof. Put $p = \mathrm{Char}(K)$. If n is a nonnegative integer such that $\alpha^{p^n} \in K$, then

$$u(\alpha)^{p^n} = u(\alpha^{p^n}) = \alpha^{p^n},$$

whence $u(\alpha) = \alpha$.   □

2.4.4. Proposition.   *Let K be a field of prime characteristic, and let L be a normal extension of K. Then an element of L is purely inseparable over K if and only if it is a fixed point of every K-automorphism of L.*

Proof.   The direct implication is a particular case of the preceding proposition. To prove the opposite, let $\alpha \in L$ and suppose that α is a fixed point of every K-automorphism of L. Then, according to 2.3.6, the minimal polynomial of α over K admits α as its only zero in L. Moreover, this polynomial splits in $L[X]$. By 2.4.2, we now conclude that α is purely inseparable over K.   □

Let K be a field of prime characteristic, and let L be an extension field of K. We say that L is **purely inseparable over** K, or that L is a **purely inseparable extension** of K, when every element of L is purely inseparable over K.

Let K be a field of prime characteristic p. If n is a nonnegative integer, the symbol K^{p^n} will denote the image of the endomorphism $\alpha \to \alpha^{p^n}$ of K. Then $(K^{p^n})_{n \geq 0}$ is a decreasing sequence of subfields of K, with $K^{p^0} = K$ and $K^{p^1} = K^p$.

It is clear that if K is a field of prime characteristic p, and if L is an extension field of K such that $L^{p^n} \subseteq K$ for some nonnegative integer n, then L is purely inseparable over K.

In particular, if K is a field of prime characteristic p and n is a nonnegative integer, then K is purely inseparable over K^{p^n}.

The following three propositions, which are stated for the sake of reference, are immediate consequences of the definitions and the preceding results.

2.4.5. Proposition.   *Let K be a field of prime characteristic, let L be an extension field of K, and let M be an extension field of L. Then M is purely inseparable over K if and only if M is purely inseparable over L and L is purely inseparable over K.*

2.4.6. Proposition.   *If K is a field of prime characteristic, then every purely inseparable extension of K is a normal extension of K.*

2.4.7. Proposition.   *Let K be a field of prime characteristic, and let L be an extension field of K. If L is purely inseparable over K, and if M is an extension field of L, then $i_{L \to M}$ is the only K-monomorphism from L to M. Conversely, if there exists an extension field M of L that is normal over K and such that $i_{L \to M}$ is the only K-monomorphism from L to M, then L is purely inseparable over K.*

We will now show how to construct finite purely inseparable extensions that are not simple. This will be our first example of a finitely generated extension that is not simple.

2.4.8. Example. Let F be a field of prime characteristic p, and let n be a positive integer. Then let $L = F(X_1, X_2, \ldots, X_n)$ and $K = F(X_1^p, X_2^p, \ldots, X_n^p)$. We claim that $L^p \subseteq K$ and $[L : K] = p^n$, so that L is purely inseparable and finite over K; but that L is not a simple extension of K when $n > 1$.

Since $L^p = F^p(X_1^p, X_2^p, \ldots, X_n^p)$, it is clear that $L^p \subseteq K$. Assuming for a moment that the equality $[L : K] = p^n$ has been established, it is possible to verify the last assertion. Indeed, if $L = K(\alpha)$ for some $\alpha \in L$, then $[K(\alpha) : K] = p^n$, whence the minimal polynomial of α over K has degree p^n; but this, in view of the relation $\alpha^p \in K$, would contradict 2.4.1 if $n > 1$.

To conclude, we shall now show, by induction on the positive integer n, that

$$[F(X_1, X_2, \ldots, X_n) : F(X_1^p, X_2^p, \ldots, X_n^p)] = p^n$$

for every field F of prime characteristic p.

If $n = 1$, this amounts to the validity of the equality $[F(X) : F(X^p)] = p$ for every field F, which is obvious because $X \notin F(X^p)$ and $X^p \in F(X^p)$.

Next assume that the result is true for some positive integer n. Consider a field F, and look at the three-term chain

$$F(X_1^p, \ldots, X_n^p, X_{n+1}^p) \subseteq F(X_1^p, \ldots, X_n^p, X_{n+1}) \subseteq F(X_1, \ldots, X_n, X_{n+1}).$$

Applying what was shown in the preceding paragraph to the field $F(X_1^p, \ldots, X_n^p)$, we see from the first inclusion that

$$[F(X_1^p, \ldots, X_n^p, X_{n+1}) : F(X_1^p, \ldots, X_n^p, X_{n+1}^p)] = p.$$

Moreover, applying the induction assumption to the field $F(X_{n+1})$, the second inclusion yields

$$[F(X_1, \ldots, X_n, X_{n+1}) : F(X_1^p, \ldots, X_n^p, X_{n+1})] = p^n.$$

Combining the last two equalities, we now obtain

$$[F(X_1, \ldots, X_n, X_{n+1}) : F(X_1^p, \ldots, X_n^p, X_{n+1}^p)] = p^{n+1},$$

which is what was needed. $\square$

In connection with the example just discussed and the next proposition, it should be mentioned that there is a simple criterion for the existence of primitive elements for finite purely inseparable extensions (problem 1).

2.4.9. Proposition. Let K be a field of prime characteristic p, and let L be a finite purely inseparable extension of K. Then $[L : K] = p^e$ for some nonnegative integer e, and $L^{p^e} \subseteq K$.

Proof. Write $L = K(\alpha_1, \alpha_2, \ldots, \alpha_n)$, where n is a positive integer and $\alpha_1, \alpha_2, \ldots, \alpha_n \in L$. Let $E_0 = K$; and for $1 \leq i \leq n$, let $E_i = K(\alpha_1, \alpha_2, \ldots, \alpha_i)$. Then $(E_i)_{0 \leq i \leq n}$ is an increasing sequence of intermediate fields between K and L such that $E_0 = K$ and $E_n = L$; therefore $[L : K] = \prod_{i=1}^{n} [E_i : E_{i-1}]$.

If $1 \le i \le n$, then $E_i = E_{i-1}(\alpha_i)$; since α_i is purely inseparable over K, it is also purely inseparable over E_{i-1}, and 2.4.1 then shows that

$$[E_i : E_{i-1}] = [E_{i-1}(\alpha_i) : E_{i-1}] = p^{e_i}$$

for some nonnegative integer e_i.

Now put $e = \sum_{i=1}^{n} e_i$. We then have

$$[L : K] = \prod_{i=1}^{n} p^{e_i} = p^e,$$

as required.

To conclude, it remains to verify that $L^{p^e} \subseteq K$. Thus, let $\alpha \in L$. By 2.4.1, we then have $[K(\alpha) : K] = p^m$, where m is a nonnegative integer such that $\alpha^{p^m} \in K$. Since $[K(\alpha) : K] | [L : K]$, we have $p^m | p^e$, whence $m \le e$. But then $\alpha^{p^e} = (\alpha^{p^m})^{p^{e-m}} \in K$.   □

It should be noted that a finite extension whose linear degree is an integer power of the characteristic is not necessarily purely inseparable (section 2.5, problem 1).

Let K be a field of prime characteristic p, and let L be an extension field of K. If n is a nonnegative integer, we shall denote by $P_n(L/K)$ the inverse image of K by the endomorphism $\alpha \to \alpha^{p^n}$ of L; alternatively, $P_n(L/K)$ can be described as the intermediate field between K and L consisting of all $\alpha \in L$ such that $\alpha^{p^n} \in K$.

It follows that the sequence $(P_n(L/K))_{n \ge 0}$ is increasing and $P_0(L/K) = K$. We shall write $P(L/K) = \cup_{n=0}^{\infty} P_n(L/K)$. Then $P(L/K)$ is also an intermediate field between K and L; it will be called the **perfect closure** of K in L. Clearly, $P(L/K)$ consists of all elements of L that are purely inseparable over K.

To say that L is purely inseparable over K amounts to saying that $P(L/K) = L$.

Now let K be a field of prime characteristic, and let L and M be algebraic closures of K. If u is a K-isomorphism from L to M, it is seen that $u(P_n(L/K)) = P_n(M/K)$ for every $n \ge 0$, and hence $u(P(L/K)) = P(M/K)$. Consequently, every K-isomorphism from L to M defines, by restriction, a K-isomorphism from $P_n(L/K)$ to $P_n(M/K)$ for every $n \ge 0$, and a K-isomorphism from $P(L/K)$ to $P(M/K)$. The "identification" of L and M by means of a K-isomorphism, therefore, can be carried over to "identifications" of the corresponding intermediate fields just defined.

In the literature on the theory of fields, the symbols $K^{p^{-n}}$ and $K^{p^{-\infty}}$ are used, respectively, to denote the fields $P_n(\overline{K}/K)$ and $P(\overline{K}/K)$ when $\overline{K}$ is "the" algebraic closure of a field K of prime characteristic p.

2.4.10. Proposition. *Let K be a field of prime characteristic, and let L be an extension field of K. Then $P(L/K)$ is the largest element of the set of all intermediate fields between K and L that are purely inseparable over K.*

Moreover, every element of L that is purely inseparable over $P(L/K)$ belongs to $P(L/K)$.

Proof. Write $P = P(L/K)$. We have already seen that P is an intermediate field between K and L, and this implies the first assertion in the conclusion.

To verify the second, put $p = \text{Char}(K)$. Let $\alpha \in L$ and suppose that α is purely inseparable over P. Then $\alpha^{p^m} \in P$ for some nonnegative integer m, whence $(\alpha^{p^m})^{p^n} \in K$ for some nonnegative integer n. Consequently, $\alpha^{p^{m+n}} = (\alpha^{p^m})^{p^n} \in K$, which implies that $\alpha \in P$. $\square$

As easy consequences of the preceding proposition, we can now obtain the "closure properties" of the class of all purely inseparable extensions. These are the exact analogs of 2.1.16, 2.1.17, and 2.1.18. Moreover, it is not necessary to give their proofs, since they are identical to those of the latter.

2.4.11. Corollary. *Let K be a field of prime characteristic, and let L be an extension field of K. If $D \subseteq P(L/K)$, then $K(D)$ is purely inseparable over K.*

2.4.12. Corollary. *Let K be a field of prime characteristic, and let M be an extension field of K. If $(L_i)_{i \in I}$ is a nonempty family of intermediate fields between K and M that are purely inseparable over K, then $\vee_{i \in I} L_i$ is purely inseparable over K.*

2.4.13. Corollary. *Let M be a field of prime characteristic, and let K and L be subfields of M. If $L \subseteq P(M/K)$, then $K \vee L$ is purely inseparable over K.*

To close this section, we would like to mention that in order to discuss pure inseparability for field extensions of arbitrary characteristic, several authors define the auxiliary notion of **characteristic exponent** of a field. This is equal to 1 or to the characteristic, according as the characteristic is 0 or a prime. With this understanding, everything discussed for purely inseparable extensions also applies in the case of characteristic 0 by taking $p = 1$; the resulting theory of pure inseparability is of course devoid of interest, all results being trivial in this case.

PROBLEMS

1. Let K be a field of prime characteristic p, and let L be a finite purely inseparable extension of K. Write $[L : K] = p^e$, where e is a nonnegative integer; and denote by d the smallest element of the set of all nonnegative integers k for which $L^{p^k} \subseteq K$. Show that $d \leq e$, and that $d = e$ if and only if L is a simple extension of K.

2. Let K be a field of prime characteristic, and let L be a perfect extension field of K. Show that $P(L/K)$ is the smallest element of the set of all perfect intermediate fields between K and L.

3. Let K be a field of prime characteristic. Prove that if K is perfect, then K is its only purely inseparable extension, so that $P_n(L/K) = K$ for every extension field L of K and every nonnegative integer n. Also prove that, conversely, if there exists a perfect extension field L of K such that $P_1(L/K) = K$, then K is perfect.

4. Let K be a field of prime characteristic p, and let L be a perfect extension field of K. Prove the following assertions:

    a. If n is a nonnegative integer, then the automorphism $\alpha \to \alpha^{p^n}$ of L defines, by restriction, an isomorphism from $P_n(L/K)$ to K.

    b. If K is not perfect, then $P_n(L/K) \subset P_{n+1}(L/K)$ for every nonnegative integer n; and $P(L/K)$ and K are not isomorphic.

5. Let K be a field of prime characteristic p, and let L be an extension field of K. Let n be a positive integer, and let $(\xi_i)_{i \in I}$ be a family of elements of L. Show that if $(\xi_i^{p^n})_{i \in I}$ is linearly independent over K, and if M is an extension field of L, then $(\xi_i)_{i \in I}$ is linearly independent over $P_n(M/K)$. Also show that, conversely, if there exists a perfect extension field M of L such that $(\xi_i)_{i \in I}$ is linearly independent over $P_n(M/K)$, then $(\xi_i^{p^n})_{i \in I}$ is linearly independent over K.

2.5. SEPARABLE EXTENSIONS

As we saw in the preceding section, purely inseparable extensions are the normal extensions in which every element is the only zero of its minimal polynomial. We shall now study the algebraic extensions that can be described by the opposite property, namely, that the minimal polynomial of every element should admit as many zeros as possible. A consequence of these properties is that these two types of algebraic extension exhibit opposite behavior in regard to embeddability: For a purely inseparable extension, it has been shown that the only embeddings are the trivial ones; for a separable extension, on the other hand, it will be seen that the largest possible number of embeddings is obtained.

Let K be a field, and let L be an extension field of K. An element of L will be said to be **separable over** K when it is algebraic over K and its minimal polynomial over K is separable. Equivalently, we could say that an element of L is separable over K when it is algebraic over K and it is a simple zero of its minimal polynomial over K.

It is clear that if K is a field, then every element of K is separable over K; and if K is a perfect field and L is an extension field of K, then 1.7.6 shows that every element of L that is algebraic over K is separable over K.

2.5.1. Proposition. *Let K be a field, and let L be an extension field of K. Then every element of L that is a simple zero of a polynomial in K[X] is separable over K.*

Proof. If $\alpha \in L$, and if α is a simple zero of a polynomial g in $K[X]$, then α is algebraic over K; if f denotes its minimal polynomial over K, then f divides g in $K[X]$, whence α is a simple zero of f. But this, as we have observed previously, means that α is separable over K. □

2.5.2. Corollary. *Let K be a field, let L be an extension field of K, and let M be an extension field of L. Then every element of M that is separable over K is separable over L.*

Proof. Indeed, every such element of M is a simple zero of its minimal polynomial over K; and this polynomial belongs to $L[X]$. □

2.5.3. Proposition. *Let K be a field of prime characteristic, and let L be an extension field of K. Then every element of L that is separable and purely inseparable over K belongs to K.*

Proof. Let $p = \mathrm{Char}(K)$. Assume that $\alpha \in L$, and that α is separable and purely inseparable over K. Then its minimal polynomial over K is separable and has the form $X^{p^n} - \alpha^{p^n}$, where n is a nonnegative integer such that $\alpha^{p^n} \in K$. But this is possible only when $n = 0$, which implies that $\alpha \in K$. □

Let K be a field, and let L be an extension field of K. We say that L is **separable over** K, or that L is a **separable extension of** K, when every element of L is separable over K.

Every field is separable over itself. If K is a field of prime characteristic, then K is the only extension field of K that is separable and purely inseparable over K. Finally, since every separable extension is an algebraic extension, an algebraically closed field possesses no proper separable extensions.

As we show next, perfect fields can be described as the fields that allow no inseparability.

2.5.4. Proposition. *Let K be a field. If K is perfect, then every algebraic extension of K is separable over K. Conversely, if there exists an algebraic closure of K that is separable over K, then K is perfect.*

Proof. Both assertions are consequences of the characterization of perfect fields given in 1.7.6.

As observed earlier, the first one is evident. To verify the second one, assume that an algebraic closure L of K is separable over K, and let f be an irreducible polynomial in $K[X]$. Then f admits a zero α in L. The monic associate of f in $K[X]$ is then the minimal polynomial of α over K, and hence it is separable. And this implies that f is separable, which is what was needed.   □

The preceding result shows, in particular, that algebraic extensions are always separable in characteristic 0. For this reason, this case can be disregarded in most arguments on separability.

2.5.5. Proposition.   *If K is a perfect field, and if L is an algebraic extension of K, then L is perfect.*

Proof.   We may assume that our fields have prime characteristic p. If $\alpha \in L$, then α admits a pth root β in an extension field of L. Since K is perfect, β is separable over K, and hence separable over L; but $\beta^p = \alpha \in L$, so that β is also purely inseparable over L. Consequently $\beta \in L$, and so $\alpha = \beta^p \in L^p$.   □

We now proceed to derive the main properties of separable extensions. We shall begin with a result stating useful necessary and sufficient conditions for separability.

2.5.6. Proposition.   *Let K be a field of prime characteristic p, and let L be an extension field of K. If L is separable over K, then $L = K \vee L^p$. Conversely, if $L = K \vee L^p$ and L is finite over K, then L is separable over K.*

Proof.   Since $L^p \subseteq K \vee L^p$, it is clear that L is purely inseparable over $K \vee L^p$. If L is separable over K, we see from 2.5.2 that it is also separable over $K \vee L^p$. It then follows that $L = K \vee L^p$.

Now suppose that L is finite over K and $L = K \vee L^p$. We claim that if a finite sequence $(\alpha_i)_{1 \le i \le m}$ of elements of L is linearly independent over K, then $(\alpha_i^p)_{1 \le i \le m}$ is also linearly independent over K. To see this, put $n = [L : K]$; we then have $m \le n$, so that $(\alpha_i)_{1 \le i \le m}$ can be extended to a linear base $(\alpha_i)_{1 \le i \le n}$ of L over K. Since $L = \sum_{i=1}^{n} K\alpha_i$, we have $L^p = \sum_{i=1}^{n} K^p \alpha_i^p$; applying 2.1.12, we now get

$$L = K \vee L^p = KL^p = \sum_{i=1}^{n} K\alpha_i^p.$$

Therefore $(\alpha_i^p)_{1 \le i \le n}$ is a generating system of the K-space L. Since the latter is n-dimensional, we conclude that $(\alpha_i^p)_{1 \le i \le n}$ is a linear base of L over K. It follows, in particular, that $(\alpha_i^p)_{1 \le i \le m}$ is linearly independent over K, and our claim is proved.

To show that L is separable over K, let $\alpha \in L$. Denote by f the minimal polynomial of α over K, and put $n = \deg(f)$. Then $(\alpha^i)_{0 \le i \le n-1}$ is linearly independent over K, and hence $(\alpha^{ip})_{0 \le i \le n-1}$ is also linearly independent over K. If f were not separable, then $f \in K[X^p]$, so that $f(X) = g(X^p)$ for some $g \in K[X]$; letting $m = \deg(g)$, it would now follow from $g(\alpha^p) = f(\alpha) = 0$ that $(\alpha^{ip})_{0 \le i \le m}$ is linearly dependent over K; but this, in view of the relations

$$n = \deg(f) = p \deg(g) = pm > m,$$

is impossible. Consequently, f is separable, which means that α is separable over K. $\qquad\square$

It is not difficult to show that the second assertion in the conclusion of the proposition just proved does not remain valid if the finiteness assumption is omitted (problem 3).

2.5.7. Proposition. *Let K be a field of prime characteristic p, and let L be an extension field of K. If $\alpha \in L$ and α is algebraic over K, then α is separable over K if and only if $K(\alpha) = K(\alpha^p)$, in which case $K(\alpha)$ is separable over K.*

Proof. If α is separable over K, then it is separable over $K(\alpha^p)$; but α is also purely inseparable over $K(\alpha^p)$, because $\alpha^p \in K(\alpha^p)$. Consequently $\alpha \in K(\alpha^p)$, which implies that $K(\alpha) = K(\alpha^p)$.

Conversely, if $K(\alpha) = K(\alpha^p)$, then

$$K \vee K(\alpha)^p = K \vee K^p(\alpha^p) = K(\alpha^p) = K(\alpha).$$

And since $K(\alpha)$ is finite over K, the separability of $K(\alpha)$ over K follows from the preceding proposition. $\qquad\square$

We shall show next that the transitivity property holds for separable extensions.

2.5.8. Proposition. *Let K be a field, let L be an extension field of K, and let M be an extension field of L. Then M is separable over K if and only if M is separable over L and L is separable over K.*

Proof. There is nothing to prove if the characteristic of our fields is 0, so that we need only consider the case in which it is a prime p. Furthermore, by 2.5.2, the direct implication is evident.

Thus, assume that M is separable over L and that L is separable over K. Let $\alpha \in M$, and denote by f the minimal polynomial of α over L. Next, let $n = \deg(f)$, and write $f(X) = \sum_{i=0}^{n-1} \gamma_i X^i + X^n$ with $\gamma_0, \gamma_1, \ldots, \gamma_{n-1} \in L$. Finally, let $E = K(\gamma_0, \gamma_1, \ldots, \gamma_{n-1})$. Since $E \subseteq L$, it is clear that E is separable over K; and since $f \in E[X]$, $f(\alpha) = 0$, and f is separable, it follows that α is separable over E.

Applying 2.5.6 and 2.5.7, we now obtain

$$E = K \vee E^p \quad \text{and} \quad E(\alpha) = E(\alpha^p),$$

whence

$$E(\alpha) = E(\alpha^p) = E \vee E(\alpha)^p = K \vee E^p \vee E(\alpha)^p = K \vee E(\alpha)^p.$$

Furthermore, since $E(\alpha) = K(\gamma_0, \gamma_1, \ldots, \gamma_{n-1}, \alpha)$, it is clear that $E(\alpha)$ is finite over K. Using 2.5.6 again, we then conclude that $E(\alpha)$ is separable over K. In particular, α is separable over K. $\qquad\square$

A natural object of study in the present discussion is the set of all separable elements in a field extension. As we shall see, this set defines an intermediate field with interesting properties.

Let K be a field, and let L be an extension field of K. We shall denote by $S(L/K)$ the set of all elements of L that are separable over K, and say that $S(L/K)$ is the **separable closure of K in L**.

It is evident from this definition that $S(L/K) = L$ if and only if L is separable over K.

2.5.9. Proposition. *Let K be a field, and let L be an extension field of K. Then $S(L/K)$ is the largest element of the set of all intermediate fields between K and L that are separable over K. Moreover, every element of L that is separable over $S(L/K)$ belongs to $S(L/K)$.*

Proof. When $\mathrm{Char}(K) = 0$, we have $S(L/K) = A(L/K)$; in this case, the conclusion follows from 2.1.15. Thus, we shall assume that K has prime characteristic, and write $S = S(L/K)$.

Since $K \subseteq S$, to prove the first assertion in the conclusion it suffices to show that S is a subfield of L. To do this, let $\alpha, \beta \in S$. Then, according to 2.5.2, β is separable over $K(\alpha)$; and by 2.5.7, we see that $K(\alpha)$ and $K(\alpha)(\beta)$ are separable over K and $K(\alpha)$, respectively. Since $K(\alpha, \beta) = K(\alpha)(\beta)$, it now follows from 2.5.8 that $K(\alpha, \beta)$ is separable over K, and so $K(\alpha, \beta) \subseteq S$. Since $\alpha - \beta, \alpha\beta \in K(\alpha, \beta)$, we have $\alpha - \beta, \alpha\beta \in S$; and if $\alpha \neq 0$, we also have $1/\alpha \in K(\alpha, \beta)$, whence $1/\alpha \in S$. We conclude that S is a subfield of L.

The second assertion in the conclusion is a consequence of 2.5.7 and 2.5.8. For if $\alpha \in L$ and α is separable over S, then $S(\alpha)$ is separable over S; since S is separable over K, it now follows that $S(\alpha)$ is separable over K. But then $S(\alpha) \subseteq S$, which implies that $\alpha \in S$. $\square$

The basic "closure properties" of the class of all separable extensions follow immediately from the proposition just proved. Just as in the case of purely inseparable extension, we shall merely state these properties, which are the exact analogs of 2.1.16, 2.1.17, and 2.1.18.

2.5.10. Corollary. *Let K be a field, and let L be an extension field of K. If $D \subseteq S(L/K)$, then $K(D)$ is separable over K.*

2.5.11. Corollary. *Let K be a field, and let M be an extension field of K. If $(L_i)_{i \in I}$ is a nonempty family of intermediate fields between K and M that are separable over K, then $\vee_{i \in I} L_i$ is separable over K.*

2.5.12. Corollary. *Let M be a field, and let K and L be subfields of M. If $L \subseteq S(M/K)$, then $K \vee L$ is separable over K.*

In the remainder of this section, we shall be concerned with concepts that have been defined for fields of prime characteristic exclusively.

With the help of the notion of "relative" separable closure, we can now state and prove a basic result on algebraic extensions in prime characteristic.

2.5.13. Proposition. *Let K be a field of prime characteristic, and let L be an algebraic extension of K. Then $S(L/K)$ is the only intermediate field* *between K and L that is separable over K and over which L is purely inseparable.*

Proof. Write $p = \text{Char}(K)$ and $S = S(L/K)$. We already know that S is separable over K. To prove that L is purely inseparable over S, let $\alpha \in L$. Let e denote the exponent of inseparability of the minimal polynomial of α over K. According to 2.1.7, the separable polynomial associated with the latter is the minimal polynomial of α^{p^e} over K; therefore α^{p^e} is separable over K, which means that $\alpha^{p^e} \in S$. Hence α is purely inseparable over S.

To conclude, let E be an intermediate field between K and L that is separable over K and over which L is purely inseparable. The separability of E over K implies that $E \subseteq S$. Therefore S is separable over E; and the pure inseparability of L over E implies that S is also purely inseparable over E. Consequently $E = S$.   □

2.5.14.   Corollary.   *Let K be a field of prime characteristic p, and let L be an extension field of K. If $\alpha \in L$ and α is algebraic over K, and if e denotes the exponent of inseparability of the minimal polynomial of α over K, then $S(K(\alpha)/K) = K(\alpha^{p^e})$.*

Proof. Since $\alpha^{p^e} \in K(\alpha^{p^e})$, it follows that $K(\alpha)$ is purely inseparable over $K(\alpha^{p^e})$. Moreover, as shown in the first part of the proof of the proposition, α^{p^e} is separable over K, and hence $K(\alpha^{p^e})$ is separable over K. The conclusion now follows immediately from the proposition.   □

2.5.15.   Corollary.   *Let K be a field of prime characteristic p, and let L be a finite extension of K. If $p \nmid [L:K]$, then L is separable over K.*

Proof. Assume, on the contrary, that L is not separable over K. Then $S(L/K) \subset L$ and L is purely inseparable over $S(L/K)$, whence $[L:S(L/K)] = p^n$ for some positive integer n. Consequently

$$[L:K] = [L:S(L/K)][S(L/K):K] = p^n[S(L/K):K],$$

which leads to the contradictory conclusion that $p \mid [L:K]$.   □

2.5.16.   Corollary.   *Let K be a field of prime characteristic, and let L be an algebraic extension of K. Then $S(L/K) = K$ if and only if L is purely inseparable over K.*

Proof. The direct implication follows immediately from the proposition. The opposite is clear, because if L is purely inseparable over K, then $S(L/K)$ is separable and purely inseparable over K.   □

The converse of corollary 2.5.15 is false. In fact, it is not difficult to show that for every prime p, there exist finite separable extensions in characteristic p having linear degree p (problem 1).

Consider a field K of prime characteristic, and an algebraic extension L of K. In light of corollary 2.5.16, it is natural to inquire about the truth of the statement "L is separable over K if and only if $P(L/K) = K$". The direct implication is true: If L is separable over K, then $P(L/K)$ is separable and purely inseparable over K, whence $P(L/K) = K$. We shall next look at an example for which the opposite implication is false. This example will also serve as an illustration of most of the ideas in the foregoing discussion.

2.5.17. Example. Let F be a field of prime characteristic p, and let $K = F(X, Y)$. Let f and g be the polynomials in $K[Z]$ defined by

$$f(Z) = X + YZ^p + Z^{p^2} \quad \text{and} \quad g(Z) = X + YZ + Z^p.$$

It is evident that $f(Z) = g(Z^p)$. Moreover, we see that f and g are irreducible in $K[Z]$: Indeed, they are seen to be irreducible in $F[X, Y, Z]$, and 1.5.13 shows that $F[X, Y]$ is a factorial domain; since K is a field of fractions of $F[X, Y]$, our assertion follows from 1.5.9.

Form a simple extension $K(\alpha)$ of K by adjunction of a zero α of f in an extension field of K. Then $[K(\alpha) : K] = \deg(f) = p^2$; and since $f' = 0$, we see that α is not separable over K, and hence $K(\alpha)$ is not separable over K. Also, note that $g' \neq 0$, so that g is the separable polynomial associated with f. In particular, by 2.5.14 and 2.1.7, we have

$$S(K(\alpha)/K) = K(\alpha^p) \quad \text{and} \quad [K(\alpha^p) : K] = \deg(g) = p.$$

We now contend that $P(K(\alpha)/K) = K$.

Assume, on the contrary, that $K \subset P(K(\alpha)/K)$. We can then choose a $\beta \in K(\alpha)$ such that $\beta \notin K$ and $\beta^p \in K$. Clearly $[K(\beta) : K] = p$; and since

$$p^2 = [K(\alpha) : K] = [K(\alpha) : K(\beta)][K(\beta) : K] = [K(\alpha) : K(\beta)] p,$$

we have $[K(\alpha) : K(\beta)] = p$. Moreover, as $f \in K(\beta)[Z]$, $f(\alpha) = 0$, and $\deg(f) = p^2$, it follows that f is reducible in $K(\beta)[Z]$.

It would now be possible to reach a contradiction if g were known to be irreducible in $K(\beta)[Z]$. For, assuming this to be true, the reducibility of f in $K(\beta)[Z]$, the relation $f(Z) = g(Z^p)$, and what was shown in 2.1.7 imply that $g \in K(\beta)^p[Z]$; that is, that $X, Y \in K(\beta)^p$. Then $X = \eta^p$ and $Y = \xi^p$, where $\eta, \xi \in K(\beta)$. We note that $\eta \notin K$, for otherwise the relations

$$X = \eta^p \quad \text{and} \quad K^p = F^p(X^p, Y^p)$$

would imply that $X \in F^p(X^p, Y^p)$, which is impossible. Similarly, we have $\xi \notin K(\eta)$, for otherwise the relations

$$Y = \xi^p \quad \text{and} \quad K(\eta)^p = K^p(\eta^p) = K^p(X) = F^p(X, Y^p)$$

would imply that $Y \in F^p(X, Y^p)$, which is also impossible. To sum up, we now have

$$\eta \notin K, \qquad \eta^p \in K, \qquad \xi \notin K(\eta), \qquad \xi^p \in K(\eta);$$

and hence

$$[K(\eta, \xi) : K] = [K(\eta, \xi) : K(\eta)][K(\eta) : K] = p^2.$$

But this, in view of the relations

$$K \subseteq K(\eta, \xi) \subseteq K(\beta) \quad \text{and} \quad [K(\beta): K] = p,$$

is a contradiction.

It remains, therefore, to establish the irreducibility of g in $K(\beta)[Z]$. To do this, we shall show first that every linear base of $K(\alpha^p)$ over K is a linear base of $K(\beta)(\alpha^p)$ over $K(\beta)$.

We know that $[K(\alpha^p): K] = p$. If $(\sigma_1, \sigma_2, \ldots, \sigma_p)$ is a linear base of $K(\alpha^p)$ over K, then $K(\alpha^p) = \sum_{i=1}^p K\sigma_i$, whence

$$K(\beta)(\alpha^p) = K(\beta) \vee K(\alpha^p) = K(\beta)K(\alpha^p) = \sum_{i=1}^p K(\beta)\sigma_i,$$

which shows that $(\sigma_1, \sigma_2, \ldots, \sigma_p)$ is a generating system of the $K(\beta)$-space $K(\beta)(\alpha^p)$. Next we see that $(\sigma_1, \sigma_2, \ldots, \sigma_p)$ is linearly independent over $K(\beta)$. For suppose that $\sum_{i=1}^p \lambda_i \sigma_i = 0$, where $\lambda_1, \lambda_2, \ldots, \lambda_p \in K(\beta)$. Since $\beta^p \in K$, it is clear that $K(\beta)^p \subseteq K$, and hence $\lambda_1^p, \lambda_2^p, \ldots, \lambda_p^p \in K$; and we have

$$\sum_{i=1}^p \lambda_i^p \sigma_i^p = \left(\sum_{i=1}^p \lambda_i \sigma_i \right)^p = 0.$$

Since $S(K(\alpha)/K) = K(\alpha^p)$, we see that $K(\alpha^p)$ is separable over K. But this, as in the proof of 2.5.6, implies that $(\sigma_1^p, \sigma_2^p, \ldots, \sigma_p^p)$ is linearly independent over K. We now conclude that $\lambda_i^p = 0$ for $1 \leq i \leq p$, whence $\lambda_i = 0$ for $1 \leq i \leq p$.

What we have shown implies that

$$[K(\beta)(\alpha^p): K(\beta)] = [K(\alpha^p): K] = p.$$

Since $g \in K(\beta)[Z]$, $g(\alpha^p) = 0$, and $\deg(g) = p$, it then follows that g is the minimal polynomial of α^p over $K(\beta)$. In particular, g is irreducible in $K(\beta)[Z]$, as was to be proved. $\qquad\square$

In chapter 4, in the context of transcendental extensions, we shall study a more general concept of separability. Roughly speaking, we can say that for fields of prime characteristic, this generalized stability consists in the requirement that linear independence over the bottom field be preserved by the Frobenius mapping of the top field. As we have seen, this condition played an essential role in both the proof of 2.5.6 and the discussion of example 2.5.17.

By proposition 2.5.13, we see that in the case of prime characteristic, the top field of an algebraic extension can be reached from the bottom field in two steps, the first consisting in a separable extension and the second in a purely inseparable extension; and furthermore, that this can be done in precisely one way.

It is natural to ask if it is possible to reverse the order of these steps. We shall now close the section by showing that in the case of a normal extension, it can be done, and again, in exactly one way. However, as the example just given shows, for an arbitrary algebraic extension such reversing may not always be possible.

2.5.18. Proposition. *Let K be a field of prime characteristic, and let L be a normal extension of K. Then $P(L/K)$ is the only intermediate field between K and L that is purely inseparable over K and over which L is separable.*

Proof. Write $P = P(L/K)$. We already know that P is purely inseparable over K. To show that L is separable over P, let $\alpha \in L$. Denoting by D the set of all zeros in L of the minimal polynomial of α over K, we see, as in the proof of 2.1.21, that $s(D) = D$ for every K-automorphism s of L. Next, define a polynomial f in $L[X]$ by $f(X) = \prod_{\rho \in D}(X - \rho)$; then

$$sf(X) = \prod_{\rho \in D}(X - s(\rho)) = \prod_{\rho \in D}(X - \rho) = f(X)$$

for every K-automorphism s of L. Consequently, the coefficients of f are fixed points of every K-automorphism of L. According to 2.4.4, this means that $f \in P[X]$. Since f is separable and $f(\alpha) = 0$, it now follows from 2.5.1 that α is separable over P.

To conclude, let E be an intermediate field between K and L that is purely inseparable over K and over which L is separable. Since E is purely inseparable over K, we then have $E \subseteq P$. This implies that P is purely inseparable over E; and in view of the separability of L over E, it also implies that P is separable over E. Therefore $E = P$. $\square$

2.5.19. Corollary. *If K is a field of prime characteristic, and if L is a normal extension of K, then $L = P(L/K) \vee S(L/K)$.*

Proof. Write $P = P(L/K)$ and $S = S(L/K)$. The proposition shows that L is separable over P, and 2.5.13 that L is purely inseparable over S. Since

$$P \subseteq P \vee S \subseteq L \quad \text{and} \quad S \subseteq P \vee S \subseteq L,$$

it then follows that L is separable and purely inseparable over $P \vee S$, whence $L = P \vee S$. $\square$

PROBLEMS

1. Let p be a prime. Prove that the polynomial $X^p - X - 1$ is irreducible in $\mathbf{Z}/p[X]$, and then deduce that there exists an extension field of $\mathbf{Z}/p$ that is separable and of linear degree p over $\mathbf{Z}/p$.

2. Let K be a field of prime characteristic, and α and β be elements of an extension field of K that are, respectively, separable and purely insep-

arable over K. Prove the following assertions:

    a. $K(\alpha, \beta) = K(\alpha + \beta)$.

    b. If $\alpha \neq 0 \neq \beta$, then $K(\alpha, \beta) = K(\alpha\beta)$.

3. Let K be a field of prime characteristic p that is not perfect, let F be an algebraic closure of K, and let $L = P(F/K)$. Show that L is an infinite purely inseparable extension of K such that $L = K \vee L^p$.

4. Show that if K is a field and L is a normal extension of K, then $S(L/K)$ is normal over K.

5. Let K be a field of prime characteristic, let L be an algebraic extension of K, and let E be an intermediate field between K and L. Verify the following assertions:

    a. If L is purely inseparable over E, then $S(L/K) \subseteq E$.

    b. If L is separable over E, then $P(L/K) \subseteq E$.

    c. If $S(L/K) \cap E = K$, then $E \subseteq P(L/K)$.

6. Let K be a field of prime characteristic, let N be an algebraic extension of K, and let L and M be intermediate fields between K and N that are, respectively, separable and purely inseparable over K. Prove that every linear base of L over K is a linear base of $L \vee M$ over M.

7. Let K be a field of prime characteristic, and let L be an algebraic extension of K. Show that L is separable over $P(L/K)$ if and only if $L = S(L/K) \vee P(L/K)$.

8. Let K be a field of prime characteristic, let L be an algebraic extension of K, and let M be a perfect extension field of L. Show that $P(M/L) = L \vee P(M/K)$.

9. Let F be a field of prime characteristic p, and let $L = F(X, Y)$ and $K = F(X^p, Y^p)$. Form a simple extension $L(\alpha)$ of L by adjunction of a zero in an extension field of L of the polynomial $X + YZ + Z^p$ in $L[Z]$. Prove the following assertions:

    a. $X^p + Y^p Z^p + Z^{p^2}$ is the minimal polynomial of α over K.

    b. $K(\alpha^p)$ is the only intermediate field strictly between K and $K(\alpha)$.

    c. $K(\alpha)$ is not separable over K.

    d. $P(K(\alpha)/K) = K$.

(This problem was communicated to the author by Flanders.)

NOTES

The first detailed study of algebraic extensions was given in Steinitz [1: §§6–21]. The terminology and notation introduced in this work have been adopted almost completely, the exceptions being the terms "radical" and "of the first kind", which have been replaced by "purely inseparable" and "separable", respectively.

Section 2.1. The proof of 2.1.3 and the lemma in the proof of 1.6.5 show that the simple algebraic extensions of a field K can be described as the fields of the form $K[X]/fK[X]$, where f is an irreducible polynomial in $K[X]$. According to the theorem on primitive elements, which will be proved in Section 3.6, every algebraic number field is a simple extension of Q, and so it is determined by an irreducible polynomial in $Q[X]$. As stated in the historical introduction, this was first observed by Kronecker, and shows that the role of C in algebraic number theory is not essential.

The results of Liouville, Hermite, and Lindemann on transcendental numbers stated in the remarks made after 2.1.19 were also mentioned in the historical introduction. The transcendence of e and π was deduced by Lindemann [1] from the following stronger theorem:

If $\alpha_1, \alpha_2, \ldots, \alpha_n$ are algebraic numbers that are linearly independent over Q, then the complex numbers

$$\exp(\alpha_1), \exp(\alpha_2), \ldots, \exp(\alpha_n)$$

are algebraically independent over Q.

This result was treated in more detail by Weierstrass [1], and is also referred to as the **Lindemann–Weierstrass theorem**. It is clear that the transcendence of e follows trivially from this theorem by taking $n = 1$. Moreover, since

$$\exp(\pi\sqrt{-1}) = \cos(\pi) + \sin(\pi)\sqrt{-1} = -1,$$

the theorem also implies that $\pi\sqrt{-1}$ is transcendental; and since $\sqrt{-1}$ is algebraic, we see that π is transcendental.

For a detailed discussion on the Lindemann–Weierstrass theorem, we refer to Jacobson [5: I, sec. 4.12]. Simple proofs of the transcendence of e and π can be found in Hadlock [1: sec. 1.7] and Stewart [1: ch. 6]. For proofs that the transcendence of π implies the impossibility of squaring the circle, the reader may consult Hadlock [1: sec. 1.4] and Stewart [1: ch. 5].

Section 2.2. Gauss gave the first complete proof of the "fundamental theorem of algebra" in his doctoral dissertation. Later on, he gave other proofs for this result; these have been collected in Gauss [1: VIII]. Many proofs have been given subsequently by other mathematicians.

The general results on algebraically closed fields discussed here are due to Steinitz [1: §§17–21]. His original exposition is no longer followed in more recent texts because the arguments are by transfinite induction (it should be noted that Zorn's lemma was not yet available when Steinitz established his theorems). Simplified proofs by transfinite induction can be found in van der Waerden [2: I, §62].

The proofs of the existence of algebraic closures sketched in problems 2 and 3 are given in Jacobson [4: sec. 4.1] and Zariski and Samuel [1: I, sec. 2.14], respectively.

Sections 2.3–2.5. It was observed in Steinitz [1: §11] that the theorems of the earlier versions of Galois theory do not remain valid for arbitrary fields. The properties of normality, pure inseparability, and separability, which define the algebraic extensions studied in these sections, were introduced by Steinitz [1: §§8, 12, 13] in order to carry out the natural extension of Galois theory to more general fields.

Chapter 3

Galois Theory

3.1. SOME VECTOR SPACES OF MAPPINGS OF FIELDS

The main objective of this section is to discuss certain very specialized notions from linear algebra that can be fruitfully applied in the theory of fields.

Let S be a set, and let F be a field. We define an F-space structure on the set of all mappings from S to F as follows: If u and v are mappings from S to F, then $u + v$ is the mapping $\rho \to u(\rho) + v(\rho)$ from S to F; and if $\alpha \in F$ and u is a mapping from S to F, then αu is the mapping $\rho \to \alpha u(\rho)$ from S to F. The F-space obtained by providing the set of all mappings from S to F with this F-space structure will be denoted by Map (S, F). The zero element of Map(S, F) is the "zero mapping", that is, the constant mapping assigning to each element of S the zero element of F; and the additive inverse in Map(S, F) of a mapping U from S to F is the mapping $\rho \to -u(\rho)$ from S to F. Finally, note that for each $\rho \in S$, the "evaluation mapping" $u \to u(\rho)$ from Map(S, F) to F is an F-linear form on Map(S, F): Indeed, the foregoing definitions imply that

$$(u + v)(\rho) = u(\rho) + v(\rho) \quad \text{and} \quad (\alpha u)(\rho) = \alpha u(\rho)$$

for all $u, v \in$ Map(S, F) and $\alpha \in F$.

No assumptions concerning the set S have been made in the definition of the F-space Map(S, F) just given. We shall be interested in particular situations where that set is provided with some algebraic structure.

Let K be a field, let V be a K-space, and let F be an extension field of K. It is evident that the K-linear mappings from V to F make up an F-subspace of $\text{Map}(V, F)$. The symbol $\text{Lin}_K(V, F)$ will be used to denote this F-subspace of $\text{Map}(V, F)$.

We begin our discussion with the following basic result on the space of linear mappings.

3.1.1. Proposition. *Let K be a field, let V be a K-space, and let F be an extension field of K.*

(i) *If $(e_i)_{i \in I}$ is a base of V, and if for each $i \in I$ the K-linear mapping from V to F such that $e_i \to 1$ and such that $e_j \to 0$ for every $j \in I - \{i\}$ is denoted by u_i, then $(u_i)_{i \in I}$ is a linearly independent family of elements of $\text{Lin}_K(V, F)$; and if I is finite, then $(u_i)_{i \in I}$ is a base of $\text{Lin}_K(V, F)$.*

(ii) *V is finite-dimensional if and only if $\text{Lin}_K(V, F)$ is finite-dimensional, in which case*

$$[V : K] = [\text{Lin}_K(V, F) : F].$$

Proof. We need only verify (i), because (ii) follows immediately from it. Thus, let $(e_i)_{i \in I}$ and $(u_i)_{i \in I}$ be as indicated in (i).

To prove the stated linear independence, suppose that $\sum_{i \in I} \alpha_i u_i = 0$, where $(\alpha_i)_{i \in I}$ is a family of elements of F such that $\alpha_i = 0$ for almost every $i \in I$. For each $j \in I$, we have

$$\alpha_j = \sum_{i \in I} \alpha_i u_i(e_j) = \left(\sum_{i \in I} \alpha_i u_i \right)(e_j) = 0,$$

as was to be shown.

To conclude, suppose now that I is finite. In order to show that $(u_i)_{i \in I}$ is a generating system of $\text{Lin}_K(V, F)$, let $u \in \text{Lin}_K(V, F)$. Write $\alpha_i = u(e_i)$ for every $i \in I$; since I is finite, it is meaningful to speak of the element $\sum_{i \in I} \alpha_i u_i$ of $\text{Lin}_K(V, F)$. For each $j \in I$, we then have

$$\left(\sum_{i \in I} \alpha_i u_i \right)(e_j) = \sum_{i \in I} \alpha_i u_i(e_j) = \alpha_j = u(e_j).$$

This means that u and $\sum_{i \in I} \alpha_i u_i$ agree at every element of the base $(e_i)_{i \in I}$ of V; therefore $u = \sum_{i \in I} \alpha_i u_i$, and so u is a linear combination of $(u_i)_{i \in I}$ with coefficients in F. □

Given fields E and F, the set of all monomorphisms from E to F will be denoted by $\text{Mon}(E, F)$; and if E and F contain a common subfield K, the set of all K-monomorphisms from E to F will be denoted by $\text{Mon}_K(E, F)$. These sets do not behave well with respect to the linear structure on $\text{Map}(E, F)$, since they are stable relative to neither the vector addition nor the scalar multiplication defining this structure. However, as we now show, an interesting property of the set of monomorphisms can be established.

3.1.2. Theorem (Dedekind). *If E and F are fields, then* $\mathrm{Mon}(E, F)$
is a linearly independent subset of $\mathrm{Map}(E, F)$.

Proof. Assume, for a contradiction, that $\mathrm{Mon}(E, F)$ is a linear dependent subset of $\mathrm{Map}(E, F)$. Then there exist nonempty finite subsets of $\mathrm{Mon}(E, F)$ that are linearly dependent subsets of $\mathrm{Map}(E, F)$; from among these, select one with the smallest possible cardinality, and denote it by Ω.

Write $n = \mathrm{Card}(\Omega)$ and $\Omega = \{u_1, u_2, \ldots, u_n\}$. Next, choose elements $\alpha_1, \alpha_2, \ldots, \alpha_n$ of F, not all equal to zero, so that $\sum_{i=1}^{n} \alpha_i u_i = 0$. The choice of Ω clearly implies that each of $\alpha_1, \alpha_2, \ldots, \alpha_n$ is different from 0. Finally, note that $n > 1$, for otherwise Ω would be a linearly dependent subset of $\mathrm{Map}(E, F)$ consisting of a single nonzero element, which is impossible.

We have, in particular $u_1 \neq u_n$; consequently, a $\sigma \in E$ can be chosen so that $u_1(\sigma) \neq u_n(\sigma)$. It is readily seen that $\sum_{i=1}^{n} \alpha_i u_i(\sigma) u_i = 0$: indeed, since $\sum_{i=1}^{n} \alpha_i u_i = 0$, for every $\rho \in E$ we have

$$\sum_{i=1}^{n} \alpha_i u_i(\sigma) u_i(\rho) = \sum_{i=1}^{n} \alpha_i u_i(\sigma \rho) = 0.$$

The two equalities

$$\sum_{i=1}^{n} \alpha_i u_i = 0 \quad \text{and} \quad \sum_{i=1}^{n} \alpha_i u_i(\sigma) u_i = 0$$

now imply that

$$\sum_{i=1}^{n-1} \alpha_i \big(u_i(\sigma) - u_n(\sigma) \big) u_i = \sum_{i=1}^{n} \alpha_i \big(u_i(\sigma) - u_n(\sigma) \big) u_i$$

$$= \sum_{i=1}^{n} \alpha_i u_i(\sigma) u_i - u_n(\sigma) \left(\sum_{i=1}^{n} \alpha_i u_i \right)$$

$$= 0.$$

Since $\alpha_1(u_1(\sigma) - u_n(\sigma)) \neq 0$, this shows that $\{u_1, u_2, \ldots, u_{n-1}\}$ is a linearly dependent subset of $\mathrm{Map}(E, F)$ having cardinality $n - 1$. This, however, is incompatible with the choice of Ω.   □

The preceding theorem, sometimes called the **Dedekind independence theorem for field monomorphisms**, was obtained by Dedekind at the end of the nineteenth century. For some reason, it remained unnoticed for many years, a circumstance that caused complications in the proofs of some results. It was after several decades that Artin rediscovered it in a slightly more general form (problem 1) and gave it the simple proof that we have reproduced here.

One of the interesting themes in Galois theory is the counting and the estimating of the number of elements of certain sets of field monomorphisms and groups of field automorphisms. The preceding results allow us to start the discussion of this topic with the following important proposition, in which an upper bound for the number of embeddings of a finite extension is given.

3.1.3. Proposition. *Let K be a field, let L be a finite extension of K, and let F be an extension field of K. Then*

$$\text{Card}(\text{Mon}_K(L, F)) \le [L : K].$$

Proof. Since $\text{Mon}_K(L, F) \subseteq \text{Lin}_K(L, F)$, it is clear from 3.1.2 that $\text{Mon}_K(L, F)$ is a linearly independent subset of $\text{Lin}_K(L, F)$; and according to 3.1.1, we have

$$[\text{Lin}_K(L, F): F] = [L : K].$$

The required inequality now follows at once. □

This proposition can be proved using only Dedekind's theorem 3.1.2 and a basic result on linear homogeneous equations with coefficients in a field (problem 4). It can also be derived from results in chapters 1 and 2 (problem 5).

3.1.4. Examples

a. Sometimes the upper bound given in 3.1.3 for the number of embeddings is not attained because the field in which the embedding takes place is not "sufficiently large".

To see this, consider the extension field $Q(\sqrt[3]{2})$ of Q. Then $[Q(\sqrt[3]{2}): Q] = 3$, so that $Q(\sqrt[3]{2})$ admits at most three Q-embeddings into any given extension field of Q.

That $\sqrt[3]{2}$ is the only real zero of the polynomial $X^3 - 2$ in $Q[X]$ implies that, for every intermediate field F between $Q(\sqrt[3]{2})$ and $\mathbf{R}$, the only Q-monomorphism from $Q(\sqrt[3]{2})$ to F is $i_{Q(\sqrt[3]{2}) \to F}$.

Suppose, on the other hand, that we now enlarge $Q(\sqrt[3]{2})$ to $Q(\sqrt[3]{2}, \omega)$, where $\omega = (-1 + \sqrt{-3})/2$. We saw in 2.3.16 that $Q(\sqrt[3]{2}, \omega)$ is a splitting field of $X^3 - 2$ over Q, whence it contains the three complex zeros of this polynomial. According to 2.1.9, it then follows that there are exactly three Q-monomorphisms from $Q(\sqrt[3]{2})$ to $Q(\sqrt[3]{2}, \omega)$.

b. Let K be a field of prime characteristic, and let L be a purely inseparable extension of K. Then, by virtue of 2.4.7, for every extension field F of L, the only K-monomorphism from L to F is $i_{L \to F}$, even if L is infinite over K and F is algebraically closed. □

It will be proved in section 3.7 that for every finite extension, the exact number of embeddings into "sufficiently large" fields is a factor of the linear degree. As we have just seen, purely inseparable extensions exhibit the most trivial behavior with respect to embeddability; it will be shown in that section that, as suggested by the terminology, exactly the opposite holds for separable extensions.

PROBLEMS

1. Let G be a group, and let F be a field. By a **character of** G **in** F we shall understand a mapping c from G to F such that $\mathrm{Im}(c) \subseteq F^*$ and such that $c(\alpha\beta) = c(\alpha)c(\beta)$ for all $\alpha, \beta \in G$. Prove the following assertions.

    a. If V is an F-space, if Ω is a finite set of characters of G in F, and if $(e_c)_{c \in \Omega}$ is a family of elements of V such that $\sum_{c \in \Omega} c(\alpha)e_c = 0$ for every $\alpha \in G$, then $e_c = 0$ for every $c \in \Omega$.

    b. The characters of G in F make up a linearly independent subset of $\mathrm{Map}(G, F)$.

2. Let G be a nontrivial subgroup of $\mathbf{Z}^+$, and let F be a field. Show that if S is a finite subset of F^*, and if $(\alpha_\sigma)_{\sigma \in S}$ is a family of elements of F such that $\sum_{\sigma \in S} \alpha_\sigma \sigma^n = 0$ for every $n \in G$, then $\alpha_\sigma = 0$ for every $\sigma \in S$.

3. Let F denote either the real field $\mathbf{R}$ or the complex field $\mathbf{C}$, and let G be a nontrivial subgroup of F^+. Show that if S is a finite subset of F^*, and if $(\alpha_\sigma)_{\sigma \in S}$ is a family of elements of F such that $\sum_{\sigma \in S} \alpha_\sigma \exp(\sigma\theta) = 0$ for every $\theta \in G$, then $\alpha_\sigma = 0$ for every $\sigma \in S$.

4. Prove 3.1.3 by using Dedekind's independence theorem, but no other result discussed previously in this book, as follows.

    Let $n = [L : K]$, and assume that there exist $n + 1$ distinct K-monomorphisms $u_1, u_2, \ldots, u_{n+1}$ from L to F. Next, choose a linear base $(\rho_1, \rho_2, \ldots, \rho_n)$ of L over K, and form the system

    $$u_1(\rho_1)\xi_1 + u_2(\rho_1)\xi_2 + \cdots + u_{n+1}(\rho_1)\xi_{n+1} = 0$$

    $$u_1(\rho_2)\xi_1 + u_2(\rho_2)\xi_2 + \cdots + u_{n+1}(\rho_2)\xi_{n+1} = 0$$

    $$\vdots \qquad \vdots \qquad \qquad \vdots \qquad \vdots$$

    $$u_1(\rho_n)\xi_1 + u_2(\rho_n)\xi_2 + \cdots + u_{n+1}(\rho_n)\xi_{n+1} = 0$$

    of n linear homogeneous equations in $n + 1$ unknowns $\xi_1, \xi_2, \ldots, \xi_{n+1}$ and with coefficients in F. Finally, obtain a contradiction to Dedekind's independence theorem by showing that if $(\alpha_1, \alpha_2, \ldots, \alpha_{n+1})$ is a nontrivial solution of this system in $F^{(n+1)}$, then $\sum_{j=1}^{n+1} \alpha_j u_j = 0$.

5. Prove the following statement, which is a generalization of 3.1.3, in each of the two ways indicated below.

    If K and F are fields, if L is a finite extension of K, and if u is a monomorphism from K to F, then the cardinality of the set of all monomorphisms from L to F extending u is less than or equal to $[L:K]$.

a. Use 0.0.1 and 3.1.3.

b. Consider first the case where $L = K(\alpha)$, where $\alpha \in L$; and show that if f denotes the minimal polynomial of α over K, then the mapping $v \rightarrow v(\alpha)$ from the set of all monomorphisms from L to F extending u to the set of all zeros of uf in F is bijective. Next, show that the general case follows from the case of a simple extension by considering a finite increasing sequence $(E_i)_{0 \leq i \leq n}$ of intermediate fields between K and L such that E_i is a simple extension of E_{i-1} for $1 \leq i \leq n$ and such that $E_0 = K$ and $E_n = L$.

6. Let K be a field, and let L be a finite extension of K. Show that if $n = [L:K]$ and $(\alpha_i)_{1 \leq i \leq n}$ is a linear base of L over K, and if $(u_i)_{1 \leq i \leq n}$ is a base of $\mathrm{Lin}_K(L, L)$, then the matrix $[u_i(\alpha_j)]_{1 \leq i, j \leq n}$ in $\mathrm{Mat}_n(L)$ is nonsingular.

One of the several approaches to the foundations of Galois theory is based on the Jacobson–Bourbaki correspondence. In the problems that follow, this topic is developed by standard techniques from linear algebra. Some of its applications to Galois theory will be presented as problems in subsequent sections.

7. Let F be a field. By an F-**algebra** we shall understand an F-subspace Φ of $\mathrm{Map}(F, F)$ enjoying the following three properties: (i) $i_F \in \Phi$; (ii) if $u, v \in \Phi$, then $u \circ v \in \Phi$; and (iii) if $u \in \Phi$, then $u(\rho + \sigma) = u(\rho) + u(\sigma)$ for all $\rho, \sigma \in F$. Prove the following assertions.
   a. If P is a subfield of F, then $\mathrm{Lin}_P(F, F)$ is an F-algebra.
   b. $\mathrm{Lin}_F(F, F) = Fi_F$, and hence $\mathrm{Lin}_F(F, F)$ is the smallest element of the set of all F-algebras.

8. Let F be a field, and let Φ be an F-algebra. We shall denote by $R(\Phi)$ the set of all $\alpha \in F$ such that $u(\alpha\rho) = \alpha u(\rho)$ for all $u \in \Phi$ and $\rho \in F$. Show that $R(\Phi)$ is the largest element of the set of all subfields P of F having the property that every element of Φ is P-linear.

9. Show that if F is a field, and if P is a subfield of F, then $R(\mathrm{Lin}_P(F, F)) = P$.

10. Let F be a field, and let n be a positive integer. Prove that if Φ is an n-dimensional F-algebra, then there exist a base $(u_i)_{1 \leq i \leq n}$ of Φ and a sequence $(\alpha_i)_{1 \leq i \leq n}$ of elements of F such that $u_i(\alpha_i) = 1$ when $1 \leq i \leq n$ and such that $u_i(\alpha_j) = 0$ when $1 \leq i, j \leq n$, and $i \neq j$.

11. Let F be a field, let n be a positive integer, and let Φ be an n-dimensional F-algebra.
   Let $(u_i)_{1 \leq i \leq n}$ and $(\alpha_i)_{1 \leq i \leq n}$ be, respectively, a base of Φ and a sequence of elements of F such that $u_i(\alpha_i) = 1$ for $1 \leq i \leq n$ and such that $u_i(\alpha_j) = 0$ when $1 \leq i, j \leq n$, and $i \neq j$. Verify the following assertions.

a. If $1 \le i, j \le n$, and $\beta \in F$, then $u_i \circ \beta u_j = u_i(\beta)u_j$.
b. If $1 \le i \le n$, then $\text{Im}(u_i) \subseteq R(\Phi)$.
c. $(\alpha_i)_{1 \le i \le n}$ is a linear base of F over $R(\Phi)$.

12. Let F be a field, and let Φ be an F-algebra. Show that Φ is finite-dimensional if and only if F is finite over $R(\Phi)$, in which case $[\Phi : F] = [F : R(\Phi)]$.

13. Let F be a field. By virtue of the results in the preceding problems, it is meaningful to speak of the mapping $P \to \text{Lin}_P(F, F)$ from the set of all fields admitting F as a finite extension to the set of all finite-dimensional F-algebras, and of the mapping $\Phi \to R(\Phi)$ from the set of all finite-dimensional F-algebras to the set of all fields admitting F as a finite extension. Prove that these mappings are mutually inverse inclusion-reversing bijections. (These bijections constitute the **Jacobson–Bourbaki correspondence** defined by the given field F.)

3.2. THE GENERAL GALOIS CORRESPONDENCES

Galois theory can be roughly described as the study of algebraic extensions by means of groups of field automorphisms. This section is devoted to the definition of these groups and the associated fields of invariants. Our principal goal is to set up the general Galois correspondences. At this preliminary stage, this can be regarded as a mere formality; we feel, however, that this will give the reader an indication of some of the central ideas of the theory.

The groups that are studied in Galois theory are groups of permutations of a special type. The general conventions concerning groups of permutations and the associated group actions will be applied in the sequel. In particular, the multiplicative notation will be used for the operations of such groups. The reader who does not feel sufficiently familiar with these concepts may wish to consult the references given in the section on prerequisites, in which the pertinent terminology and notation are explained.

Let F be a field. It is clear that the automorphisms of F form a subgroup of the symmetric group $\text{Sym}(F)$. This subgroup of $\text{Sym}(F)$ will be denoted by $\text{Aut}(F)$, and its subgroups will be called **groups of automorphisms** of F.

If Ω is a nonempty subset of $\text{Aut}(F)$, the $\alpha \in F$ such that $s(\alpha) = \alpha$ for every $s \in \Omega$ form a subfield of F, as is readily verified. This subfield of F will be denoted by $\text{Inv}(\Omega)$, and will be called the **field of invariants of** Ω. An element of F belongs to $\text{Inv}(\Omega)$ if and only if it is a fixed point of every automorphism belonging to Ω.

Next consider a field K and an extension field L of K. The K-automorphisms of L make up a group of automorphisms of L. This group will be denoted by Gal(L/K), and will be called the **Galois group of** L **over** K.

If F is a field, and if P denotes the prime subfield of F, then every automorphism of F is a P-automorphism, and hence Aut(F) = Gal(F/P).

It also follows that the Galois group of a field over itself is trivial. It will soon be seen, on the other hand, that the Galois group of a field over a proper subfield may be trivial.

Before introducing the concepts required for the general Galois correspondences, we shall collect some immediate consequences of the preceding definitions.

3.2.1. Proposition. *Let F be a field.*

(i) *If P and Q are subfields of F such that $P \subseteq Q$, then* Gal(F/P) $\supseteq$ Gal(F/Q).

(ii) *If Γ and Δ are groups of automorphisms of F such that $\Gamma \subseteq \Delta$, then* Inv(Γ) $\supseteq$ Inv(Δ).

(iii) *If Γ and P are, respectively, a group of automorphisms of F and a subfield of F, then $\Gamma \subseteq$* Gal(E/P) *if and only if $P \subseteq$* Inv(Γ).

(iv) *If P is a subfield of F, then $P \subseteq$* Inv(Gal(F/P)).

(v) *If Γ is a group of automorphisms of F, then $\Gamma \subseteq$* Gal($F/$Inv(Γ)).

Let F be a field. By a **Galois group on** F we shall understand a group Γ of automorphisms of F such that $\Gamma =$ Gal($F/$Inv(Γ)); and by an **invariant field in** F we shall understand a subfield P of F such that $P =$ Inv(Gal(F/P)).

Thus, we have introduced special terminology for the groups of automorphisms and subfields for which the inclusions in the last two assertions in the preceding proposition become equalities. Alternative terminology has been used by several authors in this context. A particularly common adjective used for this purpose is "closed", so that one speaks of "closed groups of automorphisms" and "closed subfields". However, since this adjective has a standard meaning in topology, and since we are going to deal with topological notions in connection with the Galois theory of infinite extensions, we prefer not to follow this usage.

It is evident that if F is a field, then Aut(F) and $\{i_F\}$ are Galois groups on F, and F is an invariant field in F. The extreme situation in which there are no other Galois groups and no other invariant fields obtains when the field under consideration admits no automorphism other than the identity mapping. Examples of such fields are the prime fields; others will be discussed later in the section.

We now are prepared to state and prove the results known as the **fundamental theorems of Galois theory**.

3.2.2. Theorem. *Let F be a field.*

(i) *If P is a subfield of F, then* Gal(F/P) *is a Galois group on F.*

(ii) *If Γ is a group of automorphisms of F, then $\mathrm{Inv}(\Gamma)$ is an invariant field in F.*

(iii) *The mapping $P \to \mathrm{Gal}(F/P)$ from the set of all invariant fields in F to the set of all Galois groups on F and the mapping $\Gamma \to \mathrm{Inv}(\Gamma)$ from the set of all Galois groups on F to the set of all invariant fields in F are mutually inverse inclusion-reversing bijections.*

Proof. The verifications of (i) and (ii) are similar; therefore, we shall consider only (i). If P is a subfield of F, we know that $P \subseteq \mathrm{Inv}(\mathrm{Gal}(F/P))$, which implies that

$$\mathrm{Gal}(F/P) \supseteq \mathrm{Gal}(F/\mathrm{Inv}(\mathrm{Gal}(F/P)));$$

on the oher hand, since $\mathrm{Gal}(F/P)$ is a group of automorphisms of F, we have

$$\mathrm{Gal}(F/P) \subseteq \mathrm{Gal}(F/\mathrm{Inv}(\mathrm{Gal}(F/P))).$$

It follows from these inclusions that

$$\mathrm{Gal}(F/P) = \mathrm{Gal}(F/\mathrm{Inv}(\mathrm{Gal}(F/P))),$$

which means that $\mathrm{Gal}(F/P)$ is a Galois group on F.

To verify (iii), we first note that by virtue of (i) and (ii), it is meaningful to speak of the two mappings in question. Moreover, we have already observed that these mappings are inclusion-reversing. Finally, to say that they are mutually inverse bijections amounts to saying that $P = \mathrm{Inv}(\mathrm{Gal}(F/P))$ for every invariant field P in F and that $\Gamma = \mathrm{Gal}(F/\mathrm{Inv}(\Gamma))$ for every Galois group Γ on F, hence to repeating the definitions of an invariant field and a Galois group.   □

3.2.3. Theorem. *Let K be a field, and let L be an extension field of K.*

(i) *If E is an intermediate field between K and L, then $\mathrm{Gal}(L/E)$ is a Galois group on L contained in $\mathrm{Gal}(L/K)$.*

(ii) *If Γ is a subgroup of $\mathrm{Gal}(L/K)$, then $\mathrm{Inv}(\Gamma)$ is an invariant field in L containing K.*

(iii) *The mapping $E \to \mathrm{Gal}(L/E)$ from the set of all invariant fields in L containing K to the set of all Galois groups on L contained in $\mathrm{Gal}(L/K)$ and the mapping $\Gamma \to \mathrm{Inv}(\Gamma)$ from the set of all Galois groups on L contained in $\mathrm{Gal}(L/K)$ to the set of all invariant fields in L containing K are mutually inverse inclusion-reversing bijections.*

Proof. It is clear that (i) and (ii) follow from 3.2.2 and 3.2.1; and (iii) is an immediate consequence of (i), (ii), and 3.2.2.   □

The pairs of bijections described in 3.2.2 and 3.2.3 are called the **Galois correspondences**. The first is regarded as being defined by a single field, and the second by a field extension; and hence they are viewed as "absolute" and "relative", respectively.

At this point in our discussion, some comments seem to be in order. It has to be apparent to the reader that the two theorems just presented leave much to be desired. After all, what has been done could be described by saying that Galois groups and invariant fields have been artificially defined so that the Galois correspondences could be set up. For these theorems to have the substantive meaning implied by their designation as the "fundamental theorems", it is essential that good group-theoretical and field-theoretical conditions be found under which the properties of Galois groups and invariant fields imply interesting consequences—indeed, one of our main tasks in the present chapter is to show that such conditions can be given.

The preceding theorems, which have only formal content, certainly cannot satisfy this requirement. The only justification for stating them here is that they lead immediately to some basic questions: Given a field F, can the Galois groups on F and the invariant fields in F be determined? And given a field K and an extension field L of K, can the Galois groups on L contained in $\text{Gal}(L/K)$ and the invariant fields in L containing K be determined?

These questions, needless to say, are of considerable importance in Galois theory. We shall show in subsequent sections that under suitable restrictions, satisfactory answers can be given.

3.2.4. Examples

By virtue of what has been shown in 2.1.23 and 3.1.5, we have:

a. $\text{Aut}(\mathbf{R})$ is trivial.

b. If n is an odd positive integer, then $\text{Aut}(\mathbf{Q}(\sqrt[n]{2}))$ is trivial.

c. If K is a field of prime characteristic, and if L is a purely inseparable extension of K, then $\text{Gal}(L/K)$ is trivial. $\qquad \square$

3.2.5. Example. Let K be a field and $\alpha \in K$, and suppose that α does not possess a square root in K. Form a simple extension $K(\rho)$ of K by adjunction of a square root ρ of α in an extension field of K. Then:

(i) If $\text{Char}(K) = 2$, then $\text{Gal}(K(\rho)/K)$ is trivial; and K is not invariant in $K(\rho)$.

(ii) If $\text{Char}(K) \neq 2$, then $\text{Gal}(K(\rho)/K)$ has order 2, its elements being $i_{K(\rho)}$ and the K-automorphism of $K(\rho)$ such that $\rho \to -\rho$; and K is invariant in $K(\rho)$.

By assumption, the polynomial $X^2 - \alpha$ in $K[X]$ has no zeros in K, and hence it is irreducible in $K[X]$. Consequently, since ρ and $-\rho$ are zeros of $X^2 - \alpha$ and $K(\rho) = K(-\rho)$, there exists a K-automorphism s of $K(\rho)$ such that $s(\rho) = -\rho$. It is seen that $i_{K(\rho)}$ and s are the only K-automorphisms of $K(\rho)$: For a K-automorphism of $K(\rho)$ sends ρ to a zero of $X^2 - \alpha$

in $K(\rho)$; therefore, it sends ρ to ρ or $-\rho$, and so it coincides with $i_{K(\rho)}$ or s.

To prove (i), suppose that $\mathrm{Char}(K)=2$. Then $2\rho=0$, so that $s(\rho)=-\rho=\rho$. Therefore $s=i_{K(\rho)}$, whence $\mathrm{Gal}(K(\rho)/K)$ is trivial; and this implies that K is not invariant in $K(\rho)$.

To prove (ii), suppose now that $\mathrm{Char}(K)\neq2$. Since $\rho\neq0$, we have $2\rho\neq0$, so that $s(\rho)=-\rho\neq\rho$. This shows that $s\neq i_{K(\rho)}$, and hence $\mathrm{Gal}(K(\rho)/K)$ has order 2. Finally, since ρ is not a fixed point of s, we see that $K\subseteq\mathrm{Inv}(\mathrm{Gal}(K(\rho)/K))\subset K(\rho)$; and since $[K(\rho):K]=2$, this implies that $K=\mathrm{Inv}(\mathrm{Gal}(K(\rho)/K))$, that is, that K is invariant in $K(\rho)$. □

3.2.6. Example. Let K be a field and $\alpha\in K$, and suppose that α does not admit a cube root in K. Form a simple extension $K(\rho)$ of K by adjunction of a cube root ρ of α in an extension field of K. Then:

(i) If the only cube root of unity in K is 1, then $\mathrm{Gal}(K(\rho)/K)$ is trivial; and K is not invariant in $K(\rho)$.

(ii) If there exists a cube root ω of unity in K such that $\omega\neq1$, then $\mathrm{Gal}(K(\rho)/K)$ has order 3, its elements being $i_{K(\rho)}$, the K-automorphism of $K(\rho)$ such that $\rho\to\omega\rho$, and the K-automorphism of $K(\rho)$ such that $\rho\to\omega^2\rho$; and K is invariant in $K(\rho)$.

Our assumption implies that the polynomial $X^3-\alpha$ in $K[X]$ possesses no zeros in K, and hence it is irreducible in $K[X]$.

To prove (i), suppose that 1 is the only cube root of unity in K. It suffices to verify that $\mathrm{Gal}(K(\rho)/K)$ is trivial. Assume, for a contradiction, that there exists a K-automorphism s of $K(\rho)$ such that $s\neq i_{K(\rho)}$, and write $\omega=s(\rho)/\rho$. Clearly $s(\rho)\neq\rho$, and so $\omega\neq1$; and since

$$s(\rho)^3=s(\rho^3)=s(\alpha)=\alpha=\rho^3,$$

we get $\omega^3=1$. Moreover, it follows from the equalities

$$(\omega-1)(\omega^2+\omega+1)=\omega^3-1=0$$

that $\omega^2+\omega+1=0$, whence $[K(\omega):K]\leq2$. Since $[K(\omega):K]\mid[K(\rho):K]$ and $[K(\rho):K]=3$, we conclude that $[K(\omega):K]=1$, which means that $K(\omega)=K$. But then $\omega\in K$, $\omega\neq1$, and $\omega^3=1$, which is impossible.

Now suppose, in order to prove (ii), that there exists an $\omega\in K$ such that $\omega\neq1$ and $\omega^3=1$. Since

$$(\omega\rho)^3=\omega^3\rho^3=\rho^3=\alpha \quad\text{and}\quad (\omega^2\rho)^3=\omega^6\rho^3=\rho^3=\alpha,$$

we see that $\rho,\omega\rho,\omega^2\rho$ are the three distinct zeros of $X^3-\alpha$ in $K(\rho)$. Since $K(\rho)=K(\omega\rho)=K(\omega^2\rho)$, we conclude that there exist K-automorphisms s and t of $K(\rho)$ such that $s(\rho)=\omega\rho$ and $t(\rho)=\omega^2\rho$. Next note that $i_{K(\rho)},s,t$ are the only K-automorphisms of $K(\rho)$: indeed, a K-automorphism of $K(\rho)$ sends ρ to a zero of $X^3-\alpha$ in $K(\rho)$, and hence to one of $\rho,\omega\rho,\omega^2\rho$; therefore, it coincides with one of $i_{K(\rho)},s,t$. Thus $\mathrm{Gal}(K(\rho)/K)$ has order 3. Finally, since ρ is a fixed point of neither s nor t, we have $K\subseteq\mathrm{Inv}(\mathrm{Gal}(K(\rho)/K))\subset K(\rho)$; and since $[K(\rho):K]=3$, this implies that $K=\mathrm{Inv}(\mathrm{Gal}(K(\rho)/K))$, which is what we wanted. □

3.2.7. Example. Let K be a field, and let $K(\xi)$ be a simple transcendental extension of K. Then:

(i) If $\sigma \in K(\xi) - K$, and if f and g are relatively prime polynomials in $K[X]$ such that $\sigma = f(\xi)/g(\xi)$, then $\sigma g(X) - f(X)$ is irreducible in $K(\sigma)[X]$ and has ξ as a zero, and

$$[K(\xi): K(\sigma)] = \sup(\deg(f), \deg(g)).$$

(ii) If E is an intermediate field between K and $K(\xi)$ different from K, then $K(\xi)$ is finite over E.

(iii) If $\begin{bmatrix} \alpha & \beta \\ \gamma & \delta \end{bmatrix} \in GL_2(K)$, then there exists a unique K-automorphism of $K(\xi)$ such that $\xi \to (\alpha\xi + \beta)/(\gamma\xi + \delta)$.

(iv) $\mathrm{Gal}(K(\xi)/K)$ is nonabelian.

(v) The mapping from $GL_2(K)$ to $\mathrm{Gal}(K(\xi)/K)$ assigning to each $\begin{bmatrix} \alpha & \beta \\ \gamma & \delta \end{bmatrix} \in GL_2(K)$ the K-automorphism of $K(\xi)$ such that $\xi \to (\alpha\xi + \beta)/(\gamma\xi + \delta)$ is a surjective antihomomorphism, and its kernel is the image of the natural embedding $\alpha \to \begin{bmatrix} \alpha & 0 \\ 0 & \alpha \end{bmatrix}$ of K^* in $GL_2(K)$.

(vi) If K is finite, then $\mathrm{Gal}(K(\xi)/K)$ is a group of order $\mathrm{Card}(K)^3 - \mathrm{Card}(K)$.

(vii) If Γ is an infinite subgroup of $\mathrm{Gal}(K(\xi)/K)$, then $\mathrm{Inv}(\Gamma) = K$.

(viii) If K is infinite, then K is invariant in $K(\xi)$.

To prove (i), let $\sigma \in K(\xi) - K$ and $\sigma = f(\xi)/g(\xi)$, where f and g are relatively prime polynomials in $K[X]$. Put $n = \sup(\deg(f), \deg(g))$; we can then write

$$f(X) = \sum_{i=0}^{n} \alpha_i X^i \quad \text{and} \quad g(X) = \sum_{i=0}^{n} \beta_i X^i,$$

where $\alpha_0, \alpha_1, \ldots, \alpha_n, \beta_0, \beta_1, \ldots, \beta_n \in K$, and $\alpha_n \neq 0$ or $\beta_n \neq 0$. Since $\sigma \notin K$, we obviously have $n > 0$ and $\sigma\beta_n - \alpha_n \neq 0$; and since

$$\sigma g(X) - f(X) = \sum_{i=0}^{n} (\sigma\beta_i - \alpha_i) X^i,$$

we see that $\sigma g(X) - f(X)$ is a polynomial in $K(\sigma)[X]$ of degree n. Next, note that $\sigma g(\xi) - f(\xi) = 0$, which means that ξ is a zero of $\sigma g(X) - f(X)$; therefore, ξ is algebraic over $K(\sigma)$. Since ξ is transcendental over K, this implies that σ is transcendental over K. According to 1.5.12 and 2.1.2, it now follows that $\sigma g(X) - f(X)$ is irreducible in $K(\sigma)[X]$, whence

$$[K(\xi): K(\sigma)] = [K(\sigma)(\xi): K(\sigma)] = \deg(\sigma g(X) - f(X)) = n,$$

which is what we wanted.

In order to prove (ii), consider a subfield E of $K(\xi)$ such that $K \subset E$, and choose a $\sigma \in E - K$. It then follows from (i) that $K(\xi)$ is finite over $K(\sigma)$; and since $K(\sigma) \subseteq E \subseteq K(\xi)$, we conclude that $K(\xi)$ is finite over E.

We proceed next to verify (iii). Let $\begin{bmatrix} \alpha & \beta \\ \gamma & \delta \end{bmatrix} \in GL_2(K)$; then $\alpha, \beta, \gamma, \delta$ $\in K$ and

$$\alpha\delta - \gamma\beta = \det\left(\begin{bmatrix} \alpha & \beta \\ \gamma & \delta \end{bmatrix}\right) \neq 0.$$

Consequently, we have

$$\sup(\deg(\alpha X + \beta), \deg(\gamma X + \delta)) = 1;$$

and furthermore

$$\gamma(\alpha X + \beta) - \alpha(\gamma X + \delta) = \gamma\beta - \alpha\delta \neq 0,$$

so that $\alpha X + \beta$ and $\gamma X + \delta$ are relatively prime polynomials in $K[X]$. Now let $\sigma = (\alpha\xi + \beta)/(\gamma\xi + \delta)$; the transcendence of ξ over K implies that $\sigma \notin K$. According to (i), we then have $[K(\xi): K(\sigma)] = 1$, whence $K(\sigma) = K(\xi)$. We see, in particular, that σ is transcendental over K; the existence of a K-automorphism of $K(\xi)$ such that $\xi \to \sigma$ is now an immediate consequence of 2.1.2.

It is easy to deduce (iv) from (iii). For since

$$\begin{bmatrix} 0 & 1 \\ 1 & 0 \end{bmatrix}, \begin{bmatrix} 1 & 1 \\ 0 & 1 \end{bmatrix} \in GL_2(K),$$

there exist K-automorphisms s and t of $K(\xi)$ such that

$$s(\xi) = 1/\xi \quad \text{and} \quad t(\xi) = \xi + 1.$$

The equality $st = ts$ would then imply that

$$1/\xi + 1 = s(t(\xi)) = t(s(\xi)) = 1/(\xi + 1),$$

which in turn implies that $1 + \xi + \xi^2 = 0$, in contradiction to the assumed transcendence of ξ over K.

To prove (v), we shall verify first that the mapping in question is an antihomomorphism. Thus, let

$$\begin{bmatrix} \alpha & \beta \\ \gamma & \delta \end{bmatrix}, \begin{bmatrix} \bar{\alpha} & \bar{\beta} \\ \bar{\gamma} & \bar{\delta} \end{bmatrix} \in GL_2(K);$$

then

$$\begin{bmatrix} \alpha & \beta \\ \gamma & \delta \end{bmatrix} \begin{bmatrix} \bar{\alpha} & \bar{\beta} \\ \bar{\gamma} & \bar{\delta} \end{bmatrix} = \begin{bmatrix} \alpha\bar{\alpha} + \beta\bar{\gamma} & \alpha\bar{\beta} + \beta\bar{\delta} \\ \gamma\bar{\alpha} + \delta\bar{\gamma} & \gamma\bar{\beta} + \delta\bar{\delta} \end{bmatrix}.$$

We have to show that if s, $\bar{s}$, and t are the K-automorphisms of $K(\xi)$ satisfying

$$s(\xi) = (\alpha\xi + \beta)/(\gamma\xi + \delta), \qquad \bar{s}(\xi) = (\bar{\alpha}\xi + \bar{\beta})/(\bar{\gamma}\xi + \bar{\delta}),$$

and

$$t(\xi) = ((\alpha\bar{\alpha} + \beta\bar{\gamma})\xi + (\alpha\bar{\beta} + \beta\bar{\delta}))/((\gamma\bar{\alpha} + \delta\bar{\gamma})\xi + (\gamma\bar{\beta} + \delta\bar{\delta})),$$

then $t(\xi) = \bar{s}(s(\xi))$; this presents no difficulty, for

$$\bar{s}(s(\xi)) = \bar{s}((\alpha\xi + \beta)/(\gamma\xi + \delta))$$

$$= (\alpha\bar{s}(\xi) + \beta)/(\gamma\bar{s}(\xi) + \delta))$$

$$= (\alpha((\bar{\alpha}\xi + \bar{\beta})/(\bar{\gamma}\xi + \bar{\delta})) + \beta)/(\gamma((\bar{\alpha}\xi + \bar{\beta})/(\bar{\gamma}\xi + \bar{\delta})) + \delta)$$

$$= ((\alpha\bar{\alpha} + \beta\bar{\gamma})\xi + (\alpha\bar{\beta} + \beta\bar{\delta}))/((\gamma\bar{\alpha} + \delta\bar{\gamma})\xi + (\gamma\bar{\beta} + \delta\bar{\delta})) = t(\xi).$$

Next, we have to show that our antihomomorphism is surjective. To this end, let $s \in \mathrm{Gal}(K(\xi)/K)$; then

$$K(\xi) = s(K(\xi)) = K(s(\xi)),$$

and hence $s(\xi) \notin K$. Write $s(\xi) = f(\xi)/g(\xi)$, where f and g are relatively prime polynomials in $K[X]$. Since

$$[K(\xi): K(s(\xi))] = [K(\xi): K(\xi)] = 1,$$

we deduce from (i) that $\sup(\deg(f), \deg(g)) = 1$; therefore

$$f(X) = \alpha X + \beta \quad \text{and} \quad g(X) = \gamma X + \delta,$$

where $\alpha, \beta, \gamma, \delta \in K$, and $\alpha \neq 0$ or $\gamma \neq 0$. Moreover, we see that $\alpha\delta - \gamma\beta \neq 0$: for otherwise, the equalities

$$\gamma f(X) - \alpha g(X) = \gamma(\alpha X + \beta) - \alpha(\gamma X + \delta)$$

$$= \gamma\beta - \alpha\delta = 0$$

would imply that f and g are not relatively prime. Thus

$$\det\left(\begin{bmatrix} \alpha & \beta \\ \gamma & \delta \end{bmatrix}\right) = \alpha\delta - \gamma\beta \neq 0,$$

and so $\begin{bmatrix} \alpha & \beta \\ \gamma & \delta \end{bmatrix} \in GL_2(K)$; and since

$$s(\xi) = f(\xi)/g(\xi) = (\alpha\xi + \beta)/(\gamma\xi + \delta),$$

this implies that s is the value at $\begin{bmatrix} \alpha & \beta \\ \gamma & \delta \end{bmatrix}$ of the antihomomorphism under consideration, as was to be shown. It now remains to verify the assertion concerning the kernel. To do this, let $\begin{bmatrix} \alpha & \beta \\ \gamma & \delta \end{bmatrix} \in GL_2(K)$. It is clear that $\begin{bmatrix} \alpha & \beta \\ \gamma & \delta \end{bmatrix}$ belongs to this kernel if and only if $\xi = (\alpha\xi + \beta)/(\gamma\xi + \delta)$, and hence if and only if $\beta + (\alpha - \delta)\xi + \gamma\xi^2 = 0$. This proves the assertion, because in view of the transcendence of ξ over K, the latter equality holds if and only if the equalities $\alpha = \delta$ and $\beta = 0 = \gamma$ hold.

To prove (vi), suppose now that K is finite, and write $q = \mathrm{Card}(K)$. Then K^* is a group of order $q - 1$; and according to 0.0.6, $GL_2(K)$ is a group of order $(q^2 - 1)(q^2 - q)$. Moreover, it follows from (v) that $\mathrm{Gal}(K(\xi)/K)$ is antiisomorphic to the quotient group of $GL_2(K)$ by a normal subgroup isomorphic to K^*. Consequently

$$\text{Card}(\text{Gal}(K(\xi)/K)) = (q^2 - 1)(q^2 - q)/(q - 1)$$
$$= (q^2 - 1)q = q^3 - q,$$

as required.

In order to verify (vii), let Γ be an infinite subgroup of $\text{Gal}(K(\xi)/K)$. Since

$$\Gamma \subseteq \text{Gal}(K(\xi)/\text{Inv}(\Gamma)) \subseteq \text{Mon}_{\text{Inv}(\Gamma)}(K(\xi), K(\xi)),$$

we see that $\text{Mon}_{\text{Inv}(\Gamma)}(K(\xi), K(\xi))$ is infinite. It then follows from 3.1.3 that $K(\xi)$ is infinite over $\text{Inv}(\Gamma)$; and according to (ii), this implies that $\text{Inv}(\Gamma) = K$.

To complete the discussion, we shall now deduce (viii) from (vii). Suppose that K is infinite. For each $\alpha \in K^*$, we have $\begin{bmatrix} \alpha & 0 \\ 0 & 1 \end{bmatrix} \in GL_2(K)$, and (iii) implies the existence of a K-automorphism s_α of $K(\xi)$ such that $s_\alpha(\xi) = \alpha\xi$; it is clear that $s_\alpha \neq s_\beta$ whenever $\alpha, \beta \in K^*$ and $\alpha \neq \beta$. Consequently, $\text{Gal}(K(\xi)/K)$ is infinite, and (vii) shows that $\text{Inv}(\text{Gal}(K(\xi)/K)) = K$. $\square$

Assertion (viii) in the last example is only one half of the truth. In effect, as we shall see in 3.4.12, the property of being invariant in simple transcendental extensions characterizes infinite fields.

Let K be a field, and let n be a positive integer. It is known that the image of the natural embedding

$$\alpha \rightarrow \begin{bmatrix} \alpha & 0 & 0 & \cdots & 0 \\ 0 & \alpha & 0 & \cdots & 0 \\ 0 & 0 & \alpha & \cdots & 0 \\ \vdots & \vdots & \vdots & & \vdots \\ 0 & 0 & 0 & \cdots & \alpha \end{bmatrix}$$

of K^* in $GL_n(K)$ coincides with the center of $GL_n(K)$; for $n = 2$, this is easily verified by direct computation. The quotient group of $GL_n(K)$ by its center is studied in projective geometry; it is usually denoted by $PGL_n(K)$.

What has been shown in the last example implies that the Galois group of a simple transcendental extension of a field K is antiisomorphic to $PGL_2(K)$. The K-automorphisms defined as in (iii) by the matrices in $GL_2(K)$ are called **linear fractional transformations** or **projective transformations**.

3.2.8. Example. Let K be a field of characteristic 0, and let $K(\xi)$ be a simple transcendental extension of K. For every integer n, let s_n denote the K-automorphism of $K(\xi)$ such that $s_n(\xi) = \xi + n$. The mapping $n \rightarrow s_n$ from $\mathbf{Z}^+$ to $\text{Gal}(K(\xi)/K)$ is an injective homomorphism, and hence its image is an infinite cyclic group. By virtue of what was shown in 3.2.7, it then follows that $\mathrm{Inv}(\Gamma) = K$ for every nontrivial subgroup Γ of this group. □

Let us pause for a moment in order to make some remarks concerning the preceding examples. First, they illustrate some of the typical ideas used in computing Galois groups and fields of invariants by elementary means. Also, they illustrate some of the pathology that may take place when suitable restrictions are not imposed: We see from 3.2.4 that there are fields that are not prime, but possess a unique automorphism; what was shown in 3.2.4, 3.2.5, and 3.2.6 implies that there are proper field extensions having trivial Galois group, and hence having a unique intermediate field invariant in the top field; and what was said in 3.2.8 implies the existence of a field extension whose Galois group contains infinitely many subgroups having the same field of invariants and making up a chain.

Throughout this chapter, we shall often deal with the problem of determining Galois groups and fields of invariants. It will not always be necessary to proceed by "brute force", as was done in the examples just discussed; we shall see that the main theorems of Galois theory can be used effectively in such computations. As the reader will quickly realize, however, the resulting techniques are of a very limited scope, and often have to be applied in conjunction with *ad hoc* tricks.

In the remainder of this section, we shall present the elementary generalities on groups of field automorphisms that will be required in the sequel.

3.2.9. Proposition. *Let F be a field, and let Ω be a nonempty subset of $\mathrm{Aut}(F)$. Then $\mathrm{Inv}(\Omega) = \mathrm{Inv}(\langle \Omega \rangle)$.*

Proof. Let $\alpha \in F$; and denote by Γ the stabilizer of α under $\mathrm{Aut}(F)$, that is, the subgroup of $\mathrm{Aut}(F)$ consisting of the automorphisms of F having α as a fixed point.

It follows from the definition of field of invariants that $\alpha \in \mathrm{Inv}(\Omega)$ if and only if $\Gamma \supseteq \Omega$, and that $\alpha \in \mathrm{Inv}(\langle \Omega \rangle)$ if and only if $\Gamma \supseteq \langle \Omega \rangle$. But since Γ is a subgroup of $\mathrm{Aut}(F)$, the conditions $\Gamma \supseteq \Omega$ and $\Gamma \supseteq \langle \Omega \rangle$ are equivalent, and we conclude that $\alpha \in \mathrm{Inv}(\Omega)$ if and only if $\alpha \in \mathrm{Inv}(\langle \Omega \rangle)$. □

This proposition shows that when considering fields of invariants, nothing is lost by restricting ourselves to groups—instead of arbitrary nonempty sets—of field automorphisms.

3.2.10. Proposition. *Let F be a field, and let $(\Gamma_i)_{i \in I}$ be a nonempty family of groups of automorphisms of F.*

(i) $\mathrm{Inv}(\vee_{i \in I} \Gamma_i) = \cap_{i \in I} \mathrm{Inv}(\Gamma_i)$.

(ii) $\mathrm{Inv}(\cap_{i \in I} \Gamma_i) \supseteq \vee_{i \in I} \mathrm{Inv}(\Gamma_i)$.

Proof. Since $\vee_{i \in I} \Gamma_i = \langle \cup_{i \in I} \Gamma_i \rangle$, we deduce from 3.2.9 that

$$\mathrm{Inv}\left(\bigvee_{i \in I} \Gamma_i \right) = \mathrm{Inv}\left(\bigcup_{i \in I} \Gamma_i \right) = \bigcap_{i \in I} \mathrm{Inv}(\Gamma_i),$$

which proves (i).

To verify (ii), note that $\cap_{i \in I} \Gamma_i \subseteq \Gamma_j$ for every $j \in I$. Consequently, $\mathrm{Inv}(\cap_{i \in I} \Gamma_i) \supseteq \mathrm{Inv}(\Gamma_j)$ for every $j \in I$, and so

$$\mathrm{Inv}\left(\bigcap_{i \in I} \Gamma_i \right) \supseteq \bigvee_{i \in I} \mathrm{Inv}(\Gamma_i),$$

as required. $\square$

3.2.11. Proposition. *Let F be a field, and let $(P_i)_{i \in I}$ be a nonempty family of subfields of F.*

(i) $\mathrm{Gal}(F/ \vee_{i \in I} P_i) = \cap_{i \in I} \mathrm{Gal}(F/P_i)$.

(ii) $\mathrm{Gal}(F/ \cap_{i \in I} P_i) \supseteq \vee_{i \in I} \mathrm{Gal}(F/P_i)$.

Proof. Let $s \in \mathrm{Aut}(F)$, and write $E = \mathrm{Inv}(\{s\})$. To say that $s \in \mathrm{Gal}(F/ \vee_{i \in I} P_i)$ means that $E \supseteq \vee_{i \in I} P_i$, and to say that $s \in \cap_{i \in I} \mathrm{Gal}(F/P_i)$ means that $E \supseteq P_i$ for every $i \in I$. Since $E \supseteq \vee_{i \in I} P_i$ if and only if $E \supseteq P_i$ for every $i \in I$, we conclude that $s \in \mathrm{Gal}(F/ \vee_{i \in I} P_i)$ if and only if $s \in \cap_{i \in I} \mathrm{Gal}(F/P_i)$. This proves (i).

Next note that $\cap_{i \in I} P_i \subseteq P_j$ for every $j \in I$, so that $\mathrm{Gal}(F/ \cap_{i \in I} P_i) \supseteq \mathrm{Gal}(F/P_j)$ for every $j \in I$. Therefore

$$\mathrm{Gal}\left(F \Big/ \bigcap_{i \in I} P_i \right) \supseteq \bigvee_{i \in I} \mathrm{Gal}(F/P_i),$$

which is (ii). $\square$

Examples can be given to show that the inclusions appearing in the preceding two propositions may be strict (problems 6 and 9).

3.2.12. Proposition. *Let F be a field, let P be a subfield of F, and let $s, t \in \mathrm{Aut}(F)$.*

(i) *$s\,\mathrm{Gal}(F/P) = t\,\mathrm{Gal}(F/P)$ if and only if s and t agree on P.*

(ii) *$\mathrm{Gal}(F/P)s = \mathrm{Gal}(F/P)t$ if and only if s^{-1} and t^{-1} agree on P.*

Proof. It suffices to verify one of the assertions. We have $s\,\mathrm{Gal}(F/P) = t\,\mathrm{Gal}(F/P)$ if and only if $t^{-1}s \in \mathrm{Gal}(F/P)$, hence if and only if $t^{-1}s(\alpha) = \alpha$ for every $\alpha \in P$; and the latter condition means that $s(\alpha) = t(\alpha)$ for every $\alpha \in P$. $\square$

3.2.13. Proposition. *Let F be a field and $s \in \mathrm{Aut}(F)$.*

(i) *If Γ is a group of automorphisms of F, then*

$$\mathrm{Inv}(s \Gamma s^{-1}) = s(\mathrm{Inv}(\Gamma)).$$

(ii) *If P is a subfield of F, then*

$$\mathrm{Gal}(F/s(P)) = s\,\mathrm{Gal}(F/P)s^{-1}.$$

Proof. To prove (i), let Γ be a group of automorphisms of F, and let $\alpha \in F$. Then $\alpha \in \mathrm{Inv}(s\Gamma s^{-1})$ if and only if $sts^{-1}(\alpha) = \alpha$ for every $t \in \Gamma$, hence if and only if $ts^{-1}(\alpha) = s^{-1}(\alpha)$ for every $t \in \Gamma$. And the latter condition means that $s^{-1}(\alpha) \in \mathrm{Inv}(\Gamma)$, which in turn means that $\alpha \in s(\mathrm{Inv}(\Gamma))$.

To prove (ii), consider a subfield P of F, and let $t \in \mathrm{Aut}(F)$. To say that $t \in \mathrm{Gal}(F/s(P))$ means that $ts(\alpha) = s(\alpha)$ for every $\alpha \in P$, and hence that $s^{-1}ts(\alpha) = \alpha$ for every $\alpha \in P$. But the latter condition means that $s^{-1}ts \in \mathrm{Gal}(F/P)$, which is equivalent to $t \in s\,\mathrm{Gal}(F/P)s^{-1}$. □

It is not accidental, of course, that the adjective "normal", which is of standard usage in the theory of groups, has been adopted for a certain type of field extension. Indeed, as we shall now see, the normality of subgroups in the Galois group of a field extension is related to the normality of intermediate fields over the bottom field.

3.2.14. Corollary. *Let K be a field, and let L be an extension field of K.*

(i) *If E is an intermediate field between K and L that is normal over K, then* $\mathrm{Gal}(L/E)$ *is normal in* $\mathrm{Gal}(L/K)$.

(ii) *If L is normal over K and Γ is a normal subgroup of* $\mathrm{Gal}(L/K)$, *then* $\mathrm{Inv}(\Gamma)$ *is normal over K.*

Proof. First, let E be as in (i). Then the proposition and 2.3.4 show that

$$s\,\mathrm{Gal}(L/E)s^{-1} = \mathrm{Gal}(L/s(E)) = \mathrm{Gal}(L/E)$$

for every $s \in \mathrm{Gal}(L/K)$, which means that $\mathrm{Gal}(L/E)$ is normal in $\mathrm{Gal}(L/K)$.

Assume next that the conditions described in (ii) obtain. By the proposition, we then have

$$s(\mathrm{Inv}(\Gamma)) = \mathrm{Inv}(s\Gamma s^{-1}) = \mathrm{Inv}(\Gamma)$$

for every $s \in \mathrm{Gal}(L/K)$. And according to 2.3.7, this implies the normality of $\mathrm{Inv}(\Gamma)$ over K. □

We shall next state, for the sake of reference, an important particular case of 3.1.3.

3.2.15. Proposition. *Let K be a field, and let L be a finite extension of K. Then*

$$\mathrm{Card}(\mathrm{Gal}(L/K)) \le [L:K].$$

Thus, for finite extensions, the linear degree is an upper bound for the order of the Galois group. It is known that the validity of this assertion depends in an essential manner on the assumed finiteness (section 3.3, problem 7). Note also that the stated inequality may be strict; this can be seen from the examples given earlier in the section.

3.2.16. Example. Let K be a field, and let $K(\xi)$ be a simple transcendental extension of K. Denote by s and t the K-automorphisms of $K(\xi)$ such that $s(\xi) = 1/\xi$ and $t(\xi) = 1 - \xi$. Then:

  (i) s and t are elements of order 2 in $\mathrm{Gal}(K(\xi)/K)$.
  (ii) The subgroup $\langle s, t \rangle$ of $\mathrm{Gal}(K(\xi)/K)$ has order 6.
  (iii) $\mathrm{Inv}(\langle s, t \rangle) = K((\xi^2 - \xi + 1)^3/\xi^2(\xi - 1)^2)$.

Since $s \neq i_{K(\xi)} \neq t$ and $s^2 = i_{K(\xi)} = t^2$, it clearly follows that s and t are elements of order 2 in $\mathrm{Gal}(K(\xi)/K)$, which is (i).

We now prove (ii) and (iii). Let us write $\sigma = (\xi^2 - \xi + 1)^3/\xi^2(\xi - 1)^2$. Then

$$s(\sigma) = \left(1/\xi^2 - 1/\xi + 1\right)^3 / \left(1/\xi^2\right)\left(1/\xi - 1\right)^2$$
$$= \left((1 - \xi + \xi^2)^3/\xi^6\right) / \left((1 - \xi)^2/\xi^4\right)$$
$$= (\xi^2 - \xi + 1)^3/\xi^2(\xi - 1)^2 = \sigma$$

and

$$t(\sigma) = \left((1 - \xi)^2 - (1 - \xi) + 1\right)^3 / (1 - \xi)^2\left((1 - \xi) - 1\right)^2$$
$$= (\xi^2 - \xi + 1)^3/\xi^2(\xi - 1)^2 = \sigma,$$

so that σ is a fixed point of s and t. Consequently, $\sigma \in \mathrm{Inv}(\langle s, t \rangle)$, and so $K(\sigma) \subseteq \mathrm{Inv}(\langle s, t \rangle)$.

Since $\sigma\xi^2(\xi - 1)^2 - (\xi^2 - \xi + 1)^3 = 0$, we see that ξ is a zero of the polynomial $\sigma X^2(X - 1)^2 - (X^2 - X + 1)^3$ in $K(\sigma)[X]$. Therefore

$$[K(\xi) : K(\sigma)] = [K(\sigma)(\xi) : K(\sigma)] \leq 6;$$

combining this inequality with the inclusion $K(\sigma) \subseteq \mathrm{Inv}(\langle s, t \rangle)$, we get

$$[K(\xi) : \mathrm{Inv}(\langle s, t \rangle)] \leq [K(\xi) : K(\sigma)] \leq 6.$$

Since $\langle s, t \rangle \subseteq \mathrm{Gal}(K(\xi)/\mathrm{Inv}(\langle s, t \rangle))$, it then follows from 3.2.15 that

$$\mathrm{Card}(\langle s, t \rangle) \leq [K(\xi) : \mathrm{Inv}(\langle s, t \rangle)]$$
$$\leq [K(\xi) : K(\sigma)] \leq 6.$$

In view of the relations established in the preceding two paragraphs, our argument will be complete if we show that $\mathrm{Card}(\langle s, t \rangle) \geq 6$. To do this, note first that

$$st(\xi) = \frac{\xi - 1}{\xi}, \qquad ts(\xi) = \frac{1}{1 - \xi}, \qquad tst(\xi) = \frac{\xi}{\xi - 1};$$

and then, since ξ is transcendental over K, that no two of

$$\xi, \quad \frac{1}{\xi}, \quad 1-\xi, \quad \frac{\xi-1}{\xi}, \quad \frac{1}{1-\xi}, \quad \frac{\xi}{\xi-1}$$

are equal. Consequently, the list

$$i_{K(\xi)}, \ s, \ t, \ st, \ ts, \ tst$$

contains six distinct elements of $\langle s, t \rangle$. □

With the help of the group-theoretical notion of index, we can now state and prove the following result (in the statement of which implicit use is being made of 2.3.5).

3.2.17. Proposition. *Let K be a field, and let F be a normal extension of K. Let L and M be intermediate fields between K and F such that $L \subseteq M$, and let v be a K-monomorphism from L to F.*

(i) If $\bar{v}$ denotes a K-automorphism of F extending v, and if Ω is a left transversal of $\mathrm{Gal}(F/M)$ in $\mathrm{Gal}(F/L)$, then the mapping $s \to \bar{v} \circ s \circ i_{M \to F}$ from Ω to the set of all K-monomorphisms from M to F extending v is bijective.

(ii) $[\mathrm{Gal}(F/L) : \mathrm{Gal}(F/M)]$ is the cardinality of the set of all K-monomorphisms from M to F extending v.

Proof. Note first that (ii) is an immediate consequence of (i), because the index $[\mathrm{Gal}(F/L) : \mathrm{Gal}(F/M)]$ is the common cardinality of the left transversals of $\mathrm{Gal}(F/M)$ in $\mathrm{Gal}(F/L)$.

To prove (i), now let $\bar{v}$ and Ω be as indicated in its statement. If $s \in \Omega$, then $\bar{v} \circ s \circ i_{M \to F}$ is a K-monomorphism from M to F extending v: Indeed, for every $\alpha \in L$ we have

$$(\bar{v} \circ s \circ i_{M \to F})(\alpha) = \bar{v}(s(\alpha)) = \bar{v}(\alpha) = v(\alpha).$$

It is meaningful, therefore, to speak of the mapping described in (i).

To prove the injectivity of this mapping, suppose that $s, t \in \Omega$ and $\bar{v} \circ s \circ i_{M \to F} = \bar{v} \circ t \circ i_{M \to F}$. For every $\alpha \in M$, we then have

$$\bar{v}(s(\alpha)) = (\bar{v} \circ s \circ i_{M \to F})(\alpha) = (\bar{v} \circ t \circ i_{M \to F})(\alpha) = \bar{v}(t(\alpha)),$$

whence $s(\alpha) = t(\alpha)$. Consequently s and t agree on M. But according to 3.2.12, this means that $s\,\mathrm{Gal}(F/M) = t\,\mathrm{Gal}(F/M)$; and since $s, t \in \Omega$, this equality implies that $s = t$.

To conclude, we have to show that our mapping is surjective. To this end, let w be a K-monomorphism from M to F extending v. By 2.3.5, w can be extended to a K-automorphism $\bar{w}$ of F. Then $\bar{v}$ and $\bar{w}$ extend v, and hence agree on L. By 3.2.12, this implies that $\bar{v}\,\mathrm{Gal}(F/L) = \bar{w}\,\mathrm{Gal}(F/L)$, and so $\bar{v}^{-1}\bar{w} \in \mathrm{Gal}(F/L)$. Now choose an $s \in \Omega$ so that $\bar{v}^{-1}\bar{w}\,\mathrm{Gal}(F/M) = s\,\mathrm{Gal}(F/M)$. Applying 3.2.12 once more, we see that $\bar{v}^{-1}\bar{w}$ and s agree on

M. For every $\alpha \in M$, we then have

$$(\bar{v} \circ s \circ i_{M \to F})(\alpha) = \bar{v}(s(\alpha)) = \bar{v}(\bar{v}^{-1}\bar{w}(\alpha))$$
$$= \bar{w}(\alpha) = w(\alpha);$$

therefore $w = \bar{v} \circ s \circ i_{M \to F}$.   □

3.2.18. Corollary. *Let K be a field, let L be an extension field of K, and let M be an extension field of L that is normal over K.*

(i) *If Ω is a left transversal of $\mathrm{Gal}(M/L)$ in $\mathrm{Gal}(M/K)$, then the mapping $s \to s \circ i_{L \to M}$ from Ω to $\mathrm{Mon}_K(L, M)$ is bijective.*

(ii) $[\mathrm{Gal}(M/K):\mathrm{Gal}(M/L)] = \mathrm{Card}(\mathrm{Mon}_K(L, M))$.

Proof. This is a particular case of the proposition: First, $i_{K \to M}$ is a K-monomorphism from K to M; second, i_M is a K-automorphism of M extending $i_{K \to M}$; and third, $\mathrm{Mon}_K(L, M)$ consists of all K-monomorphisms from L to M extending $i_{K \to M}$.   □

3.2.19. Corollary. *Let K be a field, let L be an extension field of K, and let M be an extension field of L that is normal over K. Let v be a K-monomorphism from L to M.*

(i) *If $\bar{v}$ denotes a K-automorphism of M extending v, then the mapping $s \to \bar{v} \circ s$ from $\mathrm{Gal}(M/L)$ to the set of all K-automorphisms of M extending v is bijective.*

(ii) $\mathrm{Gal}(M/L)$ *and the set of all K-automorphisms of M extending v are equipotent.*

Proof. This is also a particular case of the proposition: First, $\mathrm{Gal}(M/L)$ is the only left transversal of $\mathrm{Gal}(M/M)$ in $\mathrm{Gal}(M/L)$; and by 2.1.21, every K-monomorphism from M to M is a K-automorphism.   □

It can be shown that the assertions in proposition 3.2.17 and its corollaries do not remain valid if the normality assumptions are omitted (problem 10).

The next proposition (in whose statement implicit use is being made of 2.3.4) describes an important homomorphism of Galois groups.

3.2.20. Proposition. *Let K be a field, let L be a normal extension of K, and let M be an extension field of L. Then the mapping $s \to s_L$ from $\mathrm{Gal}(M/K)$ to $\mathrm{Gal}(L/K)$ is a homomorphism having $\mathrm{Gal}(M/L)$ as its kernel.*

Proof. It is clear that the mapping in question is a homomorphism. And to say that an $s \in \mathrm{Gal}(M/K)$ belongs to its kernel amounts to saying that s_L is the identity mapping on L, hence to saying that $s \in \mathrm{Gal}(M/L)$. □

An additional normality assumption yields the following supplement to the preceding proposition.

3.2.21. Proposition. *Let K be a field, let L be a normal extension of K, and let M be an extension field of L that is normal over K.*

    (i) *The homomorphism $s \rightarrow s_L$ from $\mathrm{Gal}(M/K)$ to $\mathrm{Gal}(L/K)$ is surjective; and hence the homomorphism from $\mathrm{Gal}(M/K)/\mathrm{Gal}(M/L)$ to $\mathrm{Gal}(L/K)$ such that $s\,\mathrm{Gal}(M/L) \rightarrow s_L$ for every $s \in \mathrm{Gal}(M/K)$ is an isomorphism.*

    (ii) $[\mathrm{Gal}(M/K) : \mathrm{Gal}(M/L)] = \mathrm{Card}(\mathrm{Gal}(L/K))$.

Proof. To verify (i) we need only note that the surjectivity of the homomorphism $s \rightarrow s_L$ from $\mathrm{Gal}(M/K)$ to $\mathrm{Gal}(L/K)$ follows directly from 2.3.5. And, since

$$[\mathrm{Gal}(M/K) : \mathrm{Gal}(M/L)] = \mathrm{Card}(\mathrm{Gal}(M/K)/\mathrm{Gal}(M/L)),$$

we see that (ii) is a consequence of (i).   □

We now close this section with a useful result stating that the elements of a field that are algebraic over the field of invariants of a group of automorphisms are characterized by the finiteness of their orbits.

3.2.22. Proposition. *Let F be a field, let Γ be a group of automorphisms of F, and let $\alpha \in F$. Then α is algebraic over $\mathrm{Inv}(\Gamma)$ if and only if $O_\Gamma(\alpha)$ is finite, in which case $\prod_{\beta \in O_\Gamma(\alpha)}(X - \beta)$ is its minimal polynomial over $\mathrm{Inv}(\Gamma)$.*

Proof. Suppose first that α is algebraic over $\mathrm{Inv}(\Gamma)$. If $s \in \Gamma$, then s is an $\mathrm{Inv}(\Gamma)$-automorphism of F, whence $s(\alpha)$ is a zero of the minimal polynomial of α over $\mathrm{Inv}(\Gamma)$. Since this means that every element of the orbit $O_\Gamma(\alpha)$ is a zero of this polynomial, the finiteness of $O_\Gamma(\alpha)$ follows at once.

Now suppose, conversely, that $O_\Gamma(\alpha)$ is finite. Define a polynomial f in $F[X]$ by $f(X) = \prod_{\beta \in O_\Gamma(\alpha)}(X - \beta)$. For every $s \in \Gamma$, we have $s(O_\Gamma(\alpha)) = O_\Gamma(\alpha)$, and hence

$$sf(X) = \prod_{\beta \in O_\Gamma(\alpha)}(X - s(\beta)) = \prod_{\beta \in O_\Gamma(\alpha)}(X - \beta) = f(X);$$

therefore, the coefficients of f belong to $\mathrm{Inv}(\Gamma)$, which means that $f \in \mathrm{Inv}(\Gamma)[X]$. Also, note that f is monic, $f(\alpha) = 0$, and $\deg(f) = \mathrm{Card}(O_\Gamma(\alpha))$; in particular, α is algebraic over $\mathrm{Inv}(\Gamma)$. Finally, let g denote the minimal polynomial of α over $\mathrm{Inv}(\Gamma)$. Then g is monic and divides f in $\mathrm{Inv}(\Gamma)[X]$; and, as shown in the first part of the proof, every element of $O_\Gamma(\alpha)$ is a zero of g, so that

$$\deg(g) \geq \mathrm{Card}(O_\Gamma(\alpha)) = \deg(f).$$

It then follows that $f = g$, which is what we wanted.   □

PROBLEMS

1. Let K be a field, let L be an extension field of K, and let M be an extension field of L. Verify the following assertions:
   a. If Ω is a left transversal of $\mathrm{Gal}(M/L)$ in $\mathrm{Gal}(M/K)$, then the mapping $s \to s \circ i_{L \to M}$ from Ω to $\mathrm{Mon}_K(L, M)$ is injective.
   b. $[\mathrm{Gal}(M/K) : \mathrm{Gal}(M/L)] \le \mathrm{Card}(\mathrm{Mon}_K(L, M))$.
   c. If L is finite over K, then

   $$[\mathrm{Gal}(M/K) : \mathrm{Gal}(M/L)] \le [L : K].$$

   (Note that 3.2.15 is a particular case of assertion c.)
2. Let K be a field, and let L be an extension field of K. Show that if $\alpha \in L$ and α is algebraic over K, and if Ω is a left transversal of $\mathrm{Gal}(L/K(\alpha))$ in $\mathrm{Gal}(L/K)$, then the mapping $s \to s(\alpha)$ from Ω to the set of all zeros in L of the minimal polynomial of α over K is injective.

   Use this in order to give a proof of assertion c in problem 1 that does not depend on results in section 3.1.
3. Let K be a field, and let L be a normal extension of K. Show that if E is an intermediate field between K and L, and if F denotes the normal closure of E over K in L, then

   $$\mathrm{Gal}(L/F) = \bigcap_{s \in \mathrm{Gal}(L/K)} s\,\mathrm{Gal}(L/E)\,s^{-1}.$$

4. Prove that if K is a finite field, and if $K(\xi)$ is a simple transcendental extension of K, then the set $\mathrm{Mon}_K(K(\xi), K(\xi))$ is infinite, and

   $$\mathrm{Card}(\mathrm{Gal}(K(\xi)/K)) < \mathrm{Card}(\mathrm{Mon}_K(K(\xi), K(\xi))).$$

   (Note that, in view of 2.1.21, this strict inequality cannot obtain when $K(\xi)$ is algebraic over K.)
5. Let K be a field, let L be an extension field of K, and let $\alpha \in L$. Prove that if α is algebraic over K, then the orbit of α under $\mathrm{Gal}(L/K)$ is finite.

   Give an example of a field K and an extension field L of K with the property that there exists an element of L that has a finite orbit under $\mathrm{Gal}(L/K)$, but is transcendental over K.
6. Let K be an infinite field, and let $K(\xi)$ be a simple transcendental extension of K. Let Γ and Δ denote, respectively, the subgroups of $\mathrm{Gal}(K(\xi)/K)$ consisting of the linear fractional transformations defined by matrices in $GL_2(K)$ of the types $\begin{bmatrix} \alpha & 0 \\ 0 & 1 \end{bmatrix}$ and $\begin{bmatrix} 1 & \beta \\ 0 & 1 \end{bmatrix}$. Prove that

   $$\mathrm{Inv}(\Gamma \cap \Delta) \supset \mathrm{Inv}(\Gamma) \vee \mathrm{Inv}(\Delta).$$

7. Let K be a field, and let $K(\xi)$ be a simple transcendental extension of K. Denote by s and t the K-automorphisms of $K(\xi)$ such that

$s(\xi) = -\xi$ and $t(\xi) = 1 - \xi$. Verify the following assertions:

a. If $\mathrm{Char}(K) \neq 2$, then s is an element of order 2 in $\mathrm{Gal}(K(\xi)/K)$, and $\mathrm{Inv}(\langle s \rangle) = K(\xi^2)$.

b. t is an element of order 2 in $\mathrm{Gal}(K(\xi)/K)$, and $\mathrm{Inv}(\langle t \rangle) = K(\xi^2 - \xi)$.

c. If $n \in \mathbf{Z}$, then $(ts)^n(\xi) = \xi - n$.

d. If $\mathrm{Char}(K) = 0$, then the subgroup $\langle s, t \rangle$ of $\mathrm{Gal}(K(\xi)/K)$ is infinite, and $\mathrm{Inv}(\langle s, t \rangle) = K$.

e. If p is a prime different from 2, and if $\mathrm{Char}(K) = p$, then the subgroup $\langle s, t \rangle$ of $\mathrm{Gal}(K(\xi)/K)$ is dihedral of order $2p$, and $\mathrm{Inv}(\langle s, t \rangle) = K((\xi^p - \xi)^2)$.

8. Let K be a field, and let $K(\xi)$ be a simple transcendental extension of K. Verify the following assertions.

a. $[K(\xi) : K(\xi^2)] = 2 = [K(\xi) : K(\xi^2 - \xi)]$.

b. If $\mathrm{Char}(K) = 0$, then

$$K(\xi^2) \cap K(\xi^2 - \xi) = K.$$

c. If K has prime characteristic p, then

$$K(\xi^2) \cap K(\xi^2 - \xi) = K\big((\xi^p - \xi)^2\big).$$

(Note, incidentally, that assertions a and b imply the existence of a field F containing subfields P and Q over which F is finite, but that is transcendental over $P \cap Q$.)

9. Let K be a field of characteristic 0, let $K(\xi)$ be a simple transcendental extension of K, and write $P = K(\xi^2)$ and $Q = K(\xi^2 - \xi)$. Show that

$$\mathrm{Gal}(K(\xi)/P \cap Q) \supset \mathrm{Gal}(K(\xi)/P) \vee \mathrm{Gal}(K(\xi)/Q).$$

10. Use the three-term chain $\mathbf{Q} \subset \mathbf{Q}(\sqrt{2}) \subset \mathbf{Q}(\sqrt[4]{2})$ to justify the claim made in the remark that follows 3.2.19 concerning the omission of the normality assumptions in 3.2.17, 3.2.18, and 3.2.19.

11. Let F be a field, and let P be an invariant field in F. Prove that every irreducible polynomial in $P[X]$ admitting a zero in L splits in $L[X]$.

12. Let F be a field, let Ω be a subgroup of $\mathrm{Aut}(F)$, and let V be an $\mathrm{Inv}(\Omega)$-subspace of F. We then denote by $\mathrm{Res}_\Omega(V, F)$ the F-subspace of $\mathrm{Map}(V, F)$ generated by the set of all mappings from V to F defined by restriction of the automorphisms belonging to Ω. Show that V is finite-dimensional if and only if $\mathrm{Res}_\Omega(V, F)$ is finite-dimensional, in which case

$$[V : \mathrm{Inv}(\Omega)] = [\mathrm{Res}_\Omega(V, F) : F].$$

(This striking result, which is known as **Artin's theorem**, can be used as the starting point of one of the possible presentations of Galois theory.)

13. Let F be a field, and let Ω be a subgroup of $\mathrm{Aut}(F)$. Verify the

following assertions concerning the F-subspace $F\Omega$ of $\mathrm{Map}(F, F)$ generated by Ω.

a. $F\Omega$ is an F-algebra admitting Ω as a base.

b. $R(F\Omega) = \mathrm{Inv}(\Omega)$.

c. $F\Omega = \mathrm{Res}_{\Omega}(F, F)$.

3.3. GALOIS EXTENSIONS

The classes of normal and separable extensions have been analyzed in detail in the preceding chapter. It has been seen in what sense normality can be said to provide "sufficient room" for the splitting of minimal polynomials and for the extendibility to automorphisms of embeddings of intermediate fields; and also in what sense separability can be said to imply an "abundance" of embeddings.

The field extensions that will be studied in this chapter are those defined when the conditions of normality and separability are imposed simultaneously. As may be expected from what was just said, these field extensions constitute the natural domain of Galois theory, that is, they are the field extensions that can be analyzed effectively by the properties of their Galois groups.

Let K be a field, and let L be an extension field of K. We say that L is **Galois over** K, or that L is a **Galois extension of** K, when L is normal and separable over K.

Normality and separability are, of course, independent properties. There are field extensions that are normal but not separable, and field extensions that are separable but not normal. Examples of the former, for every possible prime characteristic, have been given in 2.4.8; and of the latter, in every possible characteristic, in 2.3.1.

For a perfect field, every algebraic extension is separable; therefore, the Galois extensions of such a field are just its normal extensions. This applies, in particular, to every field of characteristic 0.

We now proceed to the discussion of the general properties of Galois extensions. We begin with three propositions that are immediate consequences of previous results on normal and separable extensions, and that are stated here explicitly for future reference.

3.3.1. Proposition. *Let K be a field, let L be an extension field of K, and let M be an extension field of L. If M is Galois over K, then M is Galois over L.*

Proof. Apply 2.3.2 and 2.5.8. □

3.3.2. Proposition. *Let K be a field, and let N be an extension field of K. If L and M are intermediate fields between K and N such that L is Galois over K, then $L \vee M$ is Galois over M.*

Proof. Apply 2.3.8, 2.5.2, and 2.5.12. □

3.3.3. Proposition. *Let K be a field, and let M be an extension field of K. If $(L_i)_{i \in I}$ is a nonempty family of intermediate fields between K and M that are Galois over K, then $\cap_{i \in I} L_i$ and $\vee_{i \in I} L_i$ are Galois over K.*

Proof. Apply 2.3.9, 2.5.8, and 2.5.11. $\qquad\qquad\qquad\qquad\qquad\qquad\square$

We continue by showing that Galois extensions can be described as splitting fields of separable polynomials.

3.3.4. Proposition. *Let K be a field. Then an extension field of K is Galois over K if and only if it is a splitting field over K of a set of separable polynomials in $K[X]$.*

Proof. Let L be an extension field of K.

If L is Galois over K, then 2.3.17 shows that L is a splitting field over K of the set of all minimal polynomials over K of the elements of L; and these polynomials are separable.

Assume next that L is a splitting field over K of a set W of separable polynomials in $K[X]$. The normality of L over K follows from 2.3.17, and so it remains to verify that L is separable over K. To this end, let D denote the set of all zeros in L of the polynomials in W. According to 2.5.1, we then have $D \subseteq S(L/K)$, every element of D being a simple zero of a polynomial in $K[X]$. Since $L = K(D)$, it now follows from 2.5.10 that L is separable over K. $\qquad\qquad\qquad\qquad\qquad\qquad\square$

3.3.5. Proposition. *Let K be a field. Then an extension field of K is Galois and finite over K if and only if it is a splitting field over K of a separable polynomial in $K[X]$.*

Proof. Let L be an extension field of K.

It is clear from the preceding proposition and 2.3.18 that if L is a splitting field over K of a separable polynomial in $K[X]$, then L is Galois and finite over K.

To conclude, now suppose that L is Galois and finite over K. Write $L = K(D)$, where D is a nonempty finite subset of L; and let W denote the set consisting of the minimal polynomials over K of the elements of D. Then L is a splitting field of W over K; writing $n = \mathrm{Card}(W)$ and $W = \{f_1, f_2, \ldots, f_n\}$, we see that L is a splitting field of $\prod_{i=1}^{n} f_i$ over K. Finally, since $f_1, f_2, \ldots, f_n$ are n distinct monic irreducible polynomials in $K[X]$, no two of them admit a common zero in L; and since they are separable, it follows that $\prod_{i=1}^{n} f_i$ is separable. $\qquad\qquad\qquad\qquad\qquad\square$

As shown by the next two propositions, every normal extension and every separable extension can be said to "generate" a Galois extension.

3.3.6. Proposition. *Let K be a field, and let L be a normal extension of K. Then $S(L/K)$ is Galois over K.*

Proof. Write $S = S(L/K)$; we have to show that S is normal over K. Thus let $\alpha \in S$, and denote by f the minimal polynomial of α over K. Our hypothesis implies that f splits in $L[X]$. Moreover, since f is separable, every zero of f in L is separable over K, and hence belongs to S. Consequently f splits in $S[X]$.   □

3.3.7. Proposition. *Let K be a field, and let L be a separable extension of K. Then every normal closure of L over K is Galois over K.*

Proof. Let M be a normal closure of L over K. We have to prove that M is separable over K. But this follows readily from 2.3.12 and 2.5.10: Denoting by T the set of all elements of M that are zeros of the minimal polynomials of the elements of L, we have $M = K(T)$; and since these polynomials are separable, we see that $T \subseteq S(M/K)$.   □

We shall now give an important characterization of Galois extensions.

3.3.8. Theorem. *Let K be a field, and let L be an extension field of K. Then L is Galois over K if and only if L is algebraic over K and K is invariant in L.*

Proof. First suppose that L is Galois over K. Clearly, L is algebraic over K. We have to show that K is invariant in L. To do this, let $\alpha \in \mathrm{Inv}(\mathrm{Gal}(L/K))$. Then $s(\alpha) = \alpha$ for every K-automorphism s of L, and we see from 2.3.6 that α is the only zero in L of its minimal polynomial over K. But since this polynomial splits in $L[X]$ and is separable, it has to be $X - \alpha$. Therefore $\alpha \in K$, as required.

Suppose, on the other hand, that L is algebraic over K and that K is invariant in L. Then by virtue of 3.2.22, the minimal polynomial over K of every element of L splits in $L[X]$ and is separable; and this means that L is normal and separable over K.   □

The results established in the foregoing discussion allow us to add some comments to those following the fundamental theorems 3.2.2 and 3.2.3.

First consider a field F. According to 3.3.8, the fields admitting F as a Galois extension are invariant in F. This naturally raises the following question: Which are the Galois groups on F related to these fields in the Galois correspondence defined by F?

Next consider a field K and a Galois extension L of K. By 3.3.1 and 3.3.8, the intermediate fields between K and L are invariant in L. Just as before, it is natural to ask the following question: Which are the Galois groups on L related to the intermediate fields in the Galois correspondence defined by K and L?

Partial answers to these questions will be given in the next section, which deals with finite Galois theory. The complete answers will be obtained—with the help of topological considerations—in the section on infinite Galois theory at the end of the chapter.

The next proposition states that in the case of a Galois extension, we can strengthen the conclusions in 3.2.14.

3.3.9.   Proposition.   *Let K be a field, and let L be a Galois extension of K.*

(i) *If E is an intermediate field between K and L, then E is Galois over K if and only if $\mathrm{Gal}(L/E)$ is normal in $\mathrm{Gal}(L/K)$.*

(ii) *If Γ is a Galois group on L contained in $\mathrm{Gal}(L/K)$, then Γ is normal in $\mathrm{Gal}(L/K)$ if and only if $\mathrm{Inv}(\Gamma)$ is Galois over K.*

Proof.   The direct implication in each part is a particular case of 3.2.14.

Now consider an intermediate field E between K and L such that $\mathrm{Gal}(L/E)$ is normal in $\mathrm{Gal}(L/K)$. By 3.2.14, it then follows that $\mathrm{Inv}(\mathrm{Gal}(L/E))$ is normal over K, and hence Galois over K. On the other hand, 3.3.1 and 3.3.8 show that $\mathrm{Inv}(\mathrm{Gal}(L/E)) = E$. Therefore E is Galois over K.

Finally, let Γ be a Galois group in L contained in $\mathrm{Gal}(L/K)$ and such that $\mathrm{Inv}(\Gamma)$ is Galois over K. Applying 3.2.14, we then see that $\mathrm{Gal}(L/\mathrm{Inv}(\Gamma))$ is normal in $\mathrm{Gal}(L/K)$. Since $\mathrm{Gal}(L/\mathrm{Inv}(\Gamma)) = \Gamma$, this means that Γ is normal in $\mathrm{Gal}(L/K)$.   □

To close the section, we shall make some brief comments concerning an important type of Galois extension.

Let K be a field, and let L be an extension field of K. We say that L is **abelian over** K, or that L is an **abelian extension of** K, when L is Galois over K and $\mathrm{Gal}(L/K)$ is abelian.

It follows from this definition and the preceding proposition that if K is a field, if L is an extension field of K, and if M is an extension field of L that is abelian over K, then M is abelian over L and L is abelian over K.

Although abelian extensions are field extensions of a special type, the description of the abelian extensions of a given field presents considerable difficulties. For example, the machinery of algebraic number theory is required in order to deal with this problem for the fields studied in number theory (i.e., the local fields and global fields). In fact, even the analysis of restricted types of abelian extensions requires auxiliary concepts that cannot be said to belong properly to the theory of fields. The only abelian extensions that we shall study in some detail are the cyclotomic extensions (in section 3.5) and the cyclic extensions (in section 3.9).

PROBLEMS

1.   Prove that if K is a field such that $\mathrm{Char}(K) \neq 2$, then every quadratic extension of K is Galois over K.

2. Prove that if K is a field of prime characteristic, and if L is a normal extension of K, then L is Galois over $P(L/K)$.

3. Let F be a field, and let Γ be a group of automorphisms of F. Show that F is algebraic over $\text{Inv}(\Gamma)$ if and only if Γ has finite orbits, in which case F is Galois over $\text{Inv}(\Gamma)$.

4. Let K be a field, and let L be a Galois extension of K. Prove that if $(\Gamma_i)_{i \in I}$ is a nonempty family of Galois groups on L contained in $\text{Gal}(L/K)$, then

$$\text{Inv}\left(\bigcap_{i \in I} \Gamma_i \right) = \bigvee_{i \in I} \text{Inv}(\Gamma_i).$$

5. Let K be a field, and let L be an extension field of K. Prove that if there exists a strictly increasing sequence $(E_n)_{n \geq 0}$ of intermediate fields between K and L that are Galois and finite over K and such that $E_0 = K$ and $\cup_{n=0}^{\infty} E_n = L$, then

$$\text{Card}(\text{Gal}(L/K)) = 2^{\aleph_0}.$$

6. Show that if K is a field, and if L is an infinite Galois extension of K, then

$$\text{Card}(\text{Gal}(L/K)) \geq 2^{\aleph_0}.$$

7. Show that if P is a prime field, and if F is an algebraic closure of P, then

$$[F:P] = \aleph_0 \quad \text{and} \quad \text{Card}(\text{Gal}(F/P)) = 2^{\aleph_0}.$$

3.4. FINITE GALOIS THEORY

We shall now take up the study of finite groups of field automorphisms and finite Galois extensions from the point of view of the Galois correspondences. This will include the results traditionally regarded as the central theorems of Galois theory.

We begin by establishing the principal property of finite groups of field automorphisms. The deep impression that Galois theory usually creates in students of algebra undoubtedly is due to its striking and elegant results. Among these, we feel, the following theorem has to be included.

3.4.1.   Theorem (Dedekind–Artin).   *Let F be a field, and let Γ be a group of automorphisms of F. Then F is finite over $\text{Inv}(\Gamma)$ if and only if Γ is finite, in which case $[F:\text{Inv}(\Gamma)] = \text{Card}(\Gamma)$ and Γ is a Galois group on F.*

Proof.   First suppose that F is finite over $\text{Inv}(\Gamma)$. According to 3.2.15, we then have

$$\text{Card}(\text{Gal}(F/\text{Inv}(\Gamma))) \leq [F:\text{Inv}(\Gamma)].$$

Since $\Gamma \subseteq \mathrm{Gal}(F/\mathrm{inv}(\Gamma))$, it clearly follows that Γ is finite and

$$\mathrm{Card}(\Gamma) \le \mathrm{Card}(\mathrm{Gal}(F/\mathrm{Inv}(\Gamma))) \le [F : \mathrm{Inv}(\Gamma)].$$

Now suppose that Γ is finite. Assume, for a contradiction, that

$$\mathrm{Card}(\Gamma) < [F : \mathrm{Inv}(\Gamma)].$$

Write $n = \mathrm{Card}(\Gamma)$ and $\Gamma = \{s_1, s_2, \ldots, s_n\}$, with $s_1 = i_F$. By the preceding inequality, we can then choose a sequence $(\rho_1, \rho_2, \ldots, \rho_{n+1})$ of elements of F that is linearly independent over $\mathrm{Inv}(\Gamma)$. Next, form the system

$$
\begin{array}{c}
s_1(\rho_1)\xi_1 + s_1(\rho_2)\xi_2 + \cdots + s_1(\rho_{n+1})\xi_{n+1} = 0 \\
s_2(\rho_1)\xi_1 + s_2(\rho_2)\xi_2 + \cdots + s_2(\rho_{n+1})\xi_{n+1} = 0 \\
\vdots \qquad \vdots \qquad \qquad \vdots \qquad \vdots \\
s_n(\rho_1)\xi_1 + s_n(\rho_2)\xi_2 + \cdots + s_n(\rho_{n+1})\xi_{n+1} = 0
\end{array}
\tag{*}
$$

of n linear homogeneous equations in $n+1$ unknowns $\xi_1, \xi_2, \ldots, \xi_{n+1}$ and with coefficients in F.

We know that $(*)$ admits nontrivial solutions in $F^{(n+1)}$. From among these we now select one with the largest possible number of coordinates equal to zero, and denote it by $(\alpha_1, \alpha_2, \ldots, \alpha_{n+1})$. Clearly, one of these coordinates may be taken to be 1.

We contend that $\alpha_1, \alpha_2, \ldots, \alpha_{n+1} \in \mathrm{Inv}(\Gamma)$. To prove this, let $t \in \Gamma$. Applying t to each of the equalities in the system

$$
\begin{array}{c}
s_1(\rho_1)\alpha_1 + s_1(\rho_2)\alpha_2 + \cdots + s_1(\rho_{n+1})\alpha_{n+1} = 0 \\
s_2(\rho_1)\alpha_1 + s_2(\rho_2)\alpha_2 + \cdots + s_2(\rho_{n+1})\alpha_{n+1} = 0 \\
\vdots \qquad \vdots \qquad \qquad \vdots \qquad \vdots \\
s_n(\rho_1)\alpha_1 + s_n(\rho_2)\alpha_2 + \cdots + s_n(\rho_{n+1})\alpha_{n+1} = 0,
\end{array}
\tag{**}
$$

we get the following system of equalities:

$$
\begin{array}{c}
ts_1(\rho_1)t(\alpha_1) + ts_1(\rho_2)t(\alpha_2) + \cdots + ts_1(\rho_{n+1})t(\alpha_{n+1}) = 0 \\
ts_2(\rho_1)t(\alpha_1) + ts_2(\rho_2)t(\alpha_2) + \cdots + ts_2(\rho_{n+1})t(\alpha_{n+1}) = 0 \\
\vdots \\
ts_n(\rho_1)t(\alpha_1) + ts_n(\rho_2)t(\alpha_2) + \cdots + ts_n(\rho_{n+1})t(\alpha_{n+1}) = 0.
\end{array}
\tag{***}
$$

Since there exists a permutation c of $\{1, 2, \ldots, n\}$ such that $ts_i = s_{c(i)}$ for $1 \le i \le n$, the equalities in $(***)$ imply that $(t(\alpha_1), t(\alpha_2), \ldots, t(\alpha_{n+1}))$ is a solution of $(*)$ in $F^{(n+1)}$. Consequently,

$$(\alpha_1 - t(\alpha_1), \alpha_2 - t(\alpha_2), \ldots, \alpha_{n+1} - t(\alpha_{n+1}))$$

is also a solution of ($*$) in $F^{(n+1)}$. Moreover, this solution has at least one more coordinate equal to 0 than our original solution: Indeed, if $1 \le i \le n$ and $\alpha_i = 0$, then $\alpha_i - t(\alpha_i) = 0 - 0 = 0$; and since $\alpha_j = 1$ for some index j, we also have $\alpha_j - t(\alpha_j) = 1 - 1 = 0$. Our choice of $(\alpha_1, \alpha_2, \ldots, \alpha_{n+1})$ now implies that

$$\left(\alpha_1 - t(\alpha_1), \alpha_2 - t(\alpha_2), \ldots, \alpha_{n+1} - t(\alpha_{n+1}) \right)$$

is the trivial solution of ($*$), that is, that $\alpha_i - t(\alpha_i) = 0$ for $1 \le i \le n+1$. Therefore $t(\alpha_i) = \alpha_i$ for $1 \le i \le n+1$, which is what was needed.

Finally, note that the first equality in ($**$) reduces to $\sum_{i=1}^{n+1} \rho_i \alpha_i = 0$, because $s_1 = i_F$. Since $\alpha_1, \alpha_2, \ldots, \alpha_{n+1} \in \text{Inv}(\Gamma)$, it then follows that $(\rho_1, \rho_2, \ldots, \rho_{n+1})$ is linearly dependent over $\text{Inv}(\Gamma)$. But this is incompatible with the choice of $(\rho_1, \rho_2, \ldots, \rho_{n+1})$ made at the beginning of the argument.

To sum up, we have proved that F is finite over $\text{Inv}(\Gamma)$ if and only if Γ is finite, in which case the inequalities

$$\text{Card}(\Gamma) \le \text{Card}(\text{Gal}(F/\text{Inv}(\Gamma))) \le [F : \text{Inv}(\Gamma)]$$

and

$$\text{Card}(\Gamma) \ge [F : \text{Inv}(\Gamma)]$$

hold. Since it is clear that these inequalities, the finiteness of the cardinalities and the dimensionality involved in them, and the inclusion $\Gamma \subseteq \text{Gal}(F/\text{Inv}(\Gamma))$ imply that

$$[F : \text{Inv}(\Gamma)] = \text{Card}(\Gamma) \quad \text{and} \quad \Gamma = \text{Gal}(F/\text{Inv}(\Gamma)),$$

the proof is complete. $\square$

The preceding proof is due to Artin. It is an ingenious application of elementary methods from linear algebra, and illustrates the elegant simplicity typical of this master's mathematical exposition.

As is usual for important results, several proofs of the Dedekind–Artin theorem have been given. These are of varied degree of difficulty and sophistication (problems 3 and 4).

We have seen that for finite extensions, the linear degree is an upper bound for the order of the Galois group. We shall now see that with the help of the Dedekind–Artin theorem, it is easy to determine the finite extensions for which this upper bound is attained.

3.4.2. Proposition. *Let K be a field, and let L be a finite extension of K. Then L is Galois over K if and only if $\text{Card}(\text{Gal}(L/K)) = [L : K]$.*

Proof. By virtue of 3.2.15, we see that $\text{Gal}(L/K)$ is finite. According to the preceding theorem, we then have

$$[L : \text{Inv}(\text{Gal}(L/K))] = \text{Card}(\text{Gal}(L/K)).$$

In view of the inclusion $K \subseteq \text{Inv}(\text{Gal}(L/K))$, this equality shows that $[L : K] = \text{Card}(\text{Gal}(L/K))$ if and only if $K = \text{Inv}(\text{Gal}(L/K))$, that is, if and only if K is invariant in L. The desired conclusion now follows at once from 3.3.8. □

Combining the preceding results, we can now derive two important equalities.

3.4.3. Proposition. *Let K be a field, and let L be a finite Galois extension of K.*

(i) *If Γ is a subgroup of $\mathrm{Gal}(L/K)$, then*

$$[\mathrm{Inv}(\Gamma):K] = [\mathrm{Gal}(L/K):\Gamma].$$

(ii) *If E is an intermediate field between K and L, then*

$$[\mathrm{Gal}(L/K):\mathrm{Gal}(L/E)] = [E:K].$$

Proof. To prove (i), let Γ be a subgroup of $\mathrm{Gal}(L/K)$. According to 3.4.1 and 3.4.2, we then have

$$[\mathrm{Inv}(\Gamma):K] = [L:K]/[L:\mathrm{Inv}(\Gamma)]$$
$$= \mathrm{Card}(\mathrm{Gal}(L/K))/\mathrm{Card}(\Gamma) = [\mathrm{Gal}(L/K):\Gamma],$$

as required.

Next note that (ii) is a consequence of (i). For if E is an intermediate field between K and L, then L is Galois over E, and the desired equality is obtained from the previous one by taking $\Gamma = \mathrm{Gal}(L/E)$. □

We now are in a position to state and prove the results known as the **fundamental theorems of finite Galois theory**. These are obtained from the general fundamental theorems in section 3.2 by specifying suitable subsets of the sets connected by the Galois correspondences.

3.4.4. Theorem. *Let F be a field.*

(i) *If P is a field admitting F as a finite Galois extension, then $\mathrm{Gal}(F/P)$ is a finite group of automorphisms of F.*

(ii) *If Γ is a finite group of automorphisms of F, then $\mathrm{Inv}(\Gamma)$ is a field admitting F as a finite Galois extension.*

(iii) *The mapping $P \to \mathrm{Gal}(F/P)$ from the set of all fields admitting F as a finite Galois extension to the set of all finite groups of automorphisms of F and the mapping $\Gamma \to \mathrm{Inv}(\Gamma)$ from the set of all finite groups of automorphisms of F to the set of all fields admitting F as a finite Galois extension are mutually inverse inclusion-reversing bijections.*

Proof. Assertion (i) follows from 3.4.2, and assertion (ii) from 3.4.1 and 3.3.8.

Also, note that by 3.4.1 and 3.3.8, every finite group of automorphisms of F is a Galois group on F, and every field admitting F as a finite Galois extension is invariant in F. Therefore, (iii) is an immediate consequence of (i), (ii), and 3.2.2. □

3.4.5. Theorem (Galois). *Let K be a field, and let L be a finite Galois extension of K.*

(i) *If E is an intermediate field between K and L, then $\mathrm{Gal}(L/E)$ is a subgroup of $\mathrm{Gal}(L/K)$.*

(ii) *If Γ is a subgroup of $\mathrm{Gal}(L/K)$, then $\mathrm{Inv}(\Gamma)$ is an intermediate field between K and L.*

(iii) *The mapping $E \to \mathrm{Gal}(L/E)$ from the set of all intermediate fields between K and L to the set of all subgroups of $\mathrm{Gal}(L/K)$ and the mapping $\Gamma \to \mathrm{Inv}(\Gamma)$ from the set of all subgroups of $\mathrm{Gal}(L/K)$ to the set of all intermediate fields between K and L are mutually inverse inclusion-reversing bijections.*

Proof. This is an easy consequence of the preceding theorem. For according to 3.4.2, every subgroup of $\mathrm{Gal}(L/K)$ is a finite group of automorphisms of L; and, by 1.4.10 and 3.3.1, every intermediate field between K and L admits L as a finite Galois extension.   □

Given a finite Galois extension, the set of all intermediate fields and the set of all subgroups of the Galois group can be regarded as ordered by the inclusion relation. Therefore, the following supplement to 3.2.10 and 3.2.11 is an immediate consequence of the inclusion-reversing character of the bijections described in Galois's theorem.

3.4.6. Corollary. *Let K be a field, and let L be a finite Galois extension of K.*

(i) *If $(\Gamma_i)_{i \in I}$ is a nonempty family of subgroups of $\mathrm{Gal}(L/K)$, then*

$$\mathrm{Inv}\left(\bigcap_{i \in I} \Gamma_i \right) = \bigvee_{i \in I} \mathrm{Inv}(\Gamma_i).$$

(ii) *If $(E_i)_{i \in I}$ is a nonempty family of intermediate fields between K and L, then*

$$\mathrm{Gal}\left(L \Big/ \bigcap_{i \in I} E_i \right) = \bigvee_{i \in I} \mathrm{Gal}(L/E_i).$$

Two general questions on the Galois correspondences were raised in the comments made immediately after Theorem 3.3.8. By virtue of the fundamental theorems 3.4.4 and 3.4.5, we can now give satisfactory answers to restricted versions of these questions.

Concerning the first, we have shown that given a field F, the Galois groups on F related to the fields admitting F as a *finite* Galois extension are precisely the *finite* groups of automorphisms of F.

As to the second, we have shown that given a field K and a *finite* Galois extension L of K, the Galois groups on L related to the intermediate fields between K and L are simply the subgroups of $\mathrm{Gal}(L/K)$.

Note that the Galois correspondence defined by a *finite* Galois extension can be said to be optimal, in the sense that it relates every subgroup of the Galois group to an intermediate field.

The preceding assertions do not remain valid if the adjective "finite" is omitted. This does not mean, however, that it is not interesting to investigate these questions in more generality. As we shall see in the last section of the chapter, it is possible to give a satisfactory theory for a special type of infinite group of field automorphisms.

To continue our discussion on the general properties of finite groups of field automorphisms, we shall now proceed to establish the existence of certain isomorphisms of Galois groups. The following two propositions (in the statements of which implicit use is being made of 3.2.20) describe the conditions under which this can be done.

3.4.7. Proposition. *Let F be a field, and let P and Q be subfields of F such that P is normal and finite over $P \cap Q$. Then the homomorphism $s \to s_P$ from $\mathrm{Gal}(P \vee Q/Q)$ to $\mathrm{Gal}(P/P \cap Q)$ is an isomorphism.*

Proof. According to 3.2.20, the normality of P over $P \cap Q$ allows us to speak of the homomorphism $s \to s_P$ from $\mathrm{Gal}(P \vee Q/P \cap Q)$ to $\mathrm{Gal}(P/P \cap Q)$; and furthermore, this homomorphism has $\mathrm{Gal}(P \vee Q/P)$ as its kernel.

Note that the homomorphism under consideration in our proposition is defined by restriction of the one described in the preceding paragraph. Therefore, its kernel is $\mathrm{Gal}(P \vee Q/P) \cap \mathrm{Gal}(P \vee Q/Q)$; and since

$$\mathrm{Gal}(P \vee Q/P) \cap \mathrm{Gal}(P \vee Q/Q) = \mathrm{Gal}(P \vee Q/P \vee Q) = \{i_{P \vee Q}\},$$

we conclude that it is injective.

In order to prove that our homomorphism is surjective, let us now denote its image by Γ; we have to verify that $\Gamma = \mathrm{Gal}(P/P \cap Q)$. Since P is finite over $P \cap Q$, it follows from 3.2.15 that $\mathrm{Gal}(P/P \cap Q)$ is finite. Therefore Γ is also finite, and the required conclusion will follow from 3.4.4 if we verify that Γ and $\mathrm{Gal}(P/P \cap Q)$ correspond to the same subfield of F in the Galois correspondence defined by F, that is, if we establish the equality

$$\mathrm{Inv}(\Gamma) = \mathrm{Inv}(\mathrm{Gal}(P/P \cap Q)).$$

Since $\Gamma \subseteq \mathrm{Gal}(P/P \cap Q)$, we have

$$\mathrm{Inv}(\Gamma) \supseteq \mathrm{Inv}(\mathrm{Gal}(P/P \cap Q)).$$

To prove the opposite inclusion, let $\alpha \in \mathrm{Inv}(\Gamma)$. Then $\alpha \in P$, and $s(\alpha) = s_P(\alpha) = \alpha$ for every $s \in \mathrm{Gal}(P \vee Q/Q)$. Thus, α is a fixed point of every Q-automorphism of $P \vee Q$. Moreover, we know from 2.3.8 that $P \vee Q$ is normal over Q. If $\mathrm{Char}(F) = 0$, then $P \vee Q$ is Galois over Q, and so $\alpha \in Q$; but then $\alpha \in P \cap Q$, and hence $\alpha \in \mathrm{Inv}(\mathrm{Gal}(P/P \cap Q))$. And if $\mathrm{Char}(F) = p$, where p is a prime, then 2.4.4 shows that $\alpha^{p^n} \in Q$ for some nonnegative integer n; but then $\alpha^{p^n} \in P \cap Q$, and a second application of 2.4.4 shows that $\alpha \in \mathrm{Inv}(\mathrm{Gal}(P/P \cap Q))$. □

3.4.8. Corollary. *Let F be a field, and let P and Q be subfields of F such that P is Galois and finite over $P \cap Q$. Then $[P \vee Q : Q] = [P : P \cap Q]$.*

Proof. By the proposition, the groups $\mathrm{Gal}(P \vee Q/Q)$ and $\mathrm{Gal}(P/P \cap Q)$ are isomorphic, and hence equipotent. The hypothesis also implies that $P \vee Q$ is Galois and finite over Q. By 3.4.2, we then have

$$[P \vee Q : Q] = \mathrm{Card}(\mathrm{Gal}(P \vee Q/Q))$$
$$= \mathrm{Card}(\mathrm{Gal}(P/P \cap Q)) = [P : P \cap Q],$$

as was to be shown. □

3.4.9. Corollary. *Let K be a field, and let N be an extension field of K. If L and M are intermediate fields between K and N such that L is Galois and finite over K, then $[L \vee M : M] \,|\, [L : K]$.*

Proof. Since L is Galois and finite over $L \cap M$, the preceding corollary shows that $[L \vee M : M] = [L : L \cap M]$. And since $[L : L \cap M] \,|\, [L : K]$, the desired conclusion follows at once. □

3.4.10. Proposition. *Let K be a field, let N be an extension field of K, and let L and M be intermediate fields between K and N that are normal and finite over K. Then the mapping $s \to (s_L, s_M)$ from $\mathrm{Gal}(L \vee M/K)$ to $\mathrm{Gal}(L/K) \times \mathrm{Gal}(M/K)$ is an injective homomorphism; and if $L \cap M = K$, then it is an isomorphism.*

Proof. It is clear from 3.2.20 that the mapping $s \to (s_L, s_M)$ from $\mathrm{Gal}(L \vee M/K)$ to $\mathrm{Gal}(L/K) \times \mathrm{Gal}(M/K)$ is a homomorphism having $\mathrm{Gal}(L \vee M/L) \cap \mathrm{Gal}(L \vee M/M)$ as its kernel. And since

$$\mathrm{Gal}(L \vee M/L) \cap \mathrm{Gal}(L \vee M/M) = \mathrm{Gal}(L \vee M/L \vee M) = \{i_{L \vee M}\},$$

we see that it is injective.

Now assume that $L \cap M = K$. We have to prove that our homomorphism is surjective. To do this, let $v \in \mathrm{Gal}(L/K)$ and $w \in \mathrm{Gal}(M/K)$. According to 3.4.7, we can write $v = s_L$ and $w = t_M$, where $s \in \mathrm{Gal}(L \vee M/M)$ and $t \in \mathrm{Gal}(L \vee M/L)$. Then $st \in \mathrm{Gal}(L \vee M/K)$, and the argument will be completed by showing that $(st)_L = v$ and $(st)_M = w$. For every $\alpha \in L$ we have

$$(st)_L(\alpha) = st(\alpha) = s(t(\alpha)) = s(\alpha) = s_L(\alpha),$$

whence $(st)_L = s_L = v$. And since $t(M) = M$, for every $\alpha \in M$ we have

$$(st)_M(\alpha) = st(\alpha) = s(t(\alpha)) = t(\alpha) = t_M(\alpha);$$

therefore $(st)_M = t_M = w$. □

The finiteness assumptions made in the preceding two propositions are superfluous. We shall be in a position to prove this at the end of the chapter, in the section on infinite Galois theory.

When the Galois group of a finite Galois extension is decomposed as the internal direct product of two subgroups, one finds that the corresponding fields of invariants are related to the top and bottom fields in the manner described in the preceding proposition. More precisely, we have the following result.

3.4.11. Proposition. *Let K be a field, let L be a finite Galois extension of K, and let Γ and Δ be subgroups of $\mathrm{Gal}(L/K)$ such that*
$$\mathrm{Gal}(L/K) = \Gamma \vee \Delta \quad and \quad \Gamma \cap \Delta = \{i_L\}.$$
Then
$$L = \mathrm{Inv}(\Gamma) \vee \mathrm{Inv}(\Delta) \quad and \quad \mathrm{Inv}(\Gamma) \cap \mathrm{Inv}(\Delta) = K.$$

Proof. By 3.4.6, we see that
$$L = \mathrm{Inv}(\{i_L\}) = \mathrm{Inv}(\Gamma \cap \Delta) = \mathrm{Inv}(\Gamma) \vee \mathrm{Inv}(\Delta).$$
And applying 3.2.10, we get
$$\mathrm{Inv}(\Gamma) \cap \mathrm{Inv}(\Delta) = \mathrm{Inv}(\Gamma \vee \Delta) = \mathrm{Inv}(\mathrm{Gal}(L/K)) = K. \qquad \square$$

The results of the foregoing discussion constitute the core of our subject. In the remainder of the section, we shall give examples illustrating how they can be used in certain special situations.

3.4.12. Example. Let K be a field, and let $K(\xi)$ be a simple transcendental extension of K. Applying the Dedekind–Artin theorem, we can now complement the analysis of $\mathrm{Gal}(K(\xi)/K)$ initiated in 3.2.7:

(i) K is invariant in $K(\xi)$ if and only if K is infinite.

(ii) The only Galois groups on $K(\xi)$ contained in $\mathrm{Gal}(K(\xi)/K)$ are $\mathrm{Gal}(K(\xi)/K)$ and its finite subgroups.

The opposite implication in (i) was proved in 3.2.7. To prove the direct implication, assume that K is invariant in $K(\xi)$. Then $\mathrm{Inv}(\mathrm{Gal}(K(\xi)/K)) = K$, and so $K(\xi)$ is infinite over $\mathrm{Inv}(\mathrm{Gal}(K(\xi)/K))$. By 3.4.1, this implies that $\mathrm{Gal}(K(\xi)/K)$ is infinite; and according to 3.2.7, this implies that K is infinite.

To verify (ii), consider a proper subgroup Γ of $\mathrm{Gal}(K(\xi)/K)$. If Γ is finite, we know from 3.4.1 that Γ is a Galois group on $K(\xi)$. On the other hand, if Γ is infinite, then $\mathrm{Gal}(K(\xi)/K)$ also is infinite, and 3.2.7 shows that
$$\mathrm{Inv}(\Gamma) = K = \mathrm{Inv}(\mathrm{Gal}(K(\xi)/K));$$
and since $\mathrm{Gal}(K(\xi)/K)$ is a Galois group on $K(\xi)$, it is clear from 3.2.3 that Γ is not a Galois group on $K(\xi)$. $\qquad \square$

3.4.13. Example. As an interesting application of the preceding results, in conjunction with Sylow's theorem, we shall prove the following assertion: If

p is a prime, and if K is a field such that the linear degree of every proper finite extension of K is divisible by p, then the linear degree over K of every finite extension of K is an integer power of p.

First, consider a finite Galois extension L of K. If Γ is a Sylow p-subgroup of $\mathrm{Gal}(L/K)$, then $p \nmid [\mathrm{Gal}(L/K) : \Gamma]$. By 3.4.3, this means that $p \nmid [\mathrm{Inv}(\Gamma) : K]$, and the hypothesis implies that $\mathrm{Inv}(\Gamma) = K$. Applying 3.4.1, we now get

$$\mathrm{Card}(\Gamma) = [L : \mathrm{Inv}(\Gamma)] = [L : K];$$

and the desired conclusion is evident, because $\mathrm{Card}(\Gamma)$ is an integer power of p.

Next, consider a finite separable extension L of K. Choose a normal closure M of L over K. According to 2.3.13 and 3.3.7, M is Galois and finite over K. It then follows from what was shown in the preceding paragraph that $[M : K]$ is an integer power of p; and since $[L : K] | [M : K]$, the same is true of $[L : K]$.

To conclude, it remains to consider the case of a finite extension L of K that is not separable over K. The existence of such a field extension implies that K is not perfect. Then, as in the beginning of the proof of 1.7.6, we see that $\mathrm{Char}(K)$ is a prime and that there exists an irreducible polynomial of degree $\mathrm{Char}(K)$ in $K[X]$. If ρ is a zero of such a polynomial in an extension field of K, then $[K(\rho) : K] = \mathrm{Char}(K)$. By hypothesis, we then have $p | \mathrm{Char}(K)$, whence $\mathrm{Char}(K) = p$. Finally, since

$$[L : K] = [L : S(L/K)][S(L/K) : K],$$

the required conclusion follows at once from 2.5.13, 2.4.9, and what was shown in the preceding paragraph.   □

3.4.14. Example. The "fundamental theorem of algebra", that is, the statement that $\mathbf{C}$ is algebraically closed, was proved in 2.2.1 using tools from complex analysis. We now are in a position to give a proof of this theorem in which the principal ingredients are the main theorems of Galois theory and Sylow's theorem. Only two properties of $\mathbf{R}$ that are not "algebraic" will be required, namely, that every nonnegative real number has a real square root and that every polynomial of odd degree in $\mathbf{R}[X]$ has a real zero. This proof is a variation, due to Artin, of one of Gauss's proofs.

Let us begin by showing that every complex number has a complex square root, that is, that given $\alpha, \beta \in \mathbf{R}$, there exist $\sigma, \tau \in \mathbf{R}$ so that $(\sigma + \tau\sqrt{-1})^2 = \alpha + \beta\sqrt{-1}$. To do this, note that

$$\alpha^2 + \beta^2 \geq 0 \quad \text{and} \quad \sqrt{\alpha^2 + \beta^2} \geq |\alpha|,$$

whence

$$\left(\alpha + \sqrt{\alpha^2 + \beta^2}\right)/2 \geq 0 \quad \text{and} \quad \left(-\alpha + \sqrt{\alpha^2 + \beta^2}\right)/2 \geq 0;$$

therefore, there exist $\sigma, \tau \in \mathbf{R}$ for which

$$\sigma^2 = \left(\alpha + \sqrt{\alpha^2 + \beta^2}\right)/2 \quad \text{and} \quad \tau^2 = \left(-\alpha + \sqrt{\alpha^2 + \beta^2}\right)/2,$$

so that

$$\sigma^2 - \tau^2 = \alpha \quad \text{and} \quad \sigma^2\tau^2 = \beta^2/4;$$

if the signs of σ and τ are chosen so that $\sigma\tau = \beta/2$, we get

$$(\sigma + \tau\sqrt{-1})^2 = (\sigma^2 - \tau^2) + 2\sigma\tau\sqrt{-1} = \alpha + \beta\sqrt{-1},$$

which is what we wanted.

It is now readily seen that $\mathbf{C}$ does not possess quadratic extensions. For assuming that a quadratic extension F of $\mathbf{C}$ exists, a $\theta \in F$ can be chosen so that $(1, \theta)$ is a linear base of F over $\mathbf{C}$. It is then clear that $F = \mathbf{C}(\theta)$; and furthermore, we can write $\theta^2 = \alpha + \beta\theta$, where $\alpha, \beta \in \mathbf{C}$. But since

$$(\theta - \beta/2)^2 = \theta^2 - \beta\theta + \beta^2/4 = \alpha + \beta^2/4 \in \mathbf{C},$$

it follows from what was shown in the preceding paragraph that $(\theta - \beta/2)^2 = \gamma^2$ for some $\gamma \in \mathbf{C}$; consequently,

$$\theta - \beta/2 = \gamma \quad \text{or} \quad \theta - \beta/2 = -\gamma,$$

which implies that $\theta \in \mathbf{C}$. And this leads to the contradictory conclusion that $F = \mathbf{C}(\theta) = \mathbf{C}$.

Next, we claim that every proper finite extension of $\mathbf{R}$ has even linear degree over $\mathbf{R}$. To see this, let F be a proper finite extension of $\mathbf{R}$, and choose an $\alpha \in F - \mathbf{R}$. Then $[\mathbf{R}(\alpha):\mathbf{R}] > 1$, and so the minimal polynomial of α over $\mathbf{R}$ has degree greater than 1. If $[F:\mathbf{R}]$ were odd, then $[\mathbf{R}(\alpha):\mathbf{R}]$ would also be odd, because $[\mathbf{R}(\alpha):\mathbf{R}]\mid[F:\mathbf{R}]$; but then the minimal polynomial of α over $\mathbf{R}$ would have a real zero, in contradiction to its irreducibility in $\mathbf{R}[X]$. This proves our claim.

Applying what was shown in 3.4.13, we now see that the linear degree over $\mathbf{R}$ of every finite extension of $\mathbf{R}$ is an integer power of 2. Since $[\mathbf{C}:\mathbf{R}] = 2$, it clearly follows that the linear degree over $\mathbf{C}$ of every finite extension of $\mathbf{C}$ is also an integer power of 2.

Finally, we can apply 2.2.3 in order to prove that $\mathbf{C}$ is algebraically closed. Assume, for a contradiction, that there exists a proper finite extension L of $\mathbf{C}$. Choose a normal closure M of L over $\mathbf{C}$. Then M is Galois and finite over $\mathbf{C}$, and

$$\mathrm{Card}(\mathrm{Gal}(M/\mathbf{C})) = [M:\mathbf{C}] = 2^e$$

for some positive integer e. By Sylow's theorem, $\mathrm{Gal}(M/\mathbf{C})$ contains a subgroup Γ of order 2^{e-1}; therefore

$$[\mathrm{Inv}(\Gamma):\mathbf{C}] = [\mathrm{Gal}(M/\mathbf{C}):\Gamma] = 2,$$

and so we have obtained a quadratic extension of **C**. This, however, is incompatible with what was proved earlier.   □

As stated in the introduction, Galois's ideas were formulated originally in terms of splitting fields and groups of permutations. In order to discuss some of the classical applications of Galois theory, we shall begin by presenting Galois's point of view. It is customary, in this context, to impose the separability condition on the polynomials under consideration. This results in no loss of generality, because every finite Galois extension is the splitting field of a separable polynomial.

Let K be a field, and let f be a separable polynomial in $K[X]$. Since every two splitting fields of f over K are K-isomorphic, they define isomorphic Galois groups over K. More precisely, if L and M are splitting fields of f over K, there exists a K-isomorphism u from L to M, and the mapping $s \to u \circ s \circ u^{-1}$ from $\mathrm{Gal}(L/K)$ to $\mathrm{Gal}(M/K)$ is an isomorphism. Thus, if the splitting fields of f over K are "identified" with each other by means of these K-isomorphisms, their Galois groups over K are "identified" with each other by means of the isomorphisms defined by the latter. From now on, whenever no ambiguity is possible, we shall speak of "the" splitting field of f over K; and its Galois group over K will be said to be "the" **Galois group of f over K**. Since the properties that will be studied are preserved by isomorphisms, this will cause no difficulty.

Our next task is to interpret the Galois group of f over K as a group of permutations. The essential point is that every K-automorphism of the splitting field of f over K permutes the zeros of f. In other words, if the splitting field of f over K is denoted by $K(\rho_1, \rho_2, \ldots, \rho_n)$, where $n = \deg(f)$ and $\rho_1, \rho_2, \ldots, \rho_n$ are the zeros of f, and if s is a K-automorphism of $K(\rho_1, \rho_2, \ldots, \rho_n)$, then there exists a unique permutation $\bar{s} \in \mathrm{Sym}(n)$ such that $s(\rho_i) = \rho_{\bar{s}(i)}$ for $1 \leq i \leq n$.

It is readily verified that the mapping $s \to \bar{s}$ from $\mathrm{Gal}(K(\rho_1, \rho_2, \ldots, \rho_n)/K)$ to $\mathrm{Sym}(n)$ is a monomorphism. Its image will also be referred to as "the" **Galois group of f over K**. Strictly speaking, this subgroup of $\mathrm{Sym}(n)$ depends on the indexing of the zeros of f; however, it is easily seen that changing the indexing will simply result in obtaining an isomorphic subgroup of $\mathrm{Sym}(n)$, which is of no consequence in our considerations.

It should be noted that, generally speaking, the Galois group of f over K is not the whole group $\mathrm{Sym}(n)$. In other words, not every permutation in $\mathrm{Sym}(n)$ is induced by a K-automorphism of the splitting field of f over K. This is the case, for example, if f admits a zero in K; for then the permutations in the Galois group of f over K admit a common fixed point.

The Galois group of a separable polynomial can therefore be interpreted in two ways: first, as the Galois group of a finite Galois extension; and second, as a subgroup of a symmetric group. This causes no confusion, since it is always clear in each situation which one of the two interpretations is intended.

As the next proposition shows, the irreducibility of the separable polynomial under consideration corresponds to the transitivity of its Galois group. Recall that if n is a positive integer, a subgroup G of $\text{Sym}(n)$ is said to be transitive when, given $1 \le i, j \le n$, there exists a permutation in G such that $i \to j$.

3.4.15. Proposition. *Let K be a field, and let f be a separable polynomial in $K[X]$. Then f is irreducible in $K[X]$ if and only if its Galois group is transitive, in which case $\deg(f)$ divides the order of this group.*

Proof. Let L be the splitting field of f over K, and write $L = K(\rho_1, \rho_2, \ldots, \rho_n)$, where $n = \deg(f)$ and $\rho_1, \rho_2, \ldots, \rho_n$ are the zeros of f. Let G denote the Galois group of f over K, identified with a subgroup of $\text{Sym}(n)$.

Suppose first that f is irreducible in $K[X]$. Then $\rho_1, \rho_2, \ldots, \rho_n$ have the same minimal polynomial over K, namely, the monic associate of f in $K[X]$. It follows that if $1 \le i, j \le n$, there exists an $s \in \text{Gal}(L/K)$ for which $s(\rho_i) = \rho_j$; therefore $\bar{s} \in G$ and $\bar{s}(i) = j$. We conclude that G is transitive. Moreover, we have

$$[\text{Gal}(L/K) : \text{Gal}(L/K(\rho_1))] = [K(\rho_1) : K] = \deg(f) = n;$$

consequently, $n \mid \text{Card}(\text{Gal}(L/K))$, which means that $n \mid \text{Card}(G)$.

Suppose now that G is transitive. Let g be an irreducible factor of f in $K[X]$. Then there exists an index i such that ρ_i is a zero of g. Furthermore, if $1 \le j \le n$, the transitivity of G implies the existence of an $s \in \text{Gal}(L/K)$ for which $\bar{s}(i) = j$, that is, for which $s(\rho_i) = \rho_j$; therefore ρ_j is also a zero of g. Since every zero of f in L is a zero of g, it follows that f divides g in $L[X]$, and hence also in $K[X]$. Thus, f and g are associates in $K[X]$, and so f is irreducible in $K[X]$. $\qquad\square$

We now proceed to derive a result that is occasionally useful in the determination of Galois groups of polynomials. To this end, it will be well to recall here that if n is a positive integer, we write $G^+ = G \cap \text{Alt}(n)$ for every subgroup G of $\text{Sym}(n)$; clearly, G^+ can also be described as the subgroup of G consisting of the even permutations belonging to G.

3.4.16. Proposition. *Let K be a field such that $\text{Char}(K) \ne 2$, and let f be a separable polynomial in $K[X]$. Denote by $K(\rho_1, \rho_2, \ldots, \rho_n)$ the splitting field of f over K, where $n = \deg(f)$ and $\rho_1, \rho_2, \ldots, \rho_n$ are the zeros of f; denote by G the Galois group of f over K; and put $\delta = \prod_{1 \le i < j < n}(\rho_i - \rho_j)$.*
   (i) $\text{Inv}(G^+) = K(\delta)$.
   (ii) $\delta^2 \in K$.
   (iii) $G \subseteq \text{Alt}(n)$ if and only if $\delta \in K$.

Proof. First, consider a transposition c in $\text{Sym}(n)$. Write $c = (hk)$, where $1 \le h < k \le n$. We then have

$$\rho_{c(h)} - \rho_{c(k)} = \rho_k - \rho_h = -(\rho_h - \rho_k),$$

$$\rho_{c(i)} - \rho_{c(j)} = \rho_i - \rho_j \quad \text{if } 1 \le i < j \le n \quad \text{and} \quad \{i, j\} \cap \{h, k\} = \phi,$$

$$\left.\begin{array}{l} \rho_{c(r)} - \rho_{c(h)} = \rho_r - \rho_k \\ \rho_{c(r)} - \rho_{c(k)} = \rho_r - \rho_h \end{array}\right\} \quad \text{if } 1 \le r < h,$$

$$\left.\begin{array}{l} \rho_{c(h)} - \rho_{c(r)} = \rho_k - \rho_r \\ \rho_{c(k)} - \rho_{c(r)} = \rho_h - \rho_r \end{array}\right\} \quad \text{if } k < r \le n,$$

and

$$\left.\begin{array}{l} \rho_{c(h)} - \rho_{c(r)} = \rho_k - \rho_r = -(\rho_r - \rho_k) \\ \rho_{c(r)} - \rho_{c(k)} = \rho_r - \rho_h = -(\rho_h - \rho_r) \end{array}\right\} \quad \text{if } h < r < k.$$

Therefore

$$\prod_{i \le i < j \le n} (\rho_{c(i)} - \rho_{c(j)}) = - \prod_{1 \le i < j \le n} (\rho_i - \rho_j) = -\delta.$$

Thus, applying a transposition in $\text{Sym}(n)$ to the indices in the product defining δ results in a change of sign. This clearly implies that

$$\prod_{1 \le i < j \le n} (\rho_{c(i)} - \rho_{c(j)}) = \text{sgn}(c)\delta$$

for every $c \in \text{Sym}(n)$. Now write $L = K(\rho_1, \rho_2, \ldots, \rho_n)$; for every $s \in \text{Gal}(L/K)$, we have

$$s(\delta) = \prod_{1 \le i < j \le n} \big(s(\rho_i) - s(\rho_j)\big)$$

$$= \prod_{1 \le i < j \le n} (\rho_{\bar{s}(i)} - \rho_{\bar{s}(j)}) = \text{sgn}(\bar{s})\delta.$$

The three assertions in the conclusion can be readily deduced from what has been shown above.

Indeed, let $s \in \text{Gal}(L/K)$; since $\text{Char}(K) \ne 2$ and $\delta \ne 0$, the equality $s(\delta) = \text{sgn}(\bar{s})\delta$ shows that $s(\delta) = \delta$ if and only if $\text{sgn}(\bar{s}) = 1$, that is, if and only if $\bar{s}$ is even. But this means that under the assumed identification of $\text{Gal}(L/K)$ with G, the subgroup $\text{Gal}(L/K(\delta))$ of the former is identified with the subgroup G^+ of the latter; therefore

$$\text{Inv}(G^+) = \text{Inv}(\text{Gal}(L/K(\delta))) = K(\delta),$$

which is (i).

In order to verify (ii) we need only note that

$$s(\delta^2) = s(\delta)^2 = (\text{sgn}(\bar{s})\delta)^2 = \delta^2$$

for every $s \in \text{Gal}(L/K)$. Consequently, $\delta^2 \in \text{Inv}(\text{Gal}(L/K)) = K$.

It only remains to prove (iii). Clearly, $G \subseteq \mathrm{Alt}(n)$ is equivalent to $G^+ = G$; and $K(\delta) = K$ is equivalent to $\delta \in K$. Moreover, according to Galois's theorem 3.4.5, $G^+ = G$ if and only if $\mathrm{Inv}(G^+) = \mathrm{Inv}(G)$. Since

$$\mathrm{Inv}(G^+) = K(\delta) \quad \text{and} \quad \mathrm{Inv}(G) = K,$$

it is now evident that $G \subseteq \mathrm{Alt}(n)$ if and only if $\delta \in K$. $\square$

In the situation described in the preceding proposition, it is customary to refer to δ^2 as the **discriminant of f over K**. The last assertion in the conclusion is frequently expressed by saying that the Galois group of f over K consists of even permutations if and only if the discriminant of f over K possesses a square root in K.

An expression for the discriminant can be obtained from the Vandermonde determinant formula:

$$\det \begin{bmatrix} 1 & 1 & 1 & \cdots & 1 \\ \rho_1 & \rho_2 & \rho_3 & \cdots & \rho_n \\ \rho_1^2 & \rho_2^2 & \rho_3^2 & \cdots & \rho_n^2 \\ \vdots & \vdots & \vdots & & \vdots \\ \rho_1^{n-1} & \rho_2^{n-1} & \rho_3^{n-1} & \cdots & \rho_n^{n-1} \end{bmatrix} = \prod_{1 \le i < j \le n} (\rho_j - \rho_i).$$

Letting $\sigma_i = \sum_{j=1}^n \rho_j^i$ for $1 \le i \le 2n - 2$, and taking the determinant of the product of the matrix appearing in the formula and its transpose, we get

$$\det \begin{bmatrix} n & \sigma_1 & \sigma_2 & \cdots & \sigma_{n-1} \\ \sigma_1 & \sigma_2 & \sigma_3 & \cdots & \sigma_n \\ \sigma_2 & \sigma_3 & \sigma_4 & \cdots & \sigma_{n+1} \\ \vdots & \vdots & & & \vdots \\ \sigma_{n-1} & \sigma_n & \sigma_{n+1} & \cdots & \sigma_{2n-2} \end{bmatrix} = \prod_{1 \le i < j \le n} (\rho_i - \rho_j)^2 = \delta^2.$$

Assuming that f is monic, write

$$f(X) = \sum_{i=0}^{n-1} \alpha_i X^i + X^n$$

with $\alpha_0, \alpha_1, \ldots, \alpha_{n-1} \in K$. Then the equality

$$\sum_{i=0}^{n-1} \alpha_i X^i + X^n = \prod_{i=1}^{n} (X - \rho_i)$$

holds in $K(\rho_1, \rho_2, \ldots, \rho_n)[X]$. Therefore, by 0.0.7, we have

$$\alpha_i = (-1)^{n-i} e_{n-i}(\rho_1, \rho_2, \ldots, \rho_n) \quad \text{for } 0 \le i \le n - 1,$$

where $e_1, e_2, \ldots, e_n$ are the elementary symmetric polynomials in $K[X_1, X_2, \ldots, X_n]$.

The fundamental theorem on symmetric polynomials shows that each of $\sigma_1, \sigma_2, \ldots, \sigma_n$ is a polynomial expression in $(\alpha_0, \alpha_1, \ldots, \alpha_{n-1})$ with integer coefficients; consequently, the same is true of δ^2, as can be seen from the determinant formula for δ^2 in terms of $\sigma_1, \sigma_2, \ldots, \sigma_n$ derived above. However, in order to perform these calculations, it is first necessary to know how to express $\sigma_1, \sigma_2, \ldots, \sigma_n$ in the required manner.

As the next two examples show, the computation of the discriminant can be carried out, not without some unpleasantness, for quadratics and cubics.

3.4.17. Example. Let K be a field, and let f be a quadratic monic polynomial in $K[X]$. Write

$$f(X) = X^2 + \alpha_1 X + \alpha_0$$

with $\alpha_0, \alpha_1 \in K$. We shall derive the familiar formula for the discriminant of f over K.

Write

$$f(X) = (X - \rho_1)(X - \rho_2),$$

where ρ_1 and ρ_2 are the zeros of f in its splitting field over K. Then

$$\rho_1 + \rho_2 = -\alpha_1 \quad \text{and} \quad \rho_1 \rho_2 = \alpha_0.$$

Using the notation of the preceding discussion, we have

$$\sigma_1 = \rho_1 + \rho_2 = -\alpha_1$$

and

$$\sigma_2 = \rho_1^2 + \rho_2^2 = (\rho_1 + \rho_2)^2 - 2\rho_1 \rho_2 = \alpha_1^2 - 2\alpha_0.$$

Consequently

$$\delta^2 = \det\left(\begin{bmatrix} 2 & \sigma_1 \\ \sigma_1 & \sigma_2 \end{bmatrix}\right) = 2\sigma_2 - \sigma_1^2;$$

and substituting in this equality the above expressions for σ_1 and σ_2, the expected formula,

$$\delta^2 = \alpha_1^2 - 4\alpha_0,$$

follows at once.   □

3.4.18. Example. Let K be a field, and let f be a cubic monic polynomial in $K[X]$. Write

$$f(X) = X^3 + \alpha_2 X^2 + \alpha_1 X + \alpha_0$$

with $\alpha_0, \alpha_1, \alpha_2 \in K$. There is also a simple—but perhaps less familiar—formula for the discriminant of f over K, which we now proceed to derive.

As before, let us begin by writing

$$f(X) = (X - \rho_1)(X - \rho_2)(X - \rho_3),$$

where ρ_1, ρ_2, ρ_3 are the zeros of f in its splitting field over K. Then

$$\rho_1 + \rho_2 + \rho_3 = -\alpha_2, \qquad \rho_1\rho_2 + \rho_1\rho_3 + \rho_2\rho_3 = \alpha_1, \qquad \rho_1\rho_2\rho_3 = -\alpha_0.$$

Continuing with the same notation, and taking into account that

$$\rho_i^3 = -\left(\alpha_2\rho_i^2 + \alpha_1\rho_i + \alpha_0\right)$$

and

$$\rho_i^4 = -\left(\alpha_2\rho_i^3 + \alpha_1\rho_i^2 + \alpha_0\rho_i\right)$$

for $1 \le i \le 3$, we get

$$\sigma_1 = \rho_1 + \rho_2 + \rho_3 = -\alpha_2,$$

$$\sigma_2 = \rho_1^2 + \rho_2^2 + \rho_3^2 = (\rho_1 + \rho_2 + \rho_3)^2 - 2(\rho_1\rho_2 + \rho_1\rho_3 + \rho_2\rho_3)$$
$$= \alpha_2^2 - 2\alpha_1,$$

$$\sigma_3 = \rho_1^3 + \rho_2^3 + \rho_3^3$$
$$= -\alpha_2\left(\rho_1^2 + \rho_2^2 + \rho_3^2\right) - \alpha_1(\rho_1 + \rho_2 + \rho_3) - 3\alpha_0$$
$$= -\alpha_2\left(\alpha_2^2 - 2\alpha_1\right) + \alpha_1\alpha_2 - 3\alpha_0$$
$$= -\alpha_2^3 + 3\alpha_1\alpha_2 - 3\alpha_0,$$

and

$$\sigma_4 = \rho_1^4 + \rho_2^4 + \rho_3^4$$
$$= -\alpha_2\left(\rho_1^3 + \rho_2^3 + \rho_3^3\right) - \alpha_1\left(\rho_1^2 + \rho_2^2 + \rho_3^2\right) - \alpha_0(\rho_1 + \rho_2 + \rho_3)$$
$$= -\alpha_2\left(-\alpha_2^3 + 3\alpha_1\alpha_2 - 3\alpha_0\right) - \alpha_1\left(\alpha_2^2 - 2\alpha_1\right) + \alpha_0\alpha_2$$
$$= \alpha_2^4 - 4\alpha_1\alpha_2^2 + 4\alpha_0\alpha_2 + 2\alpha_1^2.$$

Since

$$\delta^2 = \det\left(\begin{bmatrix} 3 & \sigma_1 & \sigma_2 \\ \sigma_1 & \sigma_2 & \sigma_3 \\ \sigma_2 & \sigma_3 & \sigma_4 \end{bmatrix}\right) = 3\sigma_2\sigma_4 + 2\sigma_1\sigma_2\sigma_3 - \sigma_2^3 - \sigma_1^2\sigma_4 - 3\sigma_3^2,$$

it now remains to substitute the above expressions for $\sigma_1, \sigma_2, \sigma_3, \sigma_4$ in this equality. The brave reader who is inclined to perform the required calculations will find that

$$\delta^2 = -4\alpha_0\alpha_2^3 + \alpha_1^2\alpha_2^2 + 18\alpha_0\alpha_1\alpha_2 - 4\alpha_1^3 - 27\alpha_0^2,$$

which is the desired formula. $\square$

Equipped with what we have shown in the foregoing discussion, we can now deal quite easily with the computation of the Galois group of a quadratic or cubic polynomial.

3.4.19. Example. Let K be a field, let f be a separable quadratic polynomial in $K[X]$, and let G denote the Galois group of f over K. Then:

(i) If f is reducible in $K[X]$, then G is trivial.

(ii) If f is irreducible in $K[X]$, then G has order 2.

The first assertion is evident: If f is reducible in $K[X]$, then it splits in $K[X]$, whence its only splitting field over K is K itself.

The second assertion follows at once from 3.4.15: If f is irreducible in $K[X]$, then G is a transitive subgroup of Sym(2), and hence $G = \mathrm{Sym}(2)$. $\square$

3.4.20. Example. Let K be a field, let f be a separable cubic polynomial in $K[X]$, and let G denote the Galois group of f over K. Then:

(i) If f is reducible in $K[X]$, then G is trivial or has order 2, according as f splits or does not split in $K[X]$.

(ii) If f is irreducible in $K[X]$, then $G = \mathrm{Alt}(3)$ or $G = \mathrm{Sym}(3)$; and if $\mathrm{Char}(K) \neq 2$, then the former or the latter obtains, according as the discriminant of f over K possesses or does not possess a square root in K.

First, suppose that f is reducible in $K[X]$. If f splits in $K[X]$, then the only splitting field of f over K is K, whence G is trivial. If f does not split in $K[X]$, then f has exactly one zero in K, and so $f(X) = (X - \rho)g(X)$, where $\rho \in K$ and g is a quadratic irreducible polynomial in $K[X]$; since it is clear that f and g have the same splitting field over K, and hence the same Galois group over K, it follows from the preceding example that G has order 2.

To conclude, suppose now that f is irreducible in $K[X]$. According to 3.4.15, we then have $G = \mathrm{Alt}(3)$ or $G = \mathrm{Sym}(3)$, because these are the only transitive subgroups of Sym(3). Finally, we need only apply 3.4.16 to verify that when $\mathrm{Char}(K) \neq 2$, the alternative is indeed decided as stated. $\square$

3.4.21. Example. Let us now look in detail at a specific case, the cubic polynomial $X^3 - 2$ in $\mathbf{Q}[X]$. We shall first determine its Galois group over $\mathbf{Q}$; and then we shall find the subgroups of this Galois group and their fields of invariants.

First, put

$$\omega = (-1 + \sqrt{-3})/2 \quad \text{and} \quad F = \mathbf{Q}\big(\sqrt[3]{2}, \omega\big).$$

Clearly, $\omega \neq 1$ and $\omega^3 = 1$, which implies that F is a splitting field of $X^3 - 2$ over $\mathbf{Q}$; in $F[X]$ we have

$$X^3 - 2 = (X - \rho_1)(X - \rho_2)(X - \rho_3),$$

where

$$\rho_1 = \sqrt[3]{2}, \qquad \rho_2 = \sqrt[3]{2}\,\omega, \qquad \rho_3 = \sqrt[3]{2}\,\omega^2.$$

Since $X^3 - 2$ is irreducible in $\mathbf{Q}[X]$ and its discriminant over $\mathbf{Q}$ is -108, it follows from 3.4.20 that the Galois group of $X^3 - 2$ over $\mathbf{Q}$ is Sym(3).

It is now meaningful, in particular, to speak of the **Q**-automorphisms s and t of F such that

$$s(\rho_1) = \rho_1, \qquad s(\rho_2) = \rho_3, \qquad s(\rho_3) = \rho_2$$

and

$$t(\rho_1) = \rho_2, \qquad t(\rho_2) = \rho_3, \qquad t(\rho_3) = \rho_1.$$

These equalities imply that s and t are elements of period 2 and 3 in $\mathrm{Gal}(F/\mathbf{Q})$, respectively. Thus $\mathrm{Gal}(F/\mathbf{Q})$ has order 6, $\langle t \rangle$ has order 3, and $s \notin \langle t \rangle$, and so we have the coset decomposition

$$\mathrm{Gal}(F/Q) = \langle t \rangle \cup s\langle t \rangle;$$

consequently the list

$$i_F, \ t, \ t^2, \ s, \ st, \ st^2$$

contains the six elements of $\mathrm{Gal}(F/\mathbf{Q})$, and $\mathrm{Gal}(F/\mathbf{Q}) = \langle s, t \rangle$.

To determine the subgroups of $\mathrm{Gal}(F/\mathbf{Q})$, note that

$$st(\rho_1) = \rho_3, \qquad st(\rho_2) = \rho_2, \qquad st(\rho_3) = \rho_1$$

and

$$st^2(\rho_1) = \rho_2, \qquad st^2(\rho_2) = \rho_1, \qquad st^2(\rho_3) = \rho_3.$$

Using these equalities, st and st^2 are readily shown to be elements of period 2 in $\mathrm{Gal}(F/\mathbf{Q})$. What we now know is certainly sufficient to assert that $\mathrm{Gal}(F/\mathbf{Q})$ contains exactly four proper nontrivial subgroups, namely,

$$\langle s \rangle, \quad \langle st \rangle, \quad \langle st^2 \rangle, \quad \langle t \rangle.$$

To complete our discussion, we still have to determine the fields of invariants of these four subgroups. To this end, note first that

$$s\left(\sqrt[3]{2}\right) = s(\rho_1) = \rho_1 = \sqrt[3]{2},$$

$$st\left(\sqrt[3]{2}\,\omega\right) = st(\rho_2) = \rho_2 = \sqrt[3]{2}\,\omega,$$

$$st^2\left(\sqrt[3]{2}\,\omega^2\right) = st^2(\rho_3) = \rho_3 = \sqrt[3]{2}\,\omega^2,$$

$$t(\omega) = t(\rho_2/\rho_1) = \rho_3/\rho_2 = \omega;$$

therefore

$$\mathbf{Q}\left(\sqrt[3]{2}\right) \subseteq \mathrm{Inv}(\langle s \rangle),$$

$$\mathbf{Q}\left(\sqrt[3]{2}\,\omega\right) \subseteq \mathrm{Inv}(\langle st \rangle),$$

$$\mathbf{Q}\left(\sqrt[3]{2}\,\omega\right) \subseteq \mathrm{Inv}(\langle st^2 \rangle),$$

$$\mathbf{Q}(\omega) \subseteq \mathrm{Inv}(\langle t \rangle).$$

Moreover, in each of these four instances, F has the same linear degree over the two subfields related by the indicated inclusion: Indeed, this is clear from the equalities

$$[F:\mathbf{Q}] = 6,$$

$$\left[\mathbf{Q}(\sqrt[3]{2}):\mathbf{Q}\right] = \left[\mathbf{Q}(\sqrt[3]{2}\,\omega):\mathbf{Q}\right] = \left[\mathbf{Q}(\sqrt[3]{2}\,\omega^2):\mathbf{Q}\right] = 3,$$

$$[\mathbf{Q}(\omega):\mathbf{Q}] = 2,$$

$$[F:\mathrm{Inv}(\langle s\rangle)] = [F:\mathrm{Inv}(\langle st\rangle)] = [F:\mathrm{Inv}(\langle st^2\rangle)] = 2,$$

$$[F:\mathrm{Inv}(\langle t\rangle)] = 3.$$

Thus, each of the above four inclusions is in fact an equality; and hence we can assert that

$$\mathbf{Q}(\sqrt[3]{2}), \quad \mathbf{Q}(\sqrt[3]{2}\,\omega), \quad \mathbf{Q}(\sqrt[3]{2}\,\omega^2), \quad \mathbf{Q}(\omega)$$

are the fields of invariants of the four proper nontrivial subgroups of $\mathrm{Gal}(F/\mathbf{Q})$, listed in the order corresponding to the order in which the latter were listed previously. Note, incidentally, that according to Galois's fundamental theorem 3.4.5, these four fields are precisely the intermediate fields strictly between $\mathbf{Q}$ and F.   □

The discussion of quartic polynomials, which is more involved than that of quadratic or cubic polynomials, will be left to the interested reader (problems 14, 15, 16). For degrees greater than 4, the computations become more complicated, and only very special situations can be handled systematically.

To close the section, we shall present the classical application of Galois theory in the study of symmetric rational functions.

3.4.22. Example. Let K be a field, and let n be a positive integer. For each $c \in \mathrm{Sym}(n)$, it is meaningful to speak of the K-automorphism of $K[X_1, X_2, \ldots, X_n]$ such that $X_i \to X_{c(i)}$ for $1 \le i \le n$; and since $K(X_1, X_2, \ldots, X_n)$ is a field of fractions of $K[X_1, X_2, \ldots, X_n]$, this is uniquely extendible to a K-automorphism of $K(X_1, X_2, \ldots, X_n)$, which will be denoted by S_c.

It is readily verified that the mapping $c \to S_c$ from $\mathrm{Sym}(n)$ to $\mathrm{Gal}(K(X_1, X_2, \ldots, X_n)/K)$ is a monomorphism. The elements of the field of invariants of its image are said to be the **symmetric rational functions in** $K(X_1, X_2, \ldots, X_n)$. Clearly, the symmetric polynomials in $K[X_1, X_2, \ldots, X_n]$ are symmetric rational functions; this is also true, in particular, of the elementary symmetric polynomials $e_1, e_2, \ldots, e_n$ in $K[X_1, X_2, \ldots, X_n]$.

Our objective here is to show, without using the fundamental theorem on symmetric polynomials, that the field consisting of the symmetric rational functions in $K(X_1, X_2, \ldots, X_n)$ is identical with $K(e_1, e_2, \ldots, e_n)$, and that

$$[K(X_1, X_2, \ldots, X_n) : K(e_1, e_2, \ldots, e_n)] = n!.$$

To do this, write

$$L = K(X_1, X_2, \ldots, X_n) \quad \text{and} \quad E = K(e_1, e_2, \ldots, e_n),$$

and denote by Ω the image of the monomorphism $c \to S_c$ from Sym(n) to Gal(L/K). We have to show that Inv$(\Omega) = E$.

First, we have $e_1, e_2, \ldots, e_n \in$ Inv(Ω), and hence $E \subseteq$ Inv(Ω). Moreover, the Dedekind–Artin theorem shows that

$$[L : \text{Inv}(\Omega)] = \text{Card}(\Omega) = n!.$$

Consequently, our conclusion will be established if it is verified that $[L : E] \leq n!$.

To this end, define a polynomial f in $E[Y]$ by

$$f(Y) = \sum_{i=0}^{n-1} (-1)^{n-i} e_{n-i}(X_1, X_2, \ldots, X_n) Y^i + Y^n.$$

According to 0.0.7, in $L[Y]$ we have the equality

$$f(Y) = \prod_{i=1}^{n} (Y - X_i);$$

and clearly

$$L = K(X_1, X_2, \ldots, X_n) = E(X_1, X_2, \ldots, X_n).$$

Therefore L is a splitting field of f over E; and since deg$(f) = n$, we conclude that $[L : E] \leq n!$, which is the required inequality. $\square$

The fact that the elementary symmetric polynomials generate the field of symmetric rational functions can be deduced directly from the fundamental theorem on symmetric polynomials (problem 12).

PROBLEMS

1. The main part of the proof of 3.4.1 is devoted to proving the following statement: If F is a field, and if Γ is a finite group of automorphisms of F, then

$$[F : \text{Inv}(\Gamma)] \leq \text{Card}(\Gamma).$$

A more general statement is the following: If F is a field, if Γ is a group of automorphisms of F, and if Ω is a subgroup of finite index of Γ, then

$$[\text{Inv}(\Omega) : \text{Inv}(\Gamma)] \leq [\Gamma : \Omega].$$

Prove the more general statement as follows: Denote by $\mathscr{L}$ the set of all left cosets of Ω in Γ; verify that there exists a mapping from $\mathscr{L} \times \mathrm{Inv}(\Omega)$ to F such that $(s\Omega, \alpha) \to s(\alpha)$ for all $s \in \Gamma$ and $\alpha \in \mathrm{Inv}(\Omega)$; and then adapt the argument used in the proof of 3.4.1 to the present situation.

2. Let F be a field. Verify the following assertions.
   a. If Ω is a Galois group on F, and if Γ is a group of automorphisms of F having Ω as a subgroup of finite index, then Γ is a Galois group on F, and
   $$[\mathrm{Inv}(\Omega):\mathrm{Inv}(\Gamma)] = [\Gamma:\Omega].$$
   b. If P is an invariant field in F, and if Q is a subfield of F that is a finite extension of P, then Q is an invariant field in F, and
   $$[\mathrm{Gal}(F/P):\mathrm{Gal}(F/Q)] = [Q:P].$$
   (These results are due to Kaplansky. They contain 3.4.3 as a particular case.)

3. Show that 3.4.1 is a consequence of 3.2.15 and the results stated in section 3.1, problem 12, and section 3.2, problem 13.

4. Show that 3.4.1 is a consequence of 3.2.15 and the results stated in section 3.2, problems 12 and 13.

5. Prove that if F is a field and Γ is a finite group of automorphisms of F, then Γ is a base of $\mathrm{Lin}_{\mathrm{Inv}(\Gamma)}(F, F)$.

6. Let F be a field, and let Γ be a finite group of automorphisms of F. Prove that if Φ is an F-algebra contained in the F-subspace of $\mathrm{Map}(F, F)$ generated by Γ, then $\Gamma \cap \Phi$ is a subgroup of Γ, and the F-subspace of $\mathrm{Map}(F, F)$ that it generates is Φ.

7. The following statement is a particular case of the result stated in problem 2.

   If F is a field, if P is an invariant field in F admitting F as a finite extension, and if Q is an intermediate field between P and F, then Q is invariant in F, and
   $$[\mathrm{Gal}(F/P):\mathrm{Gal}(F/Q)] = [Q:P].$$

   Prove this statement, without using results from the preceding or present section other than 3.4.1, in each of the two ways indicated here:
   a. Combine 3.4.1 with the result stated in section 3.2, problem 1.
   b. Combine 3.4.1 with the results stated in the preceding two problems and section 3.1, problem 9.

8. Give an example of a field F such that $\mathrm{Aut}(F)$ is infinite and such that no proper infinite subgroup of $\mathrm{Aut}(F)$ is a Galois group on F.

9. Give an example showing that 3.4.9 does not remain valid without the assumption that L be Galois over K.

10. Let K be a field, let N be an extension field of K, and let L and M be intermediate fields between K and N one of which is Galois and finite

over K and such that $L \cap M = K$. Prove that $[L \vee M : K] = [L : K][M : K]$.

11. Let K be a field such that $\text{Char}(K) \neq 3$, let f be a monic cubic polynomial in $K[X]$, and write

$$f(X) = X^3 + \alpha_2 X^2 + \alpha_1 X + \alpha_0$$

with $\alpha_0, \alpha_1, \alpha_2 \in K$. Show that if g is the polynomial in $K[X]$ defined by $g(X) = f(X - \alpha_2/3)$, then f and g have the same discriminant over K, and the coefficient of X^2 in g is 0.

12. Let K be a field, and let n be a positive integer. Use the fundamental theorem on symmetric polynomials in order to prove that if $e_1, e_2, \ldots, e_n$ denote the elementary symmetric polynomials in $K[X_1, X_2, \ldots, X_n]$, then the symmetric rational functions in $K(X_1, X_2, \ldots, X_n)$ make up the subfield $K(e_1, e_2, \ldots, e_n)$ of $K(X_1, X_2, \ldots, X_n)$.

13. A classical result in the theory of groups states that every finite group of order n is embeddable in $\text{Sym}(n)$. Use this in order to prove that every finite group is isomorphic to the Galois group of a finite Galois extension.

14. Let K be a field, let f be a separable quartic monic polynomial in $K[X]$, and write

$$f(X) = X^4 + \alpha_3 X^3 + \alpha_2 X^2 + \alpha_1 X + \alpha_0$$

with $\alpha_0, \alpha_1, \alpha_2, \alpha_3 \in K$. Let L be a splitting field of f over K, and denote by $\rho_1, \rho_2, \rho_3, \rho_4$ the zeros of f in L, so that $L = K(\rho_1, \rho_2, \rho_3, \rho_4)$ and

$$f(X) = (X - \rho_1)(X - \rho_2)(X - \rho_3)(X - \rho_4)$$

in $L[X]$. Finally, put

$$\sigma_1 = \rho_1\rho_2 + \rho_3\rho_4, \qquad \sigma_2 = \rho_1\rho_3 + \rho_2\rho_4, \qquad \sigma_3 = \rho_1\rho_4 + \rho_2\rho_3;$$

the polynomial g in $L[X]$ defined by

$$g(X) = (X - \sigma_1)(X - \sigma_2)(X - \sigma_3)$$

is said to be the **resolvent cubic** of f. Prove the following assertions.

a. g is a separable monic polynomial in $K[X]$; and furthermore

$$g(X) = X^3 - \alpha_2 X^2 + (\alpha_1\alpha_3 - 4\alpha_0)X - \alpha_0\alpha_3^2 + 4\alpha_0\alpha_2 - \alpha_1^2.$$

b. f and g have the same discriminant over K.

c. If G denotes the Galois group of f over K, then $\text{Inv}(G \cap \mathbf{V}) = K(\sigma_1, \sigma_2, \sigma_3)$, and the Galois group of g over K is isomorphic to $G/G \cap \mathbf{V}$.

15. Let K be a field, let f be separable quartic monic irreducible polynomial in $K[X]$, and let G denote the Galois group of f over K. Prove first that the order of $G/G \cap \mathbf{V}$ is one of $1, 2, 3, 6$; and then prove the following assertions.

a. If $G/G \cap \mathbf{V}$ has order 6, then $G = \text{Sym}(4)$.

b. If $G/G \cap \mathbf{V}$ has order 3, then $G = \text{Alt}(4)$.

c. If $G/G \cap \mathbf{V}$ has order 1, then $G = \mathbf{V}$.

    d. If $G/G \cap V$ has order 2, then G is dihedral of order 8 or cyclic of order 4; and furthermore, denoting by $\sigma_1, \sigma_2, \sigma_3$ the zeros of the resolvent cubic of f, the former or the latter obtains according as f is irreducible or reducible in $K(\sigma_1, \sigma_2, \sigma_3)[X]$.

16.   Complete descriptions of the Galois groups of separable quadratic and cubic polynomials have been given in 3.4.19 and 3.4.20, respectively. Give such a description of the Galois group of a separable quartic polynomial.

17.   A detailed analysis of the Galois group of $X^3 - 2$ over $\mathbf{Q}$ has been given in 3.4.21. Give such an analysis of the Galois group of $X^4 - 2$ over $\mathbf{Q}$.

3.5. ROOTS OF UNITY

Recall that the roots of unity in a field K are the elements of finite order in the multiplicative group K^*; and more precisely, that if n is a positive integer, an nth root of unity in K is an element α of K^* such that $\alpha^n = 1$.

For a fixed positive integer n, the nth roots of unity in K obviously form a multiplicative group in K. This group is finite and of order less than or equal to n, because its elements are the zeros in K of the polynomial $X^n - 1$ in $K[X]$.

On the other hand, if G is a finite multiplicative group in K of order n, then $\alpha^n = 1$ for every $\alpha \in G$; and hence G is identical with the multiplicative group consisting of the nth roots of unity in K.

In order to study the roots of unity in fields, we shall begin with a fundamental result on finite multiplicative groups in fields.

    3.5.1.   Theorem. *If K is a field, then every finite multiplicative group in K is cyclic.*

    Proof. We shall require two lemmas. The first states a property of the Euler function, and the second is a special result on finite groups. To prove the lemmas, we shall use elementary properties of cyclic groups.

    Lemma 1. *If n is a positive integer, than $n = \sum_{d \mid n} \varphi(d)$.*

    Proof. Let C be a cyclic group of order n. If d is a positive divisor of n, then C possesses a unique subgroup of order d; since such a subgroup is cyclic, it has exactly $\varphi(d)$ generators, and these are precisely the elements of C having order d. Since the order of every element of C is a positive divisor of n, the lemma now follows at once.

    Lemma 2. *Let G be a finite group having the property that for every positive divisor d of the order of G, the cardinality of the set of all $\sigma \in G$ such that $\sigma^d = 1$ is less than or equal to d. Then G is cyclic.*

Proof. Put $n = \text{Card}(G)$. For each positive divisor d of n, let S_d denote the set of all elements of G of order d. It is evident that $n = \sum_{d|n} \text{Card}(S_d)$; and by the preceding lemma, we also have $n = \sum_{d|n} \varphi(d)$.

We now claim that for every positive divisor d of n, either S_d is empty or $\text{Card}(S_d) = \varphi(d)$. Indeed, note first that if D is a subgroup of G of order d, then $\sigma^d = 1$ for every $\sigma \in D$; and, in view of our hypothesis on G, this implies that D is the set of all $\sigma \in G$ with $\sigma^d = 1$, whence $S_d \subseteq D$. This being said, assume that S_d is nonempty, and choose a $\delta \in S_d$; then the subgroup $\langle \delta \rangle$ of G generated by δ is of order d, whence $S_d \subseteq \langle \delta \rangle$; since $\langle \delta \rangle$ has exactly $\varphi(d)$ generators, and since each of these belongs to S_d, we conclude that $\text{Card}(S_d) = \varphi(d)$.

Consequently, we have $\text{Card}(S_d) \leq \varphi(d)$ for every positive divisor d of n. Since $n = \sum_{d|n} \text{Card}(S_d)$ and $n = \sum_{d|n} \varphi(d)$, it now follows that $\text{Card}(S_d) = \varphi(d)$ for every positive divisor d of n. In particular, we have $\text{Card}(S_n) = \varphi(n) > 0$, so that S_n is nonempty. Therefore, there exists an element of G of order n, which means that G is cyclic. The lemma is proved.

To prove the theorem, consider a finite multiplicative group in K. If d is a positive divisor of the order of G, every element σ of G with $\sigma^d = 1$ is a zero of the polynomial $X^d - 1$ in $K[X]$; since $\deg(X^d - 1) = d$, this implies that the set of all $\sigma \in G$ such that $\sigma^d = 1$ has cardinality less than or equal to d. By the second lemma, G is cyclic. □

The proof just given depends only on the most elementary facts about cyclic groups. The reader familiar with the structure theorems for finite abelian groups will have little difficulty in providing an alternative proof (problem 1).

Let n be a positive integer, and let K be a field. By a **primitive nth root of unity in K** we shall understand an element of order n in K^*.

As observed at the beginning of the section, the multiplicative group consisting of the nth roots of unity in K has order less than or equal to n; and in particular, by the preceding theorem, it is cyclic. Therefore, it has order n if and only if K contains a primitive nth root of unity; in this case, its generators are precisely the primitive nth roots of unity in K.

Let p be a prime, and let K be a field. If $\text{Char}(K) = p$, then an element of K admits at most one pth root in K, so that the conditions $\alpha \in K$ and $\alpha^p = 1$ imply $\alpha = 1$; therefore, K contains no primitive pth roots of unity. On the other hand, if $\text{Char}(K) \neq p$, it is seen that a $\zeta \in K$ is a primitive pth root of unity in K if and only if $\zeta \neq 1$ and $\zeta^p = 1$, since these conditions characterize the elements of order p in K^*.

3.5.2. Proposition. *Let n be a positive integer, and let K be a field containing a primitive nth root of unity.*

(i) *The set of all primitive nth roots of unity in K has cardinality $\varphi(n)$.*

(ii) *If d is a positive divisor of n, then K contains a primitive dth root of unity; in fact, if ζ is a primitive nth root of unity in K, then $\zeta^{n/d}$ is a primitive dth root of unity in K.*

(iii) $\text{Char}(K) \nmid n$.

Proof. As to the first assertion, note that the multiplicative group consisting of all nth roots of unity in K is cyclic and has order n. Therefore, it has exactly $\varphi(n)$ generators, and these are the primitive nth roots of unity in K.

To prove the second assertion, we need only recall that if an element ζ of a group has finite order n, and if d is a positive divisor of n, then the element $\zeta^{n/d}$ has order d.

The last assertion is evident when $\text{Char}(K) = 0$. Thus, assume that $\text{Char}(K) = p$, where p is a prime. If $p \mid n$, and if ζ is a primitive nth root of unity in K, then $(\zeta^{n/p})^p = \zeta^n = 1$; therefore, $\zeta^{n/p}$ would be a pth root of 1 in K, whence $\zeta^{n/p} = 1$. But this is impossible, because $1 \le n/p < n$ and ζ is an element of order n in K^*. $\qquad\qquad\square$

3.5.3. Example. Let n be an integer such that $n > 1$, let K be a field containing a primitive nth root ζ of unity, and let $K(\xi)$ be a simple transcendental extension of K. Denote by s and t the K-automorphisms of $K(\xi)$ such that $s(\xi) = 1/\xi$ and $t(\xi) = \zeta\xi$. Then:

(i) s and t are, respectively, elements of order 2 and n in $\text{Gal}(K(\xi)/K)$.

(ii) $ts = st^{n-1}$.

(iii) The subgroup $\langle s, t \rangle$ of $\text{Gal}(K(\xi)/K)$ is dihedral of order $2n$.

(iv) $\text{Inv}(\langle s, t \rangle) = K(\xi^n + 1/\xi^n)$.

To prove (i), note first that $s \ne i_{K(\xi)} \ne t$. Since it is evident that $s^2 = i_{K(\xi)}$, we see that s has order 2. To show that t has order n, it suffices to check that if k is a positive integer, then $t^k = i_{K(\xi)}$ if and only if $n \mid k$, which presents no difficulty: Indeed, we know that $\zeta^k = 1$ if and only if $n \mid k$; and since $t^k(\xi) = \zeta^k\xi$, it follows that $t^k = i_{K(\xi)}$ if and only if $\zeta^k\xi = \xi$, hence if and only if $\zeta^k = 1$.

Since

$$ts(\xi) = t(1/\xi) = 1/t(\xi) = 1/\zeta\xi$$
$$= \zeta^{n-1}/\xi = s(\zeta^{n-1}\xi) = st^{n-1}(\xi),$$

we see that $ts = st^{n-1}$, which is (ii).

Next, we shall prove (iii) and (iv). Put $\sigma = \xi^n + 1/\xi^n$; we then have $1 - \sigma\xi^n + \xi^{2n} = 0$, which means that ξ is a zero of the polynomial $1 - \sigma X^n + X^{2n}$ in $K(\sigma)[X]$, whence

$$[K(\xi) : K(\sigma)] = [K(\sigma)(\xi) : K(\sigma)] \le 2n.$$

On the other hand, the equalities

$$s(\sigma) = s(\xi)^n + 1/s(\xi)^n = 1/\xi^n + \xi^n = \sigma$$

and

$$t(\sigma) = t(\xi)^n + 1/t(\xi)^n = \zeta^n \xi^n + 1/\zeta^n \xi^n$$
$$= \xi^n + 1/\xi^n = \sigma$$

show that σ is a fixed point of s and t; therefore $\sigma \in \mathrm{Inv}(\langle s, t \rangle)$, and so $K(\sigma) \subseteq \mathrm{Inv}(\langle s, t \rangle)$.

The inequality $[K(\xi) : K(\sigma)] \le 2n$ and the inclusion $K(\sigma) \subseteq \mathrm{Inv}(\langle s, t \rangle)$ clearly imply that $[K(\xi) : \mathrm{Inv}(\langle s, t \rangle)] \le 2n$. By the Dedekind–Artin theorem, it now follows that $\langle s, t \rangle$ is finite and

$$\mathrm{Card}(\langle s, t \rangle) = [K(\xi) : \mathrm{Inv}(\langle s, t \rangle)] \le 2n.$$

We have to show that $\mathrm{Card}(\langle s, t \rangle) = 2n$ and $\mathrm{Inv}(\langle s, t \rangle) = K(\sigma)$; in view of the relations above, it suffices to establish the inequality $\mathrm{Card}(\langle s, t \rangle) \ge 2n$. To this end, we simply observe that the transcendence of ξ over K implies that no two of

$$\xi, \zeta\xi, \zeta^2\xi, \ldots, \zeta^{n-1}\xi, 1/\xi, \zeta/\xi, \zeta^2/\xi, \ldots, \zeta^{n-1}/\xi$$

are equal; and that these are the values at ξ of

$$i_{K(\xi)}, t, t^2, \ldots, t^{n-1}, s, st, st^2, \ldots, st^{n-1}.$$

Consequently, the latter list contains $2n$ distinct elements of $\langle s, t \rangle$. □

A polynomial equation of the form $X^n - \alpha = 0$ is said to be a **binomial equation** or **pure equation**. A number of results on binomial equations are required for the proof of Galois's theorem on the solvability of polynomial equations by radicals. As we show next, the existence of primitive nth roots of unity is strongly related to the splitting of a polynomial of the type $X^n - \alpha$.

3.5.4. Proposition. *Let n be a positive integer, let K be a field, and let $\alpha \in K^*$. Then $X^n - \alpha$ is separable and splits in $K[X]$ if and only if α admits an nth root in K and K contains a primitive nth root of unity; in which case, if ρ is an nth root of α in K and ζ is a primitive nth root of unity in K, then $\rho, \zeta\rho, \zeta^2\rho, \ldots, \zeta^{n-1}\rho$ are n distinct nth roots of α in K, and*

$$X^n - \alpha = \prod_{i=0}^{n-1} (X - \zeta^i \rho).$$

Proof. First suppose that $X^n - \alpha$ is separable and splits in $K[X]$. Then

$$X^n - \alpha = \prod_{i=1}^{n} (X - \rho_i),$$

where $\rho_1, \rho_2, \ldots, \rho_n$ are n distinct elements of K. Clearly, these are nth roots of α in K, whence

$$\left(\frac{\rho_i}{\rho_1}\right)^n = \frac{\rho_i^n}{\rho_1^n} = \frac{\alpha}{\alpha} = 1$$

for $1 \le i \le n$. It follows that $\rho_1/\rho_1, \rho_2/\rho_1, \ldots, \rho_n/\rho_1$ are n distinct nth roots of unity in K, and so K contains a primitive nth root of unity.

Conversely, suppose now that $\rho^n = \alpha$ for some $\rho \in K$ and that there exists a primitive nth root ζ of unity in K. For $0 \le i \le n-1$, we have

$$\left(\zeta^i \rho\right)^n = \zeta^{in}\rho^n = \rho^n = \alpha;$$

therefore $\rho, \zeta\rho, \zeta^2\rho, \ldots, \zeta^{n-1}\rho$ are n distinct nth roots of α in K, that is, n distinct zeros of $X^n - \alpha$ in K. We conclude that $X^n - \alpha$ is separable and

$$X^n - \alpha = \prod_{i=0}^{n-1} (X - \zeta^i\rho),$$

as required. $\qquad\qquad\qquad\qquad\qquad\qquad\qquad\qquad\qquad\qquad\square$

The study of splitting fields of polynomials of the type $X^n - \alpha$ will be taken up at a later time, in the section on cyclic extensions. In the rest of the present section, we shall be concerned with an important particular case.

Let n be a positive integer, and let K be a field. By a **cyclotomic field of order** n **over** K we shall understand a splitting field over K of the polynomial $X^n - 1$ in $K[X]$.

The case where $\mathrm{Char}(K)|n$ can be disregarded in the discussion of cyclotomic fields of order n over K. In effect, in this situation we have $\mathrm{Char}(K) = p$ for some prime p; and if we write $n = p^k m$, where k and m are positive integers such that $p \nmid m$, then the equalities

$$X^n - 1 = X^{p^k m} - 1 = \left(X^m - 1\right)^{p^k}$$

hold in $K[X]$. Consequently, an extension field of K is cyclotomic of order n over K if and only if it is cyclotomic of order m over K.

3.5.5. Proposition. *Let n be a positive integer, let K be a field such that $\mathrm{Char}(K) \nmid n$, and let L be a cyclotomic field of order n over K.*

(i) *L contains a primitive nth root of unity.*

(ii) *If ζ is a primitive nth root of unity in L, then $L = K(\zeta)$.*

(iii) *$\mathrm{Gal}(L/K)$ is embeddable in $(\mathbf{Z}/n)^*$.*

(iv) *L is abelian and finite over K.*

(v) *$[L:K] | \varphi(n)$.*

Proof. Since $\mathrm{Char}(K) \nmid n$, the polynomial $X^n - 1$ in $K[X]$ is separable. Assertions (i) and (ii) are then immediate consequences of 3.5.4. Moreover, according to 3.3.5, L is Galois and finite over K. Finally, note that it suffices to verify (iii); for since $(\mathbf{Z}/n)^*$ is an abelian group of order $\varphi(n)$, the embeddability of $\mathrm{Gal}(L/K)$ in $(\mathbf{Z}/n)^*$ implies (iv) and (v).

To prove (iii), choose a primitive nth root of unity ζ in L. If s is a K-automorphism of L, then $s(\zeta)$ is also a primitive nth root of unity in L, and hence an integer k_s can be chosen so that k_s and n are relatively prime, $1 \le k_s < n$, and $s(\zeta) = \zeta^{k_s}$; in particular, the coset $k_s + n\mathbf{Z}$ is an invertible element of $\mathbf{Z}/n$. We shall now complete the proof by showing that the mapping $s \to k_s + n\mathbf{Z}$ from $\mathrm{Gal}(L/K)$ to $(\mathbf{Z}/n)^*$ is an injective homomorphism.

If $s, t \in \mathrm{Gal}(L/K)$, then

$$\zeta^{k_{ts}} = ts(\zeta) = t(s(\zeta)) = t(\zeta^{k_s}) = t(\zeta)^{k_s} = \zeta^{k_t k_s};$$

therefore $k_{ts} \equiv k_t k_s \pmod{n}$, which means that the equality

$$k_{ts} + n\mathbf{Z} = (k_t + n\mathbf{Z})(k_s + n\mathbf{Z})$$

holds in $(\mathbf{Z}/n)^*$. Thus, our mapping is a homomorphism. And its injectivity is readily established: If s belongs to its kernel, then s is a K-automorphism of L such that $k_s + n\mathbf{Z} = 1 + n\mathbf{Z}$; but then $k_s = 1$ and $s(\zeta) = \zeta^{k_s} = \zeta$, and the equality $L = K(\zeta)$ shows that $s = i_L$.   □

An evident consequence of this proposition is that every intermediate field between a field K and a cyclotomic field over K is abelian over K.

For the rational field, a famous theorem of Kronecker states that the converse of this assertion is true: Every finite abelian extension of $\mathbf{Q}$ is a subfield of a cyclotomic field over $\mathbf{Q}$. However, this is not an elementary result; its proof requires number-theoretical tools that will not be developed in this book.

To conclude our discussion, we shall now proceed to derive some classical results concerning the cyclotomic polynomials. These will then be applied to the determination of the Galois groups of the cyclotomic fields over the rational field.

Let n be a positive integer. The nth roots of unity in $\mathbf{C}$ are the n complex numbers of the form $\exp(2\pi i\sqrt{-1}\,/n)$, where $1 \le i \le n$; and furthermore, exactly $\varphi(n)$ of these are primitive nth roots of unity in $\mathbf{C}$, namely, those for which i and n are relatively prime.

Denote by U_n the set consisting of all primitive nth roots of unity in $\mathbf{C}$, and let Φ_n be the monic polynomial in $\mathbf{C}[X]$ defined by

$$\Phi_n(X) = \prod_{\zeta \in U_n} (X - \zeta);$$

clearly, $\deg(\Phi_n) = \mathrm{Card}(U_n) = \varphi(n)$. This polynomial is said to be the nth **cyclotomic polynomial**.

The terminology originates from the fact that the nth roots of unity in $\mathbf{C}$ divide the unit circle into n arcs of equal length.

3.5.6. Proposition. *If n is a positive integer, then*

$$X^n - 1 = \prod_{d \mid n} \Phi_d(X).$$

Proof. Let U denote the multiplicative group consisting of the nth roots of unity in $\mathbf{C}$; then

$$X^n - 1 = \prod_{\zeta \in U} (X - \zeta).$$

And for every positive divisor d of n, let U_d denote the set of all primitive dth roots of unity in $\mathbf{C}$, that is, the set of all elements of order d in U; then

$$\Phi_d(X) = \prod_{\zeta \in U_d} (X - \zeta).$$

Arguing as in the first lemma in the proof of 3.5.1, we see that $(U_d)_{d|n}$ is a partition of U. Consequently

$$X^n - 1 = \prod_{\zeta \in U} (X - \zeta) = \prod_{d|n} \left(\prod_{\zeta \in U_d} (X - \zeta) \right) = \prod_{d|n} \Phi_d(X),$$

which is what we wanted.   □

The preceding result makes it possible to compute the cyclotomic polynomials recursively by division. For example, to compute the first six, write

$$X - 1 = \Phi_1(X),$$
$$X^2 - 1 = \Phi_1(X)\Phi_2(X),$$
$$X^3 - 1 = \Phi_1(X)\Phi_3(X),$$
$$X^4 - 1 = \Phi_1(X)\Phi_2(X)\Phi_4(X),$$
$$X^5 - 1 = \Phi_1(X)\Phi_5(X),$$
$$X^6 - 1 = \Phi_1(X)\Phi_2(X)\Phi_3(X)\Phi_6(X);$$

from these relations we then get

$$\Phi_1(X) = X - 1,$$
$$\Phi_2(X) = X + 1,$$
$$\Phi_3(X) = X^2 + X + 1,$$
$$\Phi_4(X) = X^2 + 1,$$
$$\Phi_5(X) = X^4 + X^3 + X^2 + X + 1,$$
$$\Phi_6(X) = X^2 - X + 1.$$

Note that 1 and -1 are the only nonzero coefficients appearing in the foregoing calculations. It will be seen presently that the cyclotomic polynomials have integer coefficients. For $1 \le n \le 104$ every nonzero coefficient in $\Phi_n(X)$ is 1 or -1; but the coefficient of X^7 in $\Phi_{105}(X)$ is -2. It was shown by Schur that there are cyclotomic polynomials having coefficients arbitrarily large in absolute value.

3.5.7. Proposition. *If n is a positive integer, then $\Phi_n(X) \in \mathbf{Z}[X]$.*

Proof. We shall prove this by induction. Since $\Phi_1(X) = X - 1$, we have $\Phi_1(X) \in \mathbf{Z}[X]$. Next, let n be an integer such that $n > 1$, and assume that $\Phi_k(X) \in \mathbf{Z}[X]$ for $1 \le k < n$. Letting

$$f(X) = \prod_{\substack{d \mid n \\ d < n}} \Phi_d(X),$$

we see from this assumption that $f(X) \in \mathbf{Z}[X]$. Thus, $X^n - 1$ and $f(X)$ are monic polynomials in $\mathbf{Z}[X]$, and the preceding proposition shows that the equalities

$$X^n - 1 = \prod_{d \mid n} \Phi_d(X) = f(X)\Phi_n(X)$$

hold in $\mathbf{C}[X]$. Applying the division algorithms in $\mathbf{Z}[X]$ and $\mathbf{C}[X]$, we now conclude that $\Phi_n(X) \in \mathbf{Z}[X]$. $\qquad\square$

If p is a prime, then 3.5.6 shows that

$$X^p - 1 = \Phi_1(X)\Phi_p(X) = (X - 1)\Phi_p(X),$$

whence

$$\Phi_p(X) = X^{p-1} + X^{p-2} + \cdots + X + 1.$$

We have proved in 2.1.7 that this polynomial is irreducible in $\mathbf{Q}[X]$. As the following proposition shows, this is true of every cyclotomic polynomial.

3.5.8. Proposition. *If n is a positive integer, then $\Phi_n(X)$ is irreducible in $\mathbf{Q}[X]$.*

Proof. Let ζ be a primitive nth root of unity in $\mathbf{C}$, and denote by f its minimal polynomial over $\mathbf{Q}$. Since ζ is a zero of $X^n - 1$, we then have $X^n - 1 = f(X)g(X)$ for some $g \in \mathbf{Q}[X]$. Hence f and g are monic polynomials in $\mathbf{Q}[X]$ such that $fg \in \mathbf{Z}[X]$; according to 1.5.10, this implies that $f, g \in \mathbf{Z}[X]$.

We now claim that the conditions $\theta \in \mathbf{C}$ and $f(\theta) = 0$ imply that $f(\theta^p) = 0$ for every prime p not dividing n. To verify this, assume, for a contradiction, that $f(\theta^p) \neq 0$. Then θ is an nth root of unity in $\mathbf{C}$; and from

$$f(\theta^p)g(\theta^p) = \theta^{pn} - 1 = 0$$

it is clear that $g(\theta^p) = 0$. Thus, θ is a zero of $g(X^p)$; and since f is the minimal polynomial of θ over $\mathbf{Q}$, it follows that $g(X^p) = f(X)h(X)$ for some $h \in \mathbf{Q}[X]$. Then f and h are monic polynomials in $\mathbf{Q}[X]$ such that $fh \in \mathbf{Z}[X]$, and a second application of 1.5.10 shows that $h \in \mathbf{Z}[X]$. Finally, let us indicate by $r(X) \to \bar{r}(X)$ the homomorphism from $\mathbf{Z}[X]$ to $\mathbf{Z}/p[X]$

defined by the natural projection from $\mathbf{Z}$ to $\mathbf{Z}/p$. The preceding polynomial equalities in $\mathbf{Z}[X]$ and the fact that $\gamma^p = \gamma$ for every $\gamma \in \mathbf{Z}/p$ then imply the validity of the polynomial equalities

$$X^n - 1 = \bar{f}(X)\bar{g}(X) \quad \text{and} \quad \bar{g}(X)^p = \bar{g}(X^p) = \bar{f}(X)\bar{h}(X)$$

in $\mathbf{Z}/p[X]$. These lead immediately to the desired contradiction: In the polynomial domain $\mathbf{Z}/p[X]$, take an irreducible factor of $\bar{f}(X)$; it divides $\bar{g}(X)^p$, whence it divides $\bar{g}(X)$, and so its square divides $X^n - 1$. But this is impossible, because the condition $p \nmid n$ implies that the polynomial $X^n - 1$ in $\mathbf{Z}/p[X]$ is separable. Our claim is proved.

By virtue of what has been just shown, it follows by induction that $f(\zeta^{p_1 p_2 \cdots p_k}) = 0$ whenever $p_1, p_2, \ldots, p_k$ are primes not dividing n. Since every integer greater than 1 and relatively prime to n is the product of a finite sequence of such primes, this amounts to saying that every primitive nth root of unity in $\mathbf{C}$ is a zero of f, or that every complex zero of Φ_n is a zero of f. On the other hand, we have $\Phi_n(\zeta) = 0$, whence f divides Φ_n in $\mathbf{Q}[X]$. Since Φ_n is separable, it is clear from these conditions that $\Phi_n = f$, which proves the irreducibility of Φ_n in $\mathbf{Q}[X]$.   □

We have seen in 3.5.5 that the Galois group of a cyclotomic extension of order n is embeddable in $(\mathbf{Z}/n)^*$. The irreducibility of the cyclotomic polynomials allows us to give a complete description of the Galois group of such a field extension when the bottom field is the rational field.

3.5.9. Proposition. *If n is a positive integer, and if F is a cyclotomic field of order n over $\mathbf{Q}$, then $[F:\mathbf{Q}] = \varphi(n)$, and $\mathrm{Gal}(F/\mathbf{Q})$ is isomorphic to $(\mathbf{Z}/n)^*$.*

Proof. Since $X^n - 1$ splits in $\mathbf{C}[X]$, we may assume that F is a subfield of $\mathbf{C}$. According to 3.5.5, we have $F = \mathbf{Q}(\zeta)$ for every primitive nth root ζ of unity in $\mathbf{C}$, and $\mathrm{Gal}(F/\mathbf{Q})$ is embeddable in $(\mathbf{Z}/n)^*$. Moreover, by the preceding proposition, Φ_n is the minimal polynomial over $\mathbf{Q}$ of every such ζ; therefore

$$[F:\mathbf{Q}] = [\mathbf{Q}(\zeta):\mathbf{Q}] = \deg(\Phi_n) = \varphi(n).$$

And since

$$\mathrm{Card}(\mathrm{Gal}(F/\mathbf{Q})) = [F:\mathbf{Q}] = \varphi(n) = \mathrm{Card}((\mathbf{Z}/n)^*),$$

we conclude that $\mathrm{Gal}(F/\mathbf{Q})$ is isosmorphic to $(\mathbf{Z}/n)^*$.   □

3.5.10. Example. Let F be a cyclotomic field of order 12 over $\mathbf{Q}$. We wish to determine the subgroups of $\mathrm{Gal}(F/\mathbf{Q})$ and the intermediate fields between $\mathbf{Q}$ and F.

We may take F to be a subfield of $\mathbf{C}$. Write $\zeta = \exp(2\pi\sqrt{-1}/12)$. Then ζ is one of the primitive twelfth roots of unity in $\mathbf{C}$; the others are $\zeta^5, \zeta^7, \zeta^{11}$. Also, Φ_{12} is the minimal polynomial over $\mathbf{Q}$ of each of $\zeta, \zeta^5, \zeta^7, \zeta^{11}$;

therefore, it is meaningful to speak of the **Q**-automorphisms s, t, u of F such that

$$s(\zeta) = \zeta^5, \quad t(\zeta) = \zeta^7, \quad u(\zeta) = \zeta^{11}.$$

Since $\varphi(12) = 4$, we know that $\mathrm{Gal}(F/\mathbf{Q})$ has order 4, and so its elements are i_F, s, t, u. Note that s, t, u have order 2, because $F = \mathbf{Q}(\zeta)$ and

$$s^2(\zeta) = \zeta^{25} = \zeta, \qquad t^2(\zeta) = \zeta^{49} = \zeta, \qquad u^2(\zeta) = \zeta^{121} = \zeta.$$

It now follows that $\mathrm{Gal}(F/\mathbf{Q})$ contains exactly three proper nontrivial subgroups, namely, $\langle s \rangle, \langle t \rangle, \langle u \rangle$.

Applying the Galois correspondence defined by **Q** and F, we see that there are exactly three intermediate fields strictly between **Q** and F, namely, the fields of invariants of $\langle s \rangle, \langle t \rangle, \langle u \rangle$. Moreover, each of these fields has linear degree 2 over **Q**. Since

$$s(\zeta^3) = s(\zeta)^3 = \zeta^{15} = \zeta^3 \quad \text{and} \quad t(\zeta^4) = t(\zeta)^4 = \zeta^{28} = \zeta^4,$$

we have

$$\mathbf{Q}(\zeta^3) \subseteq \mathrm{Inv}(\langle s \rangle) \quad \text{and} \quad \mathbf{Q}(\zeta^4) \subseteq \mathrm{Inv}(\langle t \rangle);$$

and since

$$\zeta^3 = \exp(\pi\sqrt{-1}/2) = \sqrt{-1}$$

and

$$\zeta^4 = \exp(2\pi\sqrt{-1}/3) = (-1 + \sqrt{-3})/2,$$

it follows that

$$\mathrm{Inv}(\langle s \rangle) = \mathbf{Q}(\zeta^3) = \mathbf{Q}(\sqrt{-1}) \quad \text{and} \quad \mathrm{Inv}(\langle t \rangle) = \mathbf{Q}(\zeta^4) = \mathbf{Q}(\sqrt{-3}).$$

Thus, we have determined two of our intermediate fields. To determine the remaining one, we now note that $\sqrt{3} \in F$, because $\sqrt{-1}, \sqrt{-3} \in F$. Consequently $\mathbf{Q} \subset \mathbf{Q}(\sqrt{3}) \subset F$; and since it is clear that

$$\mathbf{Q}(\sqrt{3}) \neq \mathbf{Q}(\sqrt{-1}) \quad \text{and} \quad \mathbf{Q}(\sqrt{3}) \neq \mathbf{Q}(\sqrt{-3}),$$

we conclude that $\mathrm{Inv}(\langle u \rangle) = \mathbf{Q}(\sqrt{3})$. $\qquad\square$

3.5.11. Example. Let n be an integer such that $n > 2$, and let ζ be a primitive nth root of unity in **C**. We shall show that

$$[\mathbf{Q}(\zeta + 1/\zeta) : \mathbf{Q}] = \varphi(n)/2.$$

Let $\eta = \zeta + 1/\zeta$. Then $\zeta^2 - \eta\zeta + 1 = 0$, and so $[\mathbf{Q}(\zeta) : \mathbf{Q}(\eta)] \leq 2$. On the other hand, since $|\zeta| = 1$, we have $1/\zeta = \bar{\zeta}$; therefore $\eta = \zeta + \bar{\zeta} \in \mathbf{R}$, which implies that $\mathbf{Q}(\eta) \subset \mathbf{Q}(\zeta)$. This shows that $[\mathbf{Q}(\zeta) : \mathbf{Q}(\eta)] = 2$.

To conclude, note now that $\mathbf{Q}(\zeta)$ is a cyclotomic field of order n over **Q**. By 3.5.9, we then have $[\mathbf{Q}(\zeta) : \mathbf{Q}] = \varphi(n)$; consequently

$$[\mathbf{Q}(\eta) : \mathbf{Q}] = [\mathbf{Q}(\zeta) : \mathbf{Q}]/[\mathbf{Q}(\zeta) : \mathbf{Q}(\eta)] = \varphi(n)/2,$$

which is what we wanted. $\qquad\square$

In algebraic number theory, it is natural to investigate the irreducibility of the cyclotomic polynomials over algebraic number fields. In other words, given a finite extension F of $\mathbf{Q}$, we would like to know which cyclotomic polynomials remain irreducible in $F[X]$. Generally speaking, of course, this irreducibility is lost when the field of coefficients is enlarged from $\mathbf{Q}$ to F. However, there are interesting particular cases in which positive results can be proved.

To conclude, let us mention that in the definition of a cyclotomic polynomial, an arbitrary algebraically closed field could be used instead of $\mathbf{C}$. In this way, we obtain a more general notion of cyclotomic polynomial, which is of interest in certain contexts. These generalized cyclotomic polynomials also have their coefficients in the prime subfield; but in general, they fail to be irreducible over it.

PROBLEMS

1. Show that theorem 3.5.1 can be deduced, without using the Euler function, from the following result on finite abelian groups.
    If G is a finite nontrivial abelian group, then G is the internal direct product of a sequence $(C_i)_{1 \le i \le n}$ of nontrivial cyclic subgroups such that $\operatorname{Card}(C_{i-1}) | \operatorname{Card}(C_i)$ for $1 < i \le n$.
2. Let n be a positive integer, and let K be a field. Prove that n is divisible by the order of the multiplicative group consisting of the nth roots of unity in K.
3. Let n be a positive integer, and let K be a field containing a primitive nth root of unity. Show that if $\rho_1, \rho_2, \dots, \rho_n$ denote the nth roots of unity in K, then

$$\prod_{\substack{1 \le i, j \le n \\ i \ne j}} (\rho_i - \rho_j) = (-1)^{n-1} n^n.$$

4. Show that if P is a subfield of $\mathbf{C}$, and if $\alpha \in \mathbf{C}$ and

$$P(\cos \alpha) = P(\sin \alpha),$$

    then

$$P(\cos(\alpha/2)) = P(\sin(\alpha/2)).$$

    Conclude that $\mathbf{Q}(\cos(\pi/2^n)) = \mathbf{Q}(\sin(\pi/2^n))$ for every nonnegative integer n.
5. Prove that the following assertions hold for every positive integer n:
    a. If ζ is a primitive 2^{n+1}th root of unity in $\mathbf{C}$, then

$$\mathbf{Q}(\zeta) = \mathbf{Q}\big(\cos(\pi/2^n), \sqrt{-1}\,\big).$$

    b. $[\mathbf{Q}(\cos(\pi/2^n)):\mathbf{Q}] = 2^{n-1}$.

       (The preceding two problems were communicated to the author by Flanders.)

6. Let n be a positive integer, let F be a field, and let s be an element of order n in $\mathrm{Aut}(F)$. Prove the following assertions:

    a. There exists a $\beta \in F$ such that $s(\beta) = 1 + \beta$.

    b. If α is an nth root of unity in F, then there exists a $\beta \in F^*$ such that $s(\beta) = \alpha\beta$.

7. Let F be an algebraically closed field of characteristic 0. Prove the following assertions:

    a. If p is a prime, and if P is a subfield of F such that $[F:P] = p$, then F is Galois over P, $\mathrm{Gal}(F/P)$ is cyclic, and every pth root of unity in F belongs to P.

    b. If p is a prime, if P is a subfield of F such that $[F:P] = p$, and if s is a generator of $\mathrm{Gal}(F/P)$, then there exist an $\alpha \in F - P$, a pth root β of α in F, and a primitive p^2th root of unity γ in F such that $\alpha^p \in P$ and $s(\beta) = \gamma\beta$.

    c. If p is an odd prime, then there exists no subfield P of F such that $[F:P] = p$.

    d. If P is a field admitting F as a proper finite extension, then $[F:P] = 2$; and there exists an $\iota \in F$ such that $\iota^2 = -1$ and $F = P(\iota)$. (This result is a special case of a theorem of Artin and Schreier. The foregoing is a sketch of a proof by Miller and Guralnick of a partial result, under the assumption that the given algebraically closed field has characteristic 0. The general theorem of Artin and Schreier, which will be presented in section 3.9, problems 17 and 18, actually shows that if an algebraically closed field is a proper finite extension of a field, then it has characteristic 0.)

8. Show that if K is a field and L is a cyclotomic field of prime order over K, then $\mathrm{Gal}(L/K)$ is cyclic.

9. Establish the following properties of the cyclotomic polynomials:

    a. If n is an odd integer such that $n > 1$, then

$$\Phi_{2n}(X) = \Phi_n(-X).$$

    b. If n is a positive integer, and if p is a prime such that $p \nmid n$, then

$$\Phi_n(X)\Phi_{pn}(X) = \Phi_n(X^p).$$

    c. If k is a positive integer, if $p_1, p_2, \ldots, p_k$ are k distinct primes, and if $e_1, e_2, \ldots, e_k$ are positive integers, then

$$\Phi_{p_1^{e_1} p_2^{e_2} \cdots p_k^{e_k}}(X) = \Phi_{p_1 p_2 \cdots p_k}\left(X^{p_1^{e_1-1} p_2^{e_2-1} \cdots p_k^{e_k-1}}\right).$$

    d. If p is a prime and e is a positive integer, then $\Phi_{p^e}(1) = p$.

e. If n is a positive integer that is not an integer power of a prime, then $\Phi_n(1) = 1$.

The remaining problems lead to the Vahlen–Capelli criterion for the reducibility of binomial equations. It is shown first that the problem can be reduced to the case where the degree is an integer power of a prime. Then, in order to deal with this particular case, two subcases are considered separately.

10. Let K be a field and $\alpha \in K$, and let m and n be relatively prime positive integers. Prove that $X^{mn} - \alpha$ is irreducible in $k[X]$ if and only if $X^m - \alpha$ and $X^n - \alpha$ are irreducible in $K[X]$.

11. Let K be a field and $\alpha \in K$, let p be a prime such that α does not possess a pth root in K, and suppose that p is odd or $\mathrm{Char}(K) = p$. Prove the following assertions:
    a. If β is a pth root of α in an extension field of K, then β does not possess a pth root in $K(\beta)$.
    b. If n is a positive integer, then $X^{p^n} - \alpha$ is irreducible in $K[X]$.

12. Let K be a field such that $\mathrm{Char}(K) \neq 2$, let $\alpha \in K$, and suppose that α does not possess a square root in K. Prove the following assertions:
    a. If β is a square root of α in an extension field of K, then β possesses a square root in $K(\beta)$ if and only if -4α possesses a fourth root in K.
    b. If n is an integer such that $n > 1$, then $X^{2^n} - \alpha$ is irreducible in $K[X]$ if and only if -4α does not possess a fourth root in K.

13. Let K be a field and $\alpha \in K$, and let n be a positive integer. Prove the following assertions:
    a. If $4 \nmid n$, then $X^n - \alpha$ is reducible in $K[X]$ if and only if there exists a prime divisor p of n such that α possesses a pth root in K.
    b. If $4 \mid n$, then $X^n - \alpha$ is reducible in $K[X]$ if and only if -4α possesses a fourth root in K or there exists a prime divisor p of n such that α possesses a pth root in K.

3.6. PRIMITIVE ELEMENTS

We now have at our disposal the background required in order to discuss the question of determining when a finite extension is simple. Here we shall prove some classical results on the existence of primitive elements. Although at one time these were regarded as important results, they can no longer be said to play a major role in the theory of fields. They are, however, of some interest *per se*, and allow certain simplifications in some arguments.

We shall first establish an interesting criterion for the existence of a primitive element in a finite extension.

3.6.1.   Theorem (Steinitz).   *Let K be a field, and let L be a finite extension of K. Then L is a simple extension of K if and only if the set of all intermediate fields between K and L is finite.*

Proof.   If K is finite, the two stated conditions hold. For in this case, L is also finite. Consequently, L admits only finitely many subfields, and so the set of all intermediate subfields between K and L is finite. Moreover, L^* is finite, hence cyclic; and every generator of L^* is a primitive element of L over K.

In the remainder of the proof we shall assume, therefore, that K is infinite. The symbol Ω will denote the set of all intermediate fields between K and L.

First, suppose that Ω is finite. Since L is finitely generated over K, an easy induction on the number of generators of L over K shows that in order to establish the existence of a primitive element of L over K, it suffices to verify that if $\alpha, \beta \in L$, then $K(\alpha, \beta)$ is a simple extension of K. To this end, note that by the finiteness of Ω and the infiniteness of K, the mapping $\rho \to K(\alpha + \rho\beta)$ from K to Ω is not injective. Consequently, there exist $\sigma, \tau \in K$ such that

$$\sigma \neq \tau \quad \text{and} \quad K(\alpha + \sigma\beta) = K(\alpha + \tau\beta).$$

If we let $\gamma = \alpha + \sigma\beta$, then we have $K(\gamma) \subseteq K(\alpha, \beta)$. But since

$$\beta = ((\alpha + \sigma\beta) - (\alpha + \tau\beta))/(\sigma - \tau),$$

and $\alpha + \sigma\beta, \alpha + \tau\beta, \sigma - \tau \in K(\gamma)$, we get $\beta \in K(\gamma)$, and hence also $\alpha = \gamma - \sigma\beta \in K(\gamma)$; therefore $K(\alpha, \beta) \subseteq K(\gamma)$. It then follows that $K(\alpha, \beta) = K(\gamma)$, as required.

Now suppose, conversely, that $L = K(\alpha)$ for some $\alpha \in L$. Denote by f the minimal polynomial of α over K, and by W the set of all monic factors of f in $L[X]$. By the unique factorization property of $L[X]$, we know that W is finite.

Let $E \in \Omega$, and let g_E denote the minimal polynomial of α over E. Since $f \in E[X]$ and $f(\alpha) = 0$, we see that g_E divides f in $E[X]$, hence also in $L[X]$, so that $g_E \in W$. Next, denote by $\overline{E}$ the subfield of E generated over K by the set of all coefficients of g_E. Since $g_E \in \overline{E}[X]$, it follows that g_E is irreducible in $\overline{E}[X]$, whence it is the minimal polynomial of α over $\overline{E}$. We then have

$$\overline{E} \subseteq E \quad \text{and} \quad [L : \overline{E}] = \deg(g_E) = [L : E],$$

which shows that $\overline{E} = E$.

Thus, it is meaningful to speak of the mapping $E \to g_E$ from Ω to W; and furthermore, this mapping is injective: if $E, F \in \Omega$ and $g_E = g_F$, then clearly $\overline{E} = \overline{F}$, which implies that $E = F$. Since W is finite, we conclude that Ω is finite, which is what we wanted. $\square$

Next we proceed to show that a suitable restriction on the separable closure of a finite extension implies the existence of a primitive element.

3.6.2.　Theorem.　*Let K be a field, and let L be a finite extension of K that is a simple extension of $S(L/K)$. Then L is a simple extension of K.*

Proof.　Exactly as in the beginning of the preceding proof, it is seen that the conclusion holds when K is finite.

Thus, assume that K is infinite, and write $S = S(L/K)$. Then S is finitely generated over K; and by induction on the number of generators of S over K, it is seen that the conclusion will follow by verifying that if $\alpha \in S$ and $\beta \in L$, then $K(\alpha, \beta)$ is a simple extension of K. To do this, let f and g denote, respectively, the minimal polynomials of α and β over K, and put $m = \deg(f)$ and $n = \deg(g)$. Then choose an extension field M of L such that f and g split in $M[X]$; thus

$$f(X) = \prod_{i=1}^{m} (X - \alpha_i) \quad \text{and} \quad g(X) = \prod_{i=1}^{n} (X - \beta_i),$$

where $\alpha_1, \alpha_2, \ldots, \alpha_m, \beta_1, \beta_2, \ldots, \beta_n \in M$. And we can take $\alpha_1 = \alpha$ and $\beta_1 = \beta$.

Since f is separable, we see that $\alpha_1, \alpha_2, \ldots, \alpha_m$ are m distinct elements of M; and since K is infinite, a $\gamma \in K$ can be chosen so that

$$\gamma \neq (\beta_j - \beta_1)/(\alpha_i - \alpha_1)$$

whenever $2 \leq i \leq m$ and $1 \leq j \leq n$. We now let $\sigma = \gamma \alpha + \beta$, and contend that σ is a primitive element of $K(\alpha, \beta)$ over K.

Let $E = K(\sigma)$. We then have $E \subseteq K(\alpha, \beta)$. Define a polynomial h in $E[X]$ by $h(X) = g(\sigma - \gamma X)$. Clearly

$$h(\alpha) = g(\sigma - \gamma \alpha) = g(\beta) = 0;$$

furthermore, if $2 \leq i \leq m$ and $1 \leq j \leq n$, then

$$\sigma - \gamma \alpha_i = \gamma \alpha + \beta - \gamma \alpha_i = \gamma \alpha_1 + \beta_1 - \gamma \alpha_i$$
$$= \gamma(\alpha_1 - \alpha_i) + \beta_1 \neq \beta_j,$$

whence

$$h(\alpha_i) = g(\sigma - \gamma \alpha_i) \neq 0.$$

Consequently, α is the only common zero of f and h in M. Moreover, these two polynomials split in $M[X]$. Therefore

$$f(X)M[X] + h(X)M[X] = (X - \alpha)M[X].$$

But since $f, h \in E[X]$, it now follows from 1.5.4 and this equality that $X - \alpha \in E[X]$. Thus $\alpha \in E$ and $\beta = \sigma - \gamma \alpha \in E$, so that $K(\alpha, \beta) \subseteq E$. We conclude that $K(\alpha, \beta) = E$, as claimed.　□

3.6.3.　Corollary.　*If K is a field, then every finite separable extension of K is a simple extension of K.*

This corollary is the result usually known as the **theorem on primitive elements**. We have obtained it as a consequence of 3.6.2. It also can be derived from 3.6.1 and some of the main results of Galois theory (problem 3).

In the comments at the end of section 2.2, it was stated that the second lemma in Artin's proof of Steinitz's theorem 2.2.14 is superfluous. The following result, which will be proved with the help of the theorem on primitive elements, provides the justification for this claim.

3.6.4. Proposition. *Let K be a field, and let L be an algebraic extension of K. If every monic irreducible polynomial in $K[X]$ admits a zero in L, then L is an algebraic closure of K.*

Proof. By virtue of 2.2.12, it suffices to prove that every nonconstant polynomial in $K[X]$ splits in $L[X]$.

Thus, let f be such a polynomial. Choose an extension field F of L so that f splits in $F[X]$, and denote by M the subfield of F generated over K by the set of all zeros of f in F. Then M is a splitting field of f over K, and hence it is normal and finite over K.

Write $S = S(M/K)$. By 3.6.3, we have $S = K(\gamma)$ for some $\gamma \in S$; and 3.3.6 shows that S is Galois over K. For every $t \in \text{Gal}(S/K)$, it is clear that $S = t(S) = K(t(\gamma))$; according to 2.3.6, this means that every zero in S of the minimal polynomial of γ over K is a primitive element of S over K. But this polynomial splits in $S[X]$, and by hypothesis it admits a zero in L. It then follows that there exists an element of L that is a primitive element of S over K, whence $S \subseteq L$.

If K has prime characteristic, write $P = P(M/K)$. It is readily verified that $P \subseteq L$: Indeed, if $\beta \in P$, then β is purely inseparable over K; therefore β is the only zero of its minimal polynomial over K, and the hypothesis implies that $\beta \in L$.

It is now easy to complete the proof. If K has characteristic 0, then $M = S$; and if K has prime characteristic, it follows from 2.5.19 that $M = P \vee S$. In either case, it is seen from what was shown in the preceding paragraphs that $M \subseteq L$; and this implies that f splits in $L[X]$. □

PROBLEMS

1. Let K be a field, let L be a finite simple extension of K, and put $n = [L:K]$. Show that there are at most 2^{n-1} intermediate fields between K and L.
2. Let K be a field, and let L be a separable extension of K. Suppose that there exists a positive integer n such that $[K(\alpha):K] \le n$ for every $\alpha \in L$. Prove that L is finite over K, and $[L:K] \le n$.

Use this result and 3.2.22, but no other result from this chapter, in order to prove the following assertion:

If F is a field, and if Γ is a finite group of automorphisms of F, then F is Galois and finite over Inv(Γ), and $[F: \text{Inv}(\Gamma)] \leq \text{Card}(\Gamma)$.

3. Use 3.3.7, 3.4.5, and 3.6.1 in order to give a proof of the theorem on primitive elements that does not depend on 3.6.2.

4. Prove that if K is a field and L is a finite simple extension of K, then every intermediate field between K and L is a simple extension of K.

5. Let K be a field, and let L be a finite extension of K. Show that L is Galois over K if and only if there exists an element of L whose orbit under Gal(L/K) has cardinality $[L: K]$, in which case the primitive elements of L over K are the elements of L enjoying this property.

3.7. SEPARABLE AND INSEPARABLE DEGREES

The notion of relative separable closure can be used to obtain an expression for the linear degree of a finite extension as the product of two positive integers. It will be seen that in a certain sense, these two factors of the linear degree measure separability and inseparability; and that the one measuring separability gives the exact number of embeddings into "sufficiently large" fields.

Let K be a field, and let L be a finite extension of K. We then write

$$[L: K]_s = [S(L/K): K] \quad \text{and} \quad [L: K]_i = [L: S(L/K)],$$

and say that $[L: K]_s$ and $[L: K]_i$ are, respectively, the **separable degree** and **inseparable degree** of L over K.

It follows from these definitions that

$$[L: K] = [L: K]_s [L: K]_i.$$

Clearly, each of $[L: K]_s = [L: K]$ and $[L: K]_i = 1$ is equivalent to $S(L/K) = L$, and hence to the separability of L over K. And in the case of prime characteristic p, each of $[L: K]_s = 1$ and $[L: K]_i = [L: K]$ is equivalent to $S(L/K) = K$, and hence to the pure inseparability of L over K; finally, since L is purely inseparable over $S(L/K)$, we know that $[L: K]_i = p^e$ for some nonnegative integer e.

In order to derive the basic properties of the separable and inseparable degrees, we shall require the following two auxiliary results (in the statement of the first of which implicit use is being made of 2.3.5).

3.7.1. Proposition. *Let K be a field, let F be a normal extension of K, and let L and M be intermediate fields between K and F such that $L \subseteq M$. For every K-monomorphism v from L to F, choose a K-automorphism $\bar{v}$ of F extending v. Then the mapping $(v, w) \to \bar{v} \circ w$ from $\text{Mon}_K(L, F) \times \text{Mon}_L(M, F)$ to $\text{Mon}_K(M, F)$ is bijective.*

Proof. To show injectivity, suppose that $(b, c), (v, w) \in$ $\mathrm{Mon}_K(L, F) \times \mathrm{Mon}_L(M, F)$ and $\bar{b} \circ c = \bar{v} \circ w$. For each $\alpha \in L$, we have

$$b(\alpha) = \bar{b}(c(\alpha)) = \bar{v}(w(\alpha)) = v(\alpha).$$

Therefore $b = v$; and since we now have $\bar{b} = \bar{v}$ and $\bar{b} \circ c = \bar{v} \circ w$, we conclude that also $c = w$.

To prove that our mapping is surjective, let $u \in \mathrm{Mon}_K(M, F)$. First put $v = u \circ i_{L \to M}$, so that $v \in \mathrm{Mon}_K(L, F)$; and then put $w = \bar{v}^{-1} \circ u$. If $\alpha \in L$, then

$$w(\alpha) = \bar{v}^{-1}(u(\alpha)) = \bar{v}^{-1}(v(\alpha)) = \bar{v}^{-1}(\bar{v}(\alpha)) = \alpha;$$

hence $w \in \mathrm{Mon}_L(M, F)$. Thus $(v, w) \in \mathrm{Mon}_K(L, F) \times \mathrm{Mon}_L(M, F)$, and it is evident that $u = \bar{v} \circ w$. $\square$

3.7.2. Proposition. *Let K be a field, and let F be an extension field of K. If $\alpha \in F$ and α is algebraic over K, then the mapping $u \to u(\alpha)$ from $\mathrm{Mon}_K(K(\alpha), F)$ to the set of all zeros in F of the minimal polynomial of α over K is bijective.*

Proof. This is a consequence of 2.1.9: for every zero β in F of the minimal polynomial of α over K, there exists a unique K-monomorphism from $K(\alpha)$ to F such that $\alpha \to \beta$. $\square$

We can now show that the number of embeddings of a finite field extension into "sufficiently large" fields coincides with the separable degree.

3.7.3. Proposition. *Let K be a field, let L be a finite extension of K, and let F be an extension field of L that is normal over K. Then*

$$\mathrm{Card}(\mathrm{Mon}_K(L, F)) = [L:K]_s.$$

Proof. By 3.6.3, $S(L/K) = K(\alpha)$ for some $\alpha \in L$. Then $[K(\alpha):K]$ $= [L:K]_s$; and since α is separable over K, its minimal polynomial possesses exactly $[L:K]_s$ zeros in F. Applying 3.7.2, we obtain $\mathrm{Card}(\mathrm{Mon}_K(K(\alpha), F)) = [L:K]_s$.

Note that $\mathrm{Mon}_{K(\alpha)}(L, F)$ consists of a single element: This is clear if $\mathrm{Char}(K) = 0$, for in this case $K(\alpha) = L$; and if $\mathrm{Char}(K) \neq 0$, it follows from 2.4.7 and the pure inseparability of L over $K(\alpha)$. Also, 3.7.1 shows that $\mathrm{Mon}_K(K(\alpha), F) \times \mathrm{Mon}_{K(\alpha)}(L, F)$ and $\mathrm{Mon}_K(L, F)$ are equipotent. Consequently, $\mathrm{Mon}_K(K(\alpha), F)$ and $\mathrm{Mon}_K(L, F)$ are equipotent, and so $\mathrm{Card}(\mathrm{Mon}_K(L, F)) = [L:K]_s$. $\square$

It can be shown that the use of the theorem on primitive elements in the preceding proof can be avoided (problem 4).

3.7.4. Corollary. *Let K be a field, let L be a finite extension of K, and let F be an algebraic closure of K. Then*

$$\mathrm{Card}(\mathrm{Mon}_K(L, F)) = [L:K]_s.$$

Proof. Choose an algebraic closure E of L. Then E and F are algebraic closures of K, and hence there exists a K-isomorphism c from E to F. Since the mapping $u \to c \circ u$ from $\text{Mon}_K(L, E)$ to $\text{Mon}_K(L, F)$ is bijective, we see that $\text{Mon}_K(L, E)$ and $\text{Mon}_K(L, F)$ are equipotent. Moreover, the proposition shows that $\text{Card}(\text{Mon}_K(L, E)) = [L : K]_s$. The desired equality now follows at once. □

3.7.5. Corollary. *Let K be a field, let L be a finite extension of K, and let M be an extension field of K. Then*

$$\text{Card}(\text{Mon}_K(L, M)) \le [L : K]_s.$$

Proof. Choose an algebraic closure F of $A(M/K)$. Then F is also an algebraic closure of K, and the preceding corollary shows that $\text{Card}(\text{Mon}_K(L, F)) = [L : K]_s$.

On the other hand, if u is a K-monomorphism from L to M, then $u(L) \subseteq A(M/K) \subseteq F$; therefore u defines, by restriction, a K-monomorphism $\bar{u}$ from L to F. Since the mapping $u \to \bar{u}$ from $\text{Mon}_K(L, M)$ to $\text{Mon}_K(L, F)$ is injective, we get $\text{Card}(\text{Mon}_K(L, M)) \le \text{Card}(\text{Mon}_K(L, F))$. □

3.7.6. Corollary. *Let K be a field, and let L be a finite extension of K. Then $\text{Card}(\text{Gal}(L/K)) \le [L : K]_s$; and $\text{Card}(\text{Gal}(L/K)) = [L : K]_s$ if and only if L is normal over K.*

Proof. We showed in 2.1.21 that $\text{Gal}(L/K) = \text{Mon}_K(L, L)$. By virtue of the preceding corollary and the proposition, we see that only the direct implication in the second assertion needs to be verified.

Thus, suppose that $\text{Card}(\text{Gal}(L/K)) = [L : K]_s$, and choose a normal closure M of L over K. According to 2.3.7, the normality of L over K will be established if we prove that $s(L) = L$ for every K-automorphism s of M.

By 3.7.3, we also have $[L : K]_s = \text{Card}(\text{Mon}_K(L, M))$, so that $\text{Gal}(L/K)$ and $\text{Mon}_K(L, M)$ are finite and equipotent. Since the mapping $t \to i_{L \to M} \circ t$ from $\text{Gal}(L/K)$ to $\text{Mon}_K(L, M)$ is injective, it is also surjective.

To conclude, let s be a K-automorphism of M. Then $s \circ i_{L \to M} \in \text{Mon}_K(L, M)$, so that $s \circ i_{L \to M} = i_{L \to M} \circ t$ for some $t \in \text{Gal}(L/K)$. Consequently

$$s(L) = s(i_{L \to M}(L)) = i_{L \to M}(t(L)) = i_{L \to M}(L) = L,$$

as required. □

It should be noted that the inequalities in 3.1.3 and 3.2.15 are weaker than the corresponding ones in the last two corollaries; and that the latter have been obtained independently of the former.

Next we show that the separable and inseparable degrees are multiplicative.

3.7.7.   Proposition.   *Let K be a field, let L be a finite extension of K, and let M be a finite extension of L. Then*

$$[M:K]_s = [M:L]_s[L:K]_s \quad \text{and} \quad [M:K]_i = [M:L]_i[L:K]_i.$$

Proof.   It suffices to verify the first equality. To do this, choose a normal closure F of M over K. By 3.7.1, the sets $\mathrm{Mon}_K(L, F) \times \mathrm{Mon}_L(M, F)$ and $\mathrm{Mon}_K(M, F)$ are equipotent, whence

$$\mathrm{Card}(\mathrm{Mon}_K(L, F))\mathrm{Card}(\mathrm{Mon}_L(M, F)) = \mathrm{Card}(\mathrm{Mon}_K(M, F)).$$

Moreover, F is normal over K and L. According to 3.7.3, we then have

$$[M:K]_s = [M:L]_s[L:K]_s,$$

which is what we wanted.   □

To conclude our discussion, we now consider the case of a simple field extension in the case of prime characteristic.

3.7.8.   Proposition.   *Let K be a field of prime characteristic, and let F be an extension field of K. If $\alpha \in F$ and α is algebraic over K, then $[K(\alpha):K]_s$ and $[K(\alpha):K]_i$ are, respectively, the reduced degree and the degree of inseparability of the minimal polynomial of α over K.*

Proof.   Let f denote the minimal polynomial of α over K. Clearly, we need to verify only that $[K(\alpha):K]_s$ is the reduced degree of f. To do this, let $p = \mathrm{Char}(K)$, and denote by e the exponent of inseparability of f. By 2.5.14, we have

$$[K(\alpha):K]_s = [S(K(\alpha)/K):K] = [K(\alpha^{p^e}):K];$$

but this, by 2.1.7, says that $[K(\alpha):K]_s$ is the reduced degree of f.   □

3.7.9.   Corollary.   *Let K be a field, and let F be a normal extension of K. If $\alpha \in F$, and if f denotes the minimal polynomial of α over K, then the equality*

$$f(X) = \prod_{u \in \mathrm{Mon}_K(K(\alpha), F)} (X - u(\alpha))^{[K(\alpha):K]_i}$$

holds in $F[X]$.

Proof.   Let D denote the set of all zeros of f in F. According to 3.7.2, the mapping $u \to u(\alpha)$ from $\mathrm{Mon}_K(K(\alpha), F)$ to D is bijective. Furthermore, by 1.7.7 and the proposition, $[K(\alpha):K]_i$ is the common multiplicity of the zeros of f in F. Since f splits in $F[X]$, our assertion now follows at once.   □

PROBLEMS

1.   Let K be a field, let L be an extension field of K, and let F be an extension field of L that is normal over K. Prove that the mapping

$v \to v \circ i_{S(L/K) \to L}$ from $\mathrm{Mon}_K(L, F)$ to $\mathrm{Mon}_K(S(L/K), F)$ is bijective.

2. Let K be a field, let F be a normal extension of K, and let L and M be intermediate fields between K and F such that $L \subseteq M$. Let v be a K-monomorphism from L to F, and let Ω denote the set of all K-monomorphisms from M to F extending v. Verify the following assertions:

    a. If $\bar{v}$ denotes a K-automorphism of F extending v, then the mapping $w \to \bar{v} \circ w$ from $\mathrm{Mon}_L(M, F)$ to Ω is bijective.

    b. If M is finite over L, then $\mathrm{Card}(\Omega) = [M : L]_s$.

3. Let K be a field, let L be an extension field of K, and let F be an extension field of L that is normal over K. Prove the following assertions.

    a. If Ω is a left transversal of $\mathrm{Gal}(F/L)$ in $\mathrm{Gal}(F/K)$, then the mapping $s \to s \circ i_{L \to F}$ from Ω to $\mathrm{Mon}_K(L, F)$ is bijective.

    b. If L is finite over K, then $[\mathrm{Gal}(F/K) : \mathrm{Gal}(F/L)] = [L : K]_s$.

4. Give a proof of 3.7.3 without using the theorem on primitive elements by considering a finite increasing sequence $(E_i)_{0 \le i \le n}$ of intermediate fields between K and $S(L/K)$ such that E_i is a simple extension of E_{i-1} for $1 \le i \le n-1$ and such that $E_0 = K$ and $E_n = S(L/K)$.

5. Show that if K is a field, and if L is a finite normal extension of K, then

$$[L : K]_s = [L : P(L/K)] \quad \text{and} \quad [L : K]_i = [P(L/K) : K].$$

3.8. NORMS AND TRACES

With every finite extension there are associated two mappings from the top field to the bottom field that play an important role in the theory of fields and in number theory. In this section, we shall define these mappings and establish their basic properties.

Let K be a field, and let L be a finite extension of K. We define mappings $N_{L/K}$ and $T_{L/K}$ from L to K, called, respectively, the **norm from L to K** and the **trace from L to K**, as follows: If $\alpha \in L$, if we write $d = [K(\alpha) : K]$ and $m = [L : K(\alpha)]$, and if $\gamma_0, \gamma_1, \ldots, \gamma_{d-1}$ denote the elements of K such that $\sum_{i=0}^{d-1} \gamma_i X^i + X^d$ is the minimal polynomial of α over K, then

$$N_{L/K}(\alpha) = \left((-1)^d \gamma_0\right)^m \quad \text{and} \quad T_{L/K}(\alpha) = -m\gamma_{d-1}.$$

It is seen from the definition just given that if K is a field and L is a finite extension of K, and if $\alpha \in L$, then $N_{L/K}(\alpha) = 0$ if and only if $\alpha = 0$; and if $\alpha \in K$, then

$$N_{L/K}(\alpha) = \alpha^{[L:K]} \quad \text{and} \quad T_{L/K}(\alpha) = [L:K]\alpha.$$

3.8.1. Examples

a. If $\alpha, \beta \in \mathbf{R}$, then

$$N_{\mathbf{C}/\mathbf{R}}(\alpha + \beta\sqrt{-1}) = \alpha^2 + \beta^2 \quad \text{and} \quad T_{\mathbf{C}/\mathbf{R}}(\alpha + \beta\sqrt{-1}) = 2\alpha.$$

For $\beta = 0$, this follows from the foregoing remarks. And when $\beta \neq 0$, we need only take into account that $X^2 - 2\alpha X + (\alpha^2 + \beta^2)$ is the minimal polynomial of $\alpha + \beta\sqrt{-1}$ over $\mathbf{R}$.

b. Let K be a field of prime characteristic p, and let n be a positive integer. If $\alpha \in K - K^p$, and if β is a p^nth root of α in an extension field of K, then $N_{K(\beta)/K}(\beta) = \alpha$ and $T_{K(\beta)/K}(\beta) = 0$.

Indeed, our assumptions imply that $X^{p^n} - \alpha$ is the minimal polynomial of β over K.   □

Because the explicit determination of the coefficients of minimal polynomials is frequently unfeasible, the definitions of the norm and trace mappings are not always easy to apply. A more convenient description of these mappings will now be given in terms of the embeddings of the given finite field extension.

3.8.2. Proposition.

Let K be a field, let L be a finite extension of K, and let F be an extension field of L that is normal over K. For every $\alpha \in L$, we then have

$$N_{L/K}(\alpha) = \left(\prod_{u \in \mathrm{Mon}_K(L, F)} u(\alpha) \right)^{[L:K]_i}$$

and

$$T_{L/K}(\alpha) = [L:K]_i \left(\sum_{u \in \mathrm{Mon}_K(L, F)} u(\alpha) \right).$$

Proof. Let us first introduce the following notation:

$$n = [L:K], \qquad n_s = [L:K]_s, \qquad n_i = [L:K]_i,$$
$$d = [K(\alpha):K], \qquad d_s = [K(\alpha):K]_s, \qquad d_i = [K(\alpha):K]_i,$$
$$m = [L:K(\alpha)], \qquad m_s = [L:K(\alpha)]_s, \qquad m_i = [L:K(\alpha)]_i,$$

and

$$\Gamma = \mathrm{Mon}_K(K(\alpha), F), \qquad \Delta = \mathrm{Mon}_{K(\alpha)}(L, F), \qquad \Omega = \mathrm{Mon}_K(L, F).$$

According to 2.3.5, the normality of F over K implies that each $v \in \Gamma$ is extendible to a K-automorphism $\bar{v}$ of F; and by 3.7.1, the mapping $(v, w) \to \bar{v} \circ w$ from $\Gamma \times \Delta$ to Ω is bijective. Since F is also normal over $K(\alpha)$, we can use 3.7.3 to obtain the equality $\mathrm{Card}(\Delta) = m_s$.

Since $(\bar{v} \circ w)(\alpha) = \bar{v}(w(\alpha)) = \bar{v}(\alpha) = v(\alpha)$ for every $(v, w) \in \Gamma \times \Delta$, it follows that

$$\prod_{u \in \Omega} u(\alpha) = \prod_{(v,w) \in \Gamma \times \Delta} (\bar{v} \circ w)(\alpha) = \left(\prod_{v \in \Gamma} v(\alpha) \right)^{m_s}$$

and

$$\sum_{u \in \Omega} u(\alpha) = \sum_{(v,w) \in \Gamma \times \Delta} (\bar{v} \circ w)(\alpha) = m_s \left(\sum_{v \in \Gamma} v(\alpha) \right);$$

and since $n_i m_s = d_i m_i m_s = d_i m$, we obtain

$$\left(\prod_{u \in \Omega} u(\alpha) \right)^{n_i} = \left(\prod_{v \in \Gamma} v(\alpha) \right)^{d_i m}$$

and

$$n_i \left(\sum_{u \in \Omega} u(\alpha) \right) = d_i m \left(\sum_{v \in \Gamma} v(\alpha) \right).$$

On the other hand, it follows from 3.7.9 that the polynomial $\prod_{v \in \Gamma}(X - v(\alpha))^{d_i}$ in $F[X]$ is the minimal polynomial of α over K; and furthermore, by 3.7.3, the normality of F over K implies that $\mathrm{Card}(\Gamma) = d_s$. Therefore

$$N_{L/K}(\alpha) = \left((-1)^d \left((-1)^{d_s} \left(\prod_{v \in \Gamma} v(\alpha) \right) \right)^{d_i} \right)^m = \left(\prod_{v \in \Gamma} v(\alpha) \right)^{d_i m}$$

and

$$T_{L/K}(\alpha) = - m \left(- d_i \left(\sum_{v \in \Gamma} v(\alpha) \right) \right) = d_i m \left(\sum_{v \in \Gamma} v(\alpha) \right).$$

Consequently

$$N_{L/K}(\alpha) = \left(\prod_{u \in \Omega} u(\alpha) \right)^{n_i} \quad \text{and} \quad T_{L/K}(\alpha) = n_i \left(\sum_{u \in \Omega} u(\alpha) \right),$$

which is the required conclusion. $\square$

The result just proved is sometimes expressed in a loose manner by saying that the norm and trace of an element are, respectively, the product and sum of the conjugates of the element, each counted as many times as its multiplicity.

Next we state two immediate consequences of the preceding proposition.

3.8.3.   Corollary.   *Let K be a field, let L be a finite extension of K, and let $\alpha \in L$.*

(i) *If L is normal over K, then*

$$N_{L/K}(\alpha) = \left(\prod_{s \in \mathrm{Gal}(L/K)} s(\alpha) \right)^{[L:K]_i}$$

and

$$T_{L/K}(\alpha) = [L:K]_i \left(\sum_{s \in \text{Gal}(L/K)} s(\alpha) \right).$$

(ii) *If K has prime characteristic, and if L is purely inseparable over* K, *then*

$$N_{L/K}(\alpha) = \alpha^{[L:K]} \quad and \quad T_{L/K}(\alpha) = [L:K]\alpha.$$

(iii) *If L is separable over K, and if F is an extension field of L that is normal over K, then*

$$N_{L/K}(\alpha) = \prod_{u \in \text{Mon}_K(L,F)} u(\alpha) \quad and \quad T_{L/K}(\alpha) = \sum_{u \in \text{Mon}_K(L,F)} u(\alpha).$$

(iv) *If L is Galois over K, then*

$$N_{L/K}(\alpha) = \prod_{s \in \text{Gal}(L/K)} s(\alpha) \quad and \quad T_{L/K}(\alpha) = \sum_{s \in \text{Gal}(L/K)} s(\alpha).$$

3.8.4. Corollary. *Let K be a field, and let L be a finite extension of K.*

(i) *If $\alpha, \beta \in L$, then*

$$N_{L/K}(\alpha\beta) = N_{L/K}(\alpha) N_{L/K}(\beta)$$

and

$$T_{L/K}(\alpha + \beta) = T_{L/K}(\alpha) + T_{L/K}(\beta).$$

(ii) *If $\alpha \in K$ and $\beta \in L$, then*

$$T_{L/K}(\alpha\beta) = \alpha T_{L/K}(\beta).$$

3.8.5. Example. Consider a complex cube root ω of unity such that $\omega \neq 1$. We shall give an explicit description of the mappings $N_{\mathbf{Q}(\omega)/\mathbf{Q}}$ and $T_{\mathbf{Q}(\omega)/\mathbf{Q}}$.

We have $X^2 - 1 = (X-1)(X^2 + X + 1)$ in $\mathbf{Q}[X]$, and so $X^2 + X + 1$ is the minimal polynomial over $\mathbf{Q}$ of ω and ω^2. Moreover, $\mathbf{Q}(\omega)$ is a splitting field of $X^2 + X + 1$ over $\mathbf{Q}$. Consequently $\mathbf{Q}(\omega)$ is Galois over $\mathbf{Q}$; and $\text{Gal}(\mathbf{Q}(\omega)/\mathbf{Q})$ is cyclic of order 2, its elements being $i_{\mathbf{Q}(\omega)}$ and the $\mathbf{Q}$-automorphism s of $\mathbf{Q}(\omega)$ such that $s(\omega) = \omega^2$.

Since $(1, \omega)$ is a linear base of $\mathbf{Q}(\omega)$ over $\mathbf{Q}$, every element of $\mathbf{Q}(\omega)$ can be written uniquely in the form $\alpha + \beta\omega$ with $\alpha, \beta \in \mathbf{Q}$. Applying 3.8.3, for every such element of $\mathbf{Q}(\omega)$ we get

$$N_{\mathbf{Q}(\omega)/\mathbf{Q}}(\alpha + \beta\omega) = (\alpha + \beta\omega)(\alpha + \beta\omega^2)$$

and

$$T_{\mathbf{Q}(\omega)/\mathbf{Q}}(\alpha + \beta\omega) = (\alpha + \beta\omega) + (\alpha + \beta\omega^2);$$

and since $\omega^3 = 1$ and $\omega^2 + \omega + 1 = 0$, this implies that

$$N_{\mathbf{Q}(\omega)/\mathbf{Q}}(\alpha + \beta\omega) = \alpha^2 - \alpha\beta + \beta^2$$

and

$$T_{Q(\omega)/Q}(\alpha + \beta\omega) = 2\alpha - \beta. \qquad \square$$

Let K be a field, and let L be a finite extension of K. It follows from the preceding discussion that it is meaningful to speak of the homomorphisms $\alpha \to N_{L/K}(\alpha)$ from L^* to K^* and $\alpha \to T_{L/K}(\alpha)$ from L^+ to K^+. As we shall see in the next section, the kernels of these homomorphisms can be determined completely for a special type of finite extension.

We shall now establish an important transitivity property of the norm and trace mappings.

3.8.6. Proposition. *Let K be a field, let L be a finite extension of K, and let M be a finite extension of L. Then*

$$N_{M/K} = N_{L/K} \circ N_{M/L} \quad and \quad T_{M/K} = T_{L/K} \circ T_{ML}.$$

Proof. Since the verifications of the two formulas are quite similar, we shall concern ourselves only with the one involving norms.

Let $\alpha \in M$, and choose a normal closure F of M over K. Next let

$$m = [L:K]_i, \qquad n = [M:L]_i, \qquad q = [M:K]_i$$

and

$$\Gamma = \mathrm{Mon}_K(L, F), \qquad \Delta = \mathrm{Mon}_L(M, F), \qquad \Omega = \mathrm{Mon}_K(M, F).$$

Since F is normal over K, we know from 2.3.5 that each $v \in \Gamma$ is extendible to a K-automorphism $\bar{v}$ of F; and then 3.7.1 shows that the mapping $(v, w) \to \bar{v} \circ w$ from $\Gamma \times \Delta$ to Ω is bijective. Finally, since F is also normal over L, it follows from 3.8.2 and 3.7.7 that

$$N_{M/K}(\alpha) = \left(\prod_{u \in \Omega} u(\alpha) \right)^q = \left(\prod_{(v,w) \in \Gamma \times \Delta} \bar{v}(w(\alpha)) \right)^{mn}$$

$$= \left(\prod_{v \in \Gamma} \bar{v}\left(\left(\prod_{w \in \Delta} w(\alpha) \right)^n \right) \right)^m = \left(\prod_{v \in \Gamma} \bar{v}(N_{M/L}(\alpha)) \right)^m$$

$$= \left(\prod_{v \in \Gamma} v(N_{M/L}(\alpha)) \right)^m = N_{L/K}(N_{M/L}(\alpha)),$$

as was to be shown. $\qquad \square$

As the next result shows, inseparability is characterized by the vanishing of the trace mapping.

3.8.7. Proposition. *Let K be a field, and let L be a finite extension of K. Then L is separable over K if and only if $T_{L/K}$ is different from the zero mapping.*

Proof. Let F be a normal closure of L over K.

Suppose first that L is not separable over K. Then K has prime characteristic p, and $[L:K]_i = p^e$ for some positive integer e. According to

3.8.2, for every $\alpha \in L$ we have

$$T_{L/K}(\alpha) = p^e\left(\sum_{u \in \mathrm{Mon}_K(L, F)} u(\alpha)\right) = 0,$$

which means that $T_{L/K}$ is the zero mapping.

To conclude, suppose that L is separable over K. By 3.8.3, we then have

$$T_{L/K}(\alpha) = \sum_{u \in \mathrm{Mon}_K(L, F)} u(\alpha)$$

for every $\alpha \in L$, and Dedekind's theorem 3.1.2 shows that $T_{L/K}$ is different from the zero mapping. $\square$

In the second part of the preceding proof, we can give an argument based on the theorem on primitive elements instead of on Dedekind's theorem (problems 4 and 10).

We also remark that in the case of a finite separable field extension, the trace mapping can be used to define a dual pairing that is of interest in number theory (problem 11).

PROBLEMS

1. Let K be a field of prime characteristic p, and let α be an element of K such that $X^p - X - \alpha$ admits no zeros in K. Prove that if β is a zero of $X^p - X - \alpha$ in an extension field of K, then $N_{K(\beta)/K}(\beta) = \alpha$; $T_{K(\beta)/K}(\beta) = 1$ when $p = 2$; and $T_{K(\beta)/K}(\beta) = 0$ when $p > 2$.

2. Let K be a field such that $\mathrm{Char}(K) \neq 2$, and let L be a quadratic extension of K. Write $L = K(\gamma)$, where $\gamma \in L$ and $\gamma^2 \in K$. Show that if $\alpha, \beta \in K$, then

$$N_{L/K}(\alpha + \beta\gamma) = \alpha^2 - \beta^2\gamma^2 \quad \text{and} \quad T_{L/K}(\alpha + \beta\gamma) = 2\alpha.$$

3. Let K be a field, let L be a finite extension of K, and put $n = [L : K]$. If $(\alpha_1, \alpha_2, \ldots, \alpha_n)$ is a linear base of L over K, the determinant of the matrix $[T_{L/K}(\alpha_i \alpha_j)]_{1 \leq i, j \leq n}$ in $\mathrm{Mat}_n(K)$ is said to be the **discriminant of L over K relative to** $(\alpha_1, \alpha_2, \ldots, \alpha_n)$, and is denoted by $\Delta_{L/K}(\alpha_1, \alpha_2, \ldots, \alpha_n)$.

   Show that if $(\alpha_1, \alpha_2, \ldots, \alpha_n)$ and $(\beta_1, \beta_2, \ldots, \beta_n)$ are linear bases of L over K, and if A is the transition matrix from $(\alpha_1, \alpha_2, \ldots, \alpha_n)$ to $(\beta_1, \beta_2, \ldots, \beta_n)$, then

$$\Delta_{L/K}(\beta_1, \beta_2, \ldots, \beta_n) = \det(A)^2 \Delta_{L/K}(\alpha_1, \alpha_2, \ldots, \alpha_n),$$

   and hence $\Delta_{L/K}(\alpha_1, \alpha_2, \ldots, \alpha_n) \neq 0$ if and only if $\Delta_{L/K}(\beta_1, \beta_2, \ldots, \beta_n) \neq 0$.

We say that L **has nonzero discriminant over** K when $\Delta_{L/K}(\alpha_1, \alpha_2, \ldots, \alpha_n) \neq 0$ for every linear base $(\alpha_1, \alpha_2, \ldots, \alpha_n)$ of L over K. According to the result just stated, if $\Delta_{L/K}(\alpha_1, \alpha_2, \ldots, \alpha_n) \neq 0$ for one linear base $(\alpha_1, \alpha_2, \ldots, \alpha_n)$ of L over K, then L has nonzero discriminant over K.

4. Show that, if K is a field, and if L is a finite extension of K, then L has nonzero discriminant over K if and only if $T_{L/K}$ is different from the zero mapping.

5. Let K be a field, let L be a finite extension of K, and let s be a K-automorphism of L. Verify the following assertions:
   a. If $\alpha \in L$, then

   $$N_{L/K}(s(\alpha)) = N_{L/K}(\alpha) \quad \text{and} \quad T_{L/K}(s(\alpha)) = T_{L/K}(\alpha).$$

   b. If $\alpha \in L^*$, then $N_{L/K}(\alpha/s(\alpha)) = 1$.
   c. If $\alpha \in L$, then $T_{L/K}(\alpha - s(\alpha)) = 0$.

6. Let p be a prime, and let ζ be a complex pth root of unity such that $\zeta \neq 1$. Prove that

   $$T_{\mathbf{Q}(\zeta)/\mathbf{Q}}\left(\sum_{i=0}^{p-2} \alpha_i \zeta^i\right) = (p-1)\alpha_0 - \sum_{i=1}^{p-2} \alpha_i$$

   for all $\alpha_0, \alpha_1, \ldots, \alpha_{p-2} \in \mathbf{Q}$. (The case where $p = 3$ has been worked out in 3.8.5.)

7. Let K be a field, and let L be a finite extension of K. Show that if $\alpha \in L$ and $m = [L : K(\alpha)]$, then the characteristic polynomial of the K-linear transformation $\rho \to \alpha\rho$ on the K-space L is the mth power of the minimal polynomial of α over K; and then deduce that $N_{L/K}(\alpha)$ and $T_{L/K}(\alpha)$ are, respectively, the determinant and the trace of this K-linear transformation.

   Use this to obtain an alternative proof of the result stated in the preceding problem.

8. Prove that if K is a field, and if L is a finite separable extension of K, then $T_{L/K}$ is surjective.

9. Let K be a field, let L be a finite separable extension of K, and let F be an extension field of L that is normal over K. Show that if $n = [L : K]$ and $u_1, u_2, \ldots, u_n$ are the K-monomorphisms from L to F, and if $(\alpha_1, \alpha_2, \ldots, \alpha_n)$ is a linear base of L over K, then

   $$\Delta_{L/K}(\alpha_1, \alpha_2, \ldots, \alpha_n) = \det\left([u_i(\alpha_j)]_{1 \leq i, j \leq n}\right)^2.$$

10. Let K be a field, let L be a finite separable extension of K, and let α be a primitive element of L over K. Put $n = [L : K]$, and let f denote the minimal polynomial of α over K. Prove the following assertions:
    a. If F is an extension field of L such that f splits in $F[X]$, and if

$\rho_1, \rho_2, \ldots, \rho_n$ are the zeros of f in F, then

$$\Delta_{L/K}(1, \alpha, \ldots, \alpha^{n-1}) = \prod_{1 \le i < j \le n} (\rho_i - \rho_j)^2.$$

b. $\Delta_{L/K}(1, \alpha, \ldots, \alpha^{n-1}) = (-1)^{n(n-1)/2} N_{L/K}(f'(\alpha))$.

11. Let K be a field, and let L be a finite separable extension of K. Verify the following assertions:

   a. The mapping $(\alpha, \beta) \to T_{L/K}(\alpha\beta)$ from $L \times L$ to K is a dual pairing on the K-space L.

   b. If $n = [L : K]$, if α is a primitive element of L over K and f denotes the minimal polynomial of α over K, and if $\beta_0, \beta_1, \ldots, \beta_{n-1}$ are the elements of L such that

   $$f(X) = (X - \alpha)\left(\sum_{i=0}^{n-1} \beta_i X^i \right)$$

   in $L[X]$, then the dual base of $(1, \alpha, \ldots, \alpha^{n-1})$ relative to the dual pairing on the K-space L described in assertion a is $(\beta_0/f'(\alpha), \beta_1/f'(\alpha), \ldots, \beta_{n-1}/f'(\alpha))$.

(Assertion b is known as **Euler's theorem**.)

   The purpose of the problems that follow is to enable us to study a special type of linear base that occurs occasionally in certain applications of Galois theory to algebraic number theory.

   Let K be a field, and let L be a finite extension of K. By a **normal base of L over K** we shall understand a linear base of L over K that is the orbit under $\text{Gal}(L/K)$ of an element of L.

   Note that if L is a simple extension of K, then $\text{Gal}(L/K)$ is equipotent to the orbit under $\text{Gal}(L/K)$ of every primitive element of L over K. However, we have not yet discussed when an orbit under $\text{Gal}(L/K)$ is linearly independent over K.

12. Let K be a field, and let L be a finite extension of K. Show that if L admits a normal base over K, then L is Galois over K.

13. Let K be a field, and let L be a finite Galois extension of K. Show that if $n = [L : K]$ and $s_1, s_2, \ldots, s_n$ denote the K-automorphisms of L, and if $\alpha \in L$, then the orbit of α under $\text{Gal}(L/K)$ is a normal base of L over K if and only if the matrix $[s_i s_j(\alpha)]_{1 \le i, j < n}$ in $\text{Mat}_n(L)$ is nonsingular.

14. Let K be a field, and let L be a finite Galois extension of K. Prove that if $\alpha \in L$ and the orbit of α under $\text{Gal}(L/K)$ is a normal base of L over K, and if E is an intermediate field between K and L, then $E = K(T_{L/E}(\alpha))$.

15. Let K be a field, let N be an extension field of K, and let L and M be intermediate fields between K and L that are Galois and finite over K and such that $L \cap M = K$. Show that if $\alpha \in L$ and $\beta \in M$, and if the orbits of α under $\text{Gal}(L/K)$ and of β under $\text{Gal}(M/K)$ are, respec-

tively, normal bases of L and M over K, then the orbit of $\alpha\beta$ under $\text{Gal}(L \vee M/K)$ is a normal base of $L \vee M$ over K.

The next three problems outline the proof of the result known as the normal base theorem. This states that every finite Galois extension possesses a normal base (in other words, this is the converse of the result stated in problem 12). When the bottom field is infinite, the argument sketched here, which is due to Artin, is based on the "algebraic independence of field automorphisms"; but when the bottom field is finite, a completely different approach is used, based on methods from linear algebra.

16. Let K be an infinite field, and let L be a finite Galois extension of K. Put $n = [L : K]$ and denote by $s_1, s_2, \ldots, s_n$ the K-automorphisms of L. Prove the following assertions:

a. If $f \in K[X_1, X_2, \ldots, X_n]$ and

$$f(s_1(\alpha), s_2(\alpha), \ldots, s_n(\alpha)) = 0$$

for every $\alpha \in L$, then $f = 0$.

b. If we write $s_{k(i,j)} = s_i s_j$ for $1 \le i, j \le n$, and if h is the polynomial in $K[X_1, X_2, \ldots, X_n]$ defined by

$$h(X_1, X_2, \ldots, X_n) = \det\left(\left[X_{k(i,j)}\right]_{1 \le i, j \le n}\right),$$

then $h \ne 0$ and

$$h(s_1(\alpha), s_2(\alpha), \ldots, s_n(\alpha)) = \det\left(\left[s_i s_j(\alpha)\right]_{1 \le i, j \le n}\right)$$

for every $\alpha \in L$.

17. Let K be a field, and let L be a finite Galois extension of K such that $\text{Gal}(L/K)$ is cyclic. Show that the minimal polynomial of every generator of $\text{Gal}(L/K)$, when regarded as a K-linear transformation on the K-space L, has degree $[L : K]$.

18. It is known that for finite fields, the Galois group of every Galois extension is cyclic. Assuming this result, which will be proved in 3.11.8, derive the **normal base theorem**: If K is a field, and if L is a finite Galois extension of K, then L admits a normal base over K.

3.9. CYCLIC EXTENSIONS

Let K be a field, and let L be an extension field of K. We say that L is **cyclic over** K, or that L is a **cyclic extension** of K, when L is Galois and finite over K and $\text{Gal}(L/K)$ is cyclic.

There is some redundancy in the definition just given. In effect, it is known that the Galois group of every infinite Galois extension is nondenumerable (section 3.3, problem 6), and so it is not cyclic. Consequently, Galois extensions with a cyclic Galois group are finite extensions.

Since finite groups of prime order are cyclic, we see that finite Galois extensions of prime linear degree are cyclic. It will also be seen that for finite fields every finite extension is cyclic.

3.9.1. Proposition. *Let K be a field, let L be an extension field of K, and let M be an extension field. If M is cyclic over K, then M is cyclic over L and L is cyclic over K.*

Proof. It is readily seen that M is cyclic over L. For M is Galois over L; and since $\mathrm{Gal}(M/L)$ is a subgroup of the cyclic group $\mathrm{Gal}(M/K)$, it also is cyclic.

To show that L is cyclic over K, we note first that $\mathrm{Gal}(M/L)$ is a normal subgroup of $\mathrm{Gal}(M/K)$; and by 3.3.9, this implies that L is Galois over K. Finally, applying 3.2.21, we see that $\mathrm{Gal}(L/K)$ is isomorphic to the cyclic group $\mathrm{Gal}(M/K)/\mathrm{Gal}(M/L)$, and hence it also is cyclic.   □

3.9.2. Proposition. *Let K be a field, and let L be a cyclic extension of K. Then the mapping $E \to [E:K]$ from the set of all intermediate fields between K and L to the set of all positive divisors of $[L:K]$ is bijective.*

Proof. Taking into account that for every positive divisor d of the order of a finite cyclic group there exists a unique subgroup of index d, we see that this is an immediate consequence of 3.4.3 and 3.4.5.   □

We have noted that given a field K and a finite extension L of K, the mappings $\alpha \to N_{L/K}(\alpha)$ from L^* to K^* and $\alpha \to T_{L/K}(\alpha)$ from L^+ to K^+ are homomorphisms. It is not difficult to show that if s is a K-automorphism of L, then

$$N_{L/K}(\alpha/s(\alpha)) = 1 \quad \text{for every } \alpha \in L^*$$

and

$$T_{L/K}(\alpha - s(\alpha)) = 0 \quad \text{for every } \alpha \in L$$

(section 3.8, problem 5). Thus, the kernel of the first homomorphism contains the elements of the form $\alpha/s(\alpha)$ with $\alpha \in L^*$ and $s \in \mathrm{Gal}(L/K)$; and that of the second contains the elements of the form $\alpha - s(\alpha)$ with $\alpha \in L$ and $s \in \mathrm{Gal}(L/K)$. As shown by the next theorem, which is frequently referred to as **Hilbert's Satz 90**, in the case of a cyclic extension the elements of the indicated forms make up the corresponding kernels.

3.9.3. Theorem (Hilbert). *Let K be a field, let L be a cyclic extension of K, and let s be a generator of $\mathrm{Gal}(L/K)$.*
(i) *If $\alpha \in L^*$ and $N_{L/K}(\alpha) = 1$, then there exists a $\beta \in L^*$ such that $\alpha = \beta/s(\beta)$.*
(ii) *If $\alpha \in L$ and $T_{L/K}(\alpha) = 0$, then there exists a $\beta \in L$ such that $\alpha = \beta - s(\beta)$.*

Proof. Write $n = [L:K]$, so that $s^n = i_L$ and $i_L, s, s^2, \ldots, s^{n-1}$ are the elements of $\mathrm{Gal}(L/K)$. Since the theorem is evident when $K = L$, we shall assume that $n > 1$.

To prove (i), suppose that $\alpha \in L^*$ and $N_{L/K}(\alpha) = 1$. By 3.8.3, we then have

$$\prod_{i=0}^{n-1} s^i(\alpha) = N_{L/K}(\alpha) = 1.$$

Define elements $\gamma_0, \gamma_1, \ldots, \gamma_{n-1}$ of L^* by

$$\gamma_0 = 1, \quad \text{and} \quad \gamma_i = \prod_{j=0}^{i-1} s^j(\alpha) \quad \text{for} \quad 1 \leq i \leq n-1;$$

then, according to Dedekind's theorem 3.1.2, a $\theta \in L$ can be chosen so that $\sum_{i=0}^{n-1} \gamma_i s^i(\theta) \neq 0$. Writing $\beta = \sum_{i=0}^{n-1} \gamma_i s^i(\theta)$, we get

$$s(\beta) = \sum_{i=0}^{n-1} s(\gamma_i) s^{i+1}(\theta)$$

$$= s(\gamma_0) s(\theta) + \sum_{i=1}^{n-2} s(\gamma_i) s^{i+1}(\theta) + s(\gamma_{n-1}) s^n(\theta)$$

$$= s(\theta) + \sum_{i=1}^{n-2} s(\gamma_i) s^{i+1}(\theta) + s(\gamma_{n-1})\theta.$$

Note next that for $1 \leq i \leq n-1$ we have

$$s(\gamma_i) = \prod_{j=0}^{i-1} s^{j+1}(\alpha) = \prod_{j=1}^{i} s^j(\alpha) = \frac{\prod_{j=0}^{i} s^j(\alpha)}{\alpha},$$

and so

$$s(\gamma_i) = \gamma_{i+1}/\alpha \quad \text{for } 1 \leq i \leq n-2, \quad \text{and} \quad s(\gamma_{n-1}) = 1/\alpha.$$

Therefore

$$s(\beta) = s(\theta) + \frac{\sum_{i=1}^{n-2} \gamma_{i+1} s^{i+1}(\theta)}{\alpha} + \frac{\theta}{\alpha}$$

$$= \frac{\theta + \alpha s(\theta) + \sum_{i=2}^{n-1} \gamma_i s^i(\theta)}{\alpha}$$

$$= \frac{\sum_{i=0}^{n-1} \gamma_i s^i(\theta)}{\alpha} = \frac{\beta}{\alpha},$$

which implies that $\alpha = \beta/s(\beta)$.

To prove (ii), now suppose that $\alpha \in L$ and $T_{L/K}(\alpha) = 0$. By 3.8.7, there exists a $\rho \in L$ such that $T_{L/K}(\rho) \neq 0$; and if we let $\theta = \rho/T_{L/K}(\rho)$, it follows from 3.8.4 that $T_{L/K}(\theta) = 1$. According to 3.8.3, we then have

$$\sum_{i=0}^{n-1} s^i(\alpha) = T_{L/K}(\alpha) = 0$$

and

$$\sum_{i=0}^{n-1} s^i(\theta) = T_{L/K}(\theta) = 1.$$

Let $\gamma_i = \sum_{j=0}^{i-1} s^j(\alpha)$ for $1 \le i \le n-1$; and let $\beta = \sum_{i=1}^{n-1} \gamma_i s^i(\theta)$. Then

$$s(\beta) = \sum_{i=1}^{n-1} s(\gamma_i) s^{i+1}(\theta)$$

$$= \sum_{i=1}^{n-2} s(\gamma_i) s^{i+1}(\theta) + s(\gamma_{n-1})\theta.$$

Moreover, for $1 \le i \le n-1$ we have

$$s(\gamma_i) = \sum_{j=0}^{i-1} s^{j+1}(\alpha) = \sum_{j=1}^{i} s^j(\alpha) = \left(\sum_{j=0}^{i} s^j(\alpha)\right) - \alpha,$$

so that

$$s(\gamma_i) = \gamma_{i+1} - \alpha \quad \text{for } 1 \le i \le n-2, \quad \text{and} \quad s(\gamma_{n-1}) = -\alpha.$$

Consequently

$$s(\beta) = \sum_{i=1}^{n-2} \gamma_{i+1} s^{i+1}(\theta) - \alpha\left(\sum_{i=1}^{n-2} s^{i+1}(\theta)\right) - \alpha\theta$$

$$= \sum_{i=2}^{n-1} \gamma_i s^i(\theta) - \alpha\left(\sum_{i=2}^{n-1} s^i(\theta)\right) - \alpha\theta,$$

whence

$$s(\beta) - \beta = -\alpha\left(\sum_{i=2}^{n-1} s^i(\theta)\right) - \alpha\theta - \gamma_1 s(\theta)$$

$$= -\alpha\theta - \alpha s(\theta) - \alpha\left(\sum_{i=2}^{n-1} s^i(\theta)\right)$$

$$= -\alpha\left(\sum_{i=0}^{n-1} s^i(\theta)\right) = -\alpha,$$

and so $\alpha = \beta - s(\beta)$. $\qquad\qquad\square$

The element β of L constructed in the first part of the preceding proof from the elements α and θ of L and the generator s of $\mathrm{Gal}(L/K)$ is commonly referred to as the **Lagrange resolvent of α and θ relative to s**.

3.9.4. Example. Let us show that the rational solutions of the equation $\alpha^2 - \alpha\beta + \beta^2 = 1$ can be described by

$$\alpha = \frac{m^2 - 2mn}{m^2 - mn + n^2} \quad \text{and} \quad \beta = \frac{n^2 - 2mn}{m^2 - mn + n^2},$$

where m and n are relatively prime integers.

To do this, we shall refer to the discussion in 3.8.5. Let ω be a complex cube root of unity such that $\omega \neq 1$; it was shown that for all $\alpha, \beta \in \mathbf{Q}$ we have

$$N_{\mathbf{Q}(\omega)/\mathbf{Q}}(\alpha + \beta\omega) = \alpha^2 - \alpha\beta + \beta^2,$$

so that the condition $\alpha^2 - \alpha\beta + \beta^2 = 1$ is equivalent to $N_{\mathbf{Q}(\omega)/\mathbf{Q}}(\alpha + \beta\omega) = 1$. On the other hand, we have $\mathrm{Gal}(\mathbf{Q}(\omega)/\mathbf{Q}) = \langle s \rangle$, where s is the $\mathbf{Q}$-automorphism of $\mathbf{Q}(\omega)$ such that $s(\omega) = \omega^2$.

Our conclusion now follows at once from Hilbert's theorem 3.9.3, the remarks preceding it, and what was shown in 3.8.5: If $\alpha, \beta \in \mathbf{Q}$, to say that $N_{\mathbf{Q}(\omega)/\mathbf{Q}}(\alpha + \beta\omega) = 1$ amounts to saying that there exist $\gamma, \delta \in \mathbf{Q}$, not both equal to 0, such that

$$\alpha + \beta\omega = \frac{s(\gamma + \delta\omega)}{\gamma + \delta\omega} = \frac{\gamma + \delta\omega^2}{\gamma + \delta\omega}$$

$$= \left(\frac{\gamma^2 - 2\gamma\delta}{\gamma^2 - \gamma\delta + \delta^2} \right) + \left(\frac{\delta^2 - 2\gamma\delta}{\gamma^2 - \gamma\delta + \delta^2} \right) \omega;$$

and this equality is equivalent to the two equalities

$$\alpha = \frac{\gamma^2 - 2\gamma\delta}{\gamma^2 - \gamma\delta + \delta^2} \quad \text{and} \quad \beta = \frac{\delta^2 - 2\gamma\delta}{\gamma^2 - \gamma\delta + \delta^2}. \qquad \square$$

Cyclic extensions occur naturally in Galois's theory of equations. This is because under a suitable hypothesis concerning roots of unity, they are the splitting fields of the polynomials defining binomial equations. The precise result is contained in the next two propositions.

3.9.5. Proposition. *Let n be a positive integer, let K be a field containing a primitive nth root of unity, and let L be a cyclic extension of K such that $[L : K] = n$. Then there exists an $\alpha \in K^*$ such that*:
   (i) $X^n - \alpha$ *is irreducible in $K[X]$.*
   (ii) L *is a splitting field of $X^n - \alpha$ over K.*
   (iii) $L = K(\rho)$ *for every nth root ρ of α in L.*

Proof. Let ζ be a primitive nth root of unity in K, and let s be a generator of $\mathrm{Gal}(L/K)$. Then

$$N_{L/K}(1/\zeta) = 1/\zeta^n = 1,$$

and 3.9.3 implies that $1/\zeta = \beta/s(\beta)$ for some $\beta \in L^*$.

It follows from the equality $s(\beta) = \zeta\beta$ that

$$s(\beta^n) = s(\beta)^n = \zeta^n \beta^n = \beta^n,$$

that is, β^n is a fixed point of s; and clearly $\beta^n \neq 0$. Since $\mathrm{Gal}(L/K) = \langle s \rangle$, we conclude that $\beta^n \in K^*$. We now put $\alpha = \beta^n$, and contend that α satisfies the three conditions stated in the conclusion.

The equality $s(\beta) = \zeta\beta$ also implies that

$$s^i(\beta) = \zeta^i\beta \quad \text{for } 0 \le i \le n-1.$$

According to 2.3.6, this shows that the zeros in L of the minimal polynomial of β over K are $\beta, \zeta\beta, \zeta^2\beta, \ldots, \zeta^{n-1}\beta$. Since this polynomial splits in $L[X]$ and is separable, it has degree n, and hence it is $X^n - \alpha$.

Thus, $X^n - \alpha$ is irreducible in $K[X]$. It also follows that $[K(\beta): K] = n = [L: K]$, whence

$$L = K(\beta) = K(\zeta^i\beta) \quad \text{for } 0 \le i \le n-1;$$

and this completes the argument, because $\beta, \zeta\beta, \zeta^2\beta, \ldots, \zeta^{n-1}\beta$ are the nth roosts of α in L. □

The use of the norm mapping $N_{L/K}$ and Hilbert's theorem in the preceding proof is not essential. For the existence of a $\beta \in L$ such that $s(\beta) = \zeta\beta$ can be established directly from the assumptions that s is an element of order n in $\text{Aut}(L)$ and that ζ is an nth root of unity in L (section 3.5, problem 6).

3.9.6. Proposition. *Let n be a positive integer, let K be a field containing a primitive nth root of unity, and let $\alpha \in K$. If L is a splitting field of $X^n - \alpha$ over K, then L is cyclic over K and $[L: K] | n$.*

Proof. We shall assume that $\alpha \in K^*$, since the conclusion is evident when $\alpha = 0$.

Since $\text{Char}(K) \nmid n$, the polynomial $X^n - \alpha$ in $K[X]$ is separable, and so L is Galois and finite over K. Choose a zero ρ of $X^n - \alpha$ in L, and denote by U the multiplicative group consisting of the nth roots of unity in K. According to 3.5.4, we have

$$L = K(\rho) \quad \text{and} \quad X^n - \alpha = \prod_{\zeta \in U} (X - \zeta\rho).$$

If $s \in \text{Gal}(L/K)$, then $s(\rho)$ is a zero of $X^n - \alpha$ in L, and hence a $\zeta_s \in U$ can be chosen so that $s(\rho) = \zeta_s\rho$.

It is readily verified that the mapping $s \to \zeta_s$ from $\text{Gal}(L/K)$ to U is a homomorphism: If $s, t \in \text{Gal}(L/K)$, then

$$\zeta_{ts}\rho = ts(\rho) = t(s(\rho)) = t(\zeta_s\rho) = \zeta_s t(\rho) = \zeta_s\zeta_t\rho,$$

whence $\zeta_{ts} = \zeta_s\zeta_t = \zeta_t\zeta_s$. Moreover, this homomorphism is injective: If $s \in \text{Gal}(L/K)$ and $\zeta_s = 1$, then $s(\rho) = \zeta_s\rho = \rho$; and since $L = K(\rho)$, this implies that $s = i_L$.

Since U is cyclic of order n, and since $\text{Gal}(L/K)$ is embeddable in U, it now follows that $\text{Gal}(L/K)$ is cyclic and $\text{Card}(\text{Gal}(L/K)) | n$. Therefore L is cyclic over K and $[L: K] | n$, as was to be shown. □

It is readily verified that the last two propositions do not remain valid if the existence of the indicated primitive roots of unity is not assumed.

It will be shown next that for a field K of prime characteristic p, a cyclic extension of linear degree p over K can be described as a splitting field over K of a polynomial in $K[X]$ of the type $X^p - X - \alpha$.

Before stating the result, however, we would like to recall that, as we saw in 1.6.6, a polynomial in $K[X]$ of this type either is irreducible in $K[X]$ or splits in $K[X]$. In the latter case, its only splitting field over K is K itself; thus, the irreducibility condition in the statement merely excludes this trivial case.

3.9.7. Proposition (Artin–Schreier). *Let K be a field of prime characteristic p, and let L be an extension field of K such that $[L:K]=p$. Then L is cyclic over K if and only if there exists an $\alpha \in K^*$ such that $X^p - X - \alpha$ is irreducible in $K[X]$ and admits L as a splitting field over K, in which case $L = K(\rho)$ for every zero ρ of $X^p - X - \alpha$ in L.*

Proof. First suppose that there is an $\alpha \in K^*$ such that $X^p - X - \alpha$ is irreducible in $K[X]$ and admits L as a splitting field over K. Since $(X^p - X - \alpha)' = -1$, we see that $X^p - X - \alpha$ is separable, and hence L is Galois and finite over K. Moreover, L has prime linear degree over K. As observed at the beginning of the section, it then follows that L is cyclic over K.

Now suppose, conversely, that L is cyclic over K, and let s be a generator of $\mathrm{Gal}(L/K)$. Then

$$T_{L/K}(-1) = -T_{L/K}(1) = -p = 0,$$

and 3.9.3 shows that $-1 = \beta - s(\beta)$ for some $\beta \in L$. Since $s(\beta) = 1 + \beta \neq \beta$, it is clear that $\beta \notin K$; but then

$$K \subset K(\beta) \subseteq L \quad \text{and} \quad [L:K] = p,$$

which implies that $L = K(\beta)$. It also follows that

$$s(\beta^p - \beta) = s(\beta)^p - s(\beta) = (1+\beta)^p - (1+\beta)$$
$$= (1+\beta^p) - (1+\beta) = \beta^p - \beta,$$

and so $\beta^p - \beta$ is a fixed point of s; since $\mathrm{Gal}(L/K) = \langle s \rangle$, this means that $\beta^p - \beta \in K$. Now let $\alpha = \beta^p - \beta$; then

$$[K(\beta):K] = p = \deg(X^p - X - \alpha),$$

whence $X^p - X - \alpha$ is the minimal polynomial of β over K. This proves the irreducibility of $X^p - X - \alpha$ in $K[X]$, which in turn implies that $\alpha \neq 0$. Finally, in view of what was shown in 1.3.8 and 1.6.2, this polynomial splits in $L[X]$, its zeros in L being $\beta + 1, \beta + 2, \ldots, \beta + p$. Since

$$L = K(\beta) = K(\beta + i) \quad \text{for } 1 \leq i \leq p,$$

it follows that L is a splitting field of $X^p - X - \alpha$ over K, and that $L = K(\rho)$ for every zero ρ of $X^p - X - \alpha$ in L.   □

It is possible to avoid the use of the trace mapping $T_{L/K}$ and Hilbert's theorem in the second part of the proof just given. Indeed, the existence of a $\beta \in L$ such that $s(\beta) = 1 + \beta$ can be deduced directly from the assumption that s is an element of order p in $\mathrm{Aut}(L)$ (section 3.5, problem 6).

Let K be a field of prime characteristic p. According to the preceding proposition and what was shown in 1.6.2 and 1.6.6, the following three conditions are equivalent:

a. K admits a cyclic extension of linear degree p over K.
b. There exists an $\alpha \in K$ such that $X^p - X - \alpha$ is irreducible in $K[X]$.
c. There exists an $\alpha \in K$ such that $\alpha \neq \beta^p - \beta$ for every $\beta \in K$.

It can now be seen (with the help of the result stated in section 2.1, problem 14) that if K admits a cyclic extension of linear degree p over K, then there exists a sequence $(L_n)_{n \geq 0}$ of extension fields of K with $L_0 = K$ and such that L_{n+1} is a cyclic extension of L_n and $[L_n : K] = p^n$ for every $n \geq 0$. We cannot assert, however, that such a sequence consists of cyclic extensions of K.

In truth, a more precise result is valid: Under the same hypothesis on K, it was proved by Albert that there exists a sequence of cyclic extensions of K enjoying the foregoing properties (problem 10).

PROBLEMS

1. Let p be a prime, and let n be a positive integer. Prove that if K is a field and L is a cyclic extension of K such that $[L : K] = p^n$, and if E denotes the intermediate field between K and L for which $[E : K] = p^{n-1}$, then $L = K(\alpha)$ for every $\alpha \in L - E$.

2. Let K be a field, and let L be a cyclic extension of K. Let d be a positive divisor of $[L : K]$, and denote by E the intermediate field between K and L having linear degree $[L : K]/d$ over K. Show that if $\alpha \in K$ and $\alpha^d \in \mathrm{Im}(N_{L/K})$, then $\alpha \in \mathrm{Im}(N_{E/K})$.

3. Let K be a field, let F be an algebraic closure of K, and let s be a K-automorphism of F. Prove that every finite extension of $\mathrm{Inv}(\{s\})$ is cyclic over $\mathrm{Inv}(\{s\})$. (This result is due to Artin.)

4. Let p be a prime, let K be a field containing a primitive pth root of unity, and let $K(\xi)$ be a simple transcendental extension of K. Form a

simple extension $K(\sigma)$ of $K(\xi)$ by adjunction of a pth root of ξ in an extension field of $K(\xi)$. Prove the following assertions:

a. $K(\sigma)$ is cyclic over $K(\xi)$.

b. If $\theta \in K(\sigma)$, and if ζ is a primitive pth root of unity in K, then $N_{K(\sigma)/K(\xi)}(\theta) = 1$ if and only if $\theta = f(\zeta\sigma)/f(\sigma)$ for some nonzero polynomial f in $K[X]$ such that $\deg(f) < p$.

5. Let n be a positive integer, let K be a field containing a primitive nth root of unity, and let L be a cyclic extension of K such that $[L:K] = n$. Write $L = K(\rho)$, where $\rho \in L$ and $\rho^n \in K$. Prove that if $\sigma \in L$, then $\sigma^n \in K$ if and only if $\sigma = \gamma\rho^k$ for some $\gamma \in K$ and $k \in \mathbf{Z}$; in which case $L = K(\sigma)$ if and only if k and n are relatively prime and $\gamma \neq 0$.

6. Let n be a positive integer, let K be a field containing a primitive nth root of unity, and let $\alpha \in K$. Prove that if L is a splitting field of $X^n - \alpha$ over K, then $[L:K]$ is the smallest element of the set of all positive integers k for which α^k possesses an nth root in K.

7. Let K be a field of prime characteristic p, and let L be a cyclic extension of K such that $[L:K] = p$. Write $L = K(\rho)$, where $\rho \in L$ and $\rho^p - \rho \in K$. Prove that if $\sigma \in L$, then $\sigma^p - \sigma \in K$ if and only if $\sigma = \gamma + k\rho$ for some $\gamma \in K$ and $k \in \mathbf{Z}$; in which case $L = K(\sigma)$ if and only if $p \nmid k$.

Let G be a finite nontrivial cyclic group, and let $\mathrm{Card}(G) = p_1^{n_1}p_2^{n_2}\cdots p_k^{n_k}$ be the prime factorization of its order. Then G is the internal direct product of subgroups $C_1, C_2, \ldots, C_k$ such that $\mathrm{Card}(C_i) = p_i^{n_i}$ for $1 \leq i \leq n$. It then follows from 3.4.11 that in order to study the structure of cyclic extensions, it suffices to consider the cyclic extensions having an integer power of a prime as linear degree.

The next seven problems, which are classical results due to Albert, are devoted to the study of this special type of cyclic extension.

8. Let K be a field of prime characteristic p, let L be a cyclic extension of K, and suppose that $[L:K] = p^n$ for some positive integer n. Let s be a generator of $\mathrm{Gal}(L/K)$, and denote by E the intermediate field between K and L for which $[E:K] = p^{n-1}$. Show that there exist $\alpha \in E$ and $\rho \in L$ with the following properties:

a. $X^p - X - \alpha$ is irreducible in $E[X]$.

b. ρ is a zero of $X^p - X - \alpha$.

c. $L = K(\rho)$.

d. $s^{p^{n-1}}(\rho) = \rho + 1$.

e. If β denotes the element $s(\rho) - \rho$ of L, then

$$\beta \in E, \qquad T_{E/K}(\beta) = 1, \quad \text{and} \quad \beta^p - \beta = s(\alpha) - \alpha.$$

9. Let K be a field of prime characteristic p, and let L be a cyclic extension of K. Suppose that $[L:K] = p^{n-1}$ for some integer $n > 1$,

and let s be a generator of $\mathrm{Gal}(L/K)$. Show that there exist $\alpha, \beta \in L$ with the following properties:

a. $T_{L/K}(\beta) = 1$ and $\beta^p - \beta = s(\alpha) - \alpha$.

b. If $\omega \in K$, then $X^p - X - \alpha - \omega$ is irreducible in $L[X]$.

c. If $\omega \in K$, and if ρ is a zero of $X^p - X - \alpha - \omega$ in an extension field of L, then $L(\rho)$ is cyclic over K and $[L(\rho): K] = p^n$; furthermore, s is uniquely extendible to a K-automorphism of $L(\rho)$ such that $\rho \to \rho + \beta$, and this K-automorphism of $L(\rho)$ is a generator of $\mathrm{Gal}(L(\rho)/K)$.

d. If M is an extension field of L that is cyclic over K and such that $[M: K] = p^n$, then there exists an $\omega \in K$ such that M is a splitting field of $X^p - X - \alpha - \omega$ over L.

10. Let K be a field of prime characteristic p. Prove that if K admits a cyclic extension of linear degree p over K, then there exists a sequence $(L_n)_{n \geq 1}$ of cyclic extensions of K such that L_{n+1} is an extension field of L_n and $[L_n: K] = p^n$ for every $n \geq 1$.

11. Let p be a prime, let K be a field containing a primitive pth root of unity, and let L be a cyclic extension of K. Suppose that $[L: K] = p^n$ for some positive integer n, and denote by E the intermediate field between K and L with $[E: K] = p^{n-1}$. Let ζ and s be, respectively, a primitive pth root of unity in K and a generator of $\mathrm{Gal}(L/K)$. Show that there exist $\alpha \in E^*$ and $\rho \in L^*$ with the following properties:

a. $X^p - \alpha$ is irreducible in $E[X]$.

b. ρ is a pth root of α.

c. $L = K(\rho)$.

d. $s^{p^{n-1}}(\rho) = \zeta\rho$.

e. If β denotes the element $s(\rho)/\rho$ of L^*, then

$$\beta \in E^*, \qquad N_{E/K}(\beta) = \zeta, \quad \text{and} \quad \beta^p = s(\alpha)/\alpha.$$

12. Let p be a prime, let K be a field containing a primitive pth root of unity, and let L be a cyclic extension of K. Suppose that $[L: K] = p^{n-1}$ for some integer $n > 1$, and let s be a generator of $\mathrm{Gal}(L/K)$. Show that if there exists a $\beta \in L^*$ such that $N_{L/K}(\beta)$ is a primitive pth root of unity in K, then there exists an $\alpha \in L^*$ with the following properties:

a. $s(\alpha)/\alpha = \beta^p$.

b. If $\omega \in K^*$, then $X^p - \alpha\omega$ is irreducible in $L[X]$.

c. If $\omega \in K^*$, and if ρ is a pth root of $\alpha\omega$ in a extension field of L, then $L(\rho)$ is cyclic over K and $[L(\rho): K] = p^n$; furthermore, s is uniquely extendible to a K-automorphism of $L(\rho)$ such that $\rho \to \beta\rho$, and this K-automorphism of $L(\rho)$ is a generator of $\mathrm{Gal}(L(\rho)/K)$.

d. If M is an extension field of L that is cyclic over K and such that $[M: K] = p^n$, then there exists an $\omega \in K^*$ such that M is a splitting field of $X^p - \alpha\omega$ over L.

13. Let p be a prime, let K be a field containing a primitive pth root of
     unity, and let L be a cyclic extension of K. Suppose that $[L:K] = p^n$
     for some positive integer n. Prove that L is a subfield of a cyclic
     extension of K having linear degree p^{n+1} over K if and only if the pth
     roots of unity in K belong to the multiplicative group $N_{L/K}(L^*)$ in K.

14. Let K be a field such that $\mathrm{Char}(K) \neq 2$, and let L be a quadratic
     extension of K. Write $L = K(\gamma)$, where $\gamma \in L$ and $\gamma^2 \in K$. Show that L
     is a subfield of a quartic cyclic extension of K if and only if there exist
     $\alpha, \beta \in K$ such that $\gamma^2 = \alpha^2 + \beta^2$.

     Deduce from this that there exists no quartic cyclic extension of
     $\mathbf{Q}$ containing $\mathbf{Q}(\sqrt{-1})$ as a subfield.

     The objective of the remaining four problems is to outline a
     discussion of a striking theorem of Artin–Schreier, to which reference
     has been made in section 3.5, problem 7. The first two, which are
     interesting in themselves, can be considered as a prelude; the last two
     contain the main results.

15. Let K be a field, let p be a prime such that K admits a finite extension
     whose linear degree over K is divisible by p, and suppose that p is odd
     or $\mathrm{Char}(K) = p$. Prove that if n is a positive integer, then K admits a
     finite extension whose linear degree over K is divisible by p^n.

16. Let K be a field admitting a finite extension whose linear degree over K
     is divisible by 4. Prove that if n is an integer and $n > 1$, then K admits a
     finite extension whose linear degree is divisible by 2^n.

17. Let K be a field admitting a quadratic extension but no quartic
     extensions. Prove the following assertions:
     a. If $\alpha \in K^*$, then exactly one of α and $-\alpha$ possesses a square root in
        K.
     b. The set of all elements of K possessing a square root in K is stable
        relative to the addition and multiplication of K.
     c. $\mathrm{Char}(K) = 0$.

18. Let F be an algebraically closed field, and let P be a field admitting F
     as a proper finite extension. Prove the following assertions.
     a. F is a quadratic extension of P.
     b. P admits no quartic extensions.
     c. There exists an $\iota \in F$ such that $\iota^2 = -1$ and $F = P(\iota)$.

3.10. SOLVABILITY BY RADICALS

This section will be devoted to the main result of Galois's investigations on
the solvability of polynomial equations by radicals. Some basic facts on
finite solvable groups are essential for the understanding of the discussion;

the reader unfamiliar with solvable groups may wish to consult one of the books on the theory of groups cited in the section on prerequisites.

It is first necessary to give a precise formulation of the problem. Indeed, what is meant by saying that a polynomial equation is solvable by radicals? It seems reasonable to say that such an equation is solvable by radicals when we can give a finite procedure for finding its roots in which every step involves a rational operation on field elements or the extraction of a root of a field element. The former can be carried out inside the field to which the elements belong; but in order to perform the latter, it may be necessary to pass to an extension field.

We are thus led to the following definitions. Let K be a field, and let L be an extension field of K. A finite sequence $(\alpha_i)_{1 \le i \le n}$ of elements of L is said to be a **radical sequence for L over K** when $L = K(\alpha_1, \alpha_2, \ldots, \alpha_n)$ and there exists a sequence $(k_i)_{1 \le i \le n}$ of positive integers such that $\alpha_1^{k_1} \in K$ and such that $\alpha_i^{k_i} \in K(\alpha_1, \alpha_2, \ldots, \alpha_{i-1})$ for $1 < i \le n$. We say that L is a **radical extension of K** when L admits a radical sequence over K.

This being said, consider a field K and a nonconstant polynomial f in $K[X]$. We say that the equation $f(X) = 0$ is **solvable by radicals over K** when there exists a radical extension L of K such that f splits in $L[X]$.

This definition expresses in a precise manner the intuitive idea formulated above. For if L is a radical extension of K such that f splits in $L[X]$, and if $(\alpha_i)_{1 \le i \le n}$ is a radical sequence for L over K, then the roots of the equation $f(X) = 0$ are contained in L, and so have the form

$$g(\alpha_1, \alpha_2, \ldots, \alpha_n)/h(\alpha_1, \alpha_2, \ldots, \alpha_n) \quad \text{with } g, h \in K[X_1, X_2, \ldots, X_n];$$

therefore, they are obtained by rational operations and extractions of roots, starting from elements in the field K.

Let us begin with the auxiliary technicalities on radical extensions that are required in the proof of Galois's theorem.

3.10.1. Proposition. *Let K be a field, let L be an extension field of K, and let M be an extension field of L.*

(i) *If L is a radical extension of K, and if M is a radical extension of L, then M is a radical extension of K.*

(ii) *If M is a radical extension of K, then M is a radical extension of L.*

Proof. To prove the first assertion, note that if $(\alpha_i)_{1 \le i \le m}$ and $(\beta_j)_{1 \le j \le n}$ are, respectively, radical sequences of L over K and of M over L, then

$$(\alpha_1, \alpha_2, \ldots, \alpha_m, \beta_1, \beta_2, \ldots, \beta_n)$$

is a radical sequence of M over K.

As to the second assertion, simply observe that every radical sequence of M over K is also a radical sequence of M over L. $\qquad \square$

3.10.2. Proposition. *Let K be a field, let N be an extension field of K, and let L and M be intermediate fields between K and N.*

(i) *If L is a radical extension of K, then $L \vee M$ is a radical extension of M.*

(ii) *If L and M are radical extensions of K, then $L \vee M$ is a radical extension of K.*

Proof. The first assertion is clear, since every radical sequence of L over K is also a radical sequence of $L \vee M$ over M. And the second follows immediately from the first and the preceding proposition.   □

3.10.3. Proposition. *Let K be a field, and let L be a radical extension of K. Then every normal closure of L over K is a radical extension of K.*

Proof. Let M be a normal closure of L over K, and write $\Gamma = \text{Gal}(M/K)$. Since L is finite over K, we know from 2.3.13 that M is also finite over K; and by 3.2.15, this implies that Γ is finite. Thus, to reach the required conclusion, it will suffice to verify that $M = \vee_{s \in \Gamma} s(L)$: indeed, for every $s \in \Gamma$, the subfield $s(L)$ of M is a radical extension of K, because it is K-isomorphic to L; and the preceding proposition would then show that M is a radical extension of K.

To prove the desired equality, denote by T the set of all elements of M that are zeros of the minimal polynomials over K of the elements of L. By 2.3.12, we then have $M = K(T)$. Moreover, we see that $T \subseteq \cup_{s \in \Gamma} s(L)$: for if $\rho \in T$, then ρ is a zero of the minimal polynomial over K of an $\alpha \in L$; therefore, according to 2.3.6, there exists a K-automorphism s of M for which $s(\alpha) = \rho$, and hence $\rho \in s(L)$. It then follows that

$$M = K(T) \subseteq \bigvee_{s \in \Gamma} s(L) \subseteq M,$$

and so $M = \vee_{s \in \Gamma} s(L)$.   □

We now are prepared to establish Galois's remarkable criterion for the solvability by radicals of a polynomial equation over a field of characteristic 0.

3.10.4. Theorem (Galois). *Let K be a field of characteristic 0, and let f be a nonconstant polynomial in $K[X]$. Then the equation $f(X) = 0$ is solvable by radicals over K if and only if the Galois group of f over K is solvable.*

Proof. First suppose that the equation $f(X) = 0$ is solvable by radicals over K. It then follows from 2.3.15 and the preceding proposition that there exists a normal radical extension L of K such that f splits in $L[X]$. Let $(\alpha_i)_{1 \le i \le n}$ be a radical sequence for L over K, and let $(k_i)_{1 \le i \le n}$ be a sequence of positive integers such that $\alpha_1^{k_1} \in K$ and such that $\alpha_i^{k_i} \in K(\alpha_1, \alpha_2, \ldots, \alpha_{i-1})$ for $1 < i \le n$. Next let $k = \prod_{i=1}^{n} k_i$, and let F be a cyclotomic field of order k over L.

By 2.3.18, we know that L is a splitting field over K of a polynomial $g(X)$ in $K[X]$; consequently, F is a splitting field over K of the polynomial $(X^k - 1)g(X)$, and a second application of 2.3.18 shows that F also is normal over K. Moreover, the hypothesis that $\mathrm{Char}(K) = 0$ implies that F is separable over K, and hence Galois over K. Finally, note that according to 3.5.5, we have $F = L(\zeta)$, where ζ is a primitive kth root of unity in F.

We now define an increasing sequence $(E_i)_{0 \le i \le n+1}$ of intermediate fields between K and F as follows:

$$E_0 = K, \quad E_1 = K(\zeta), \quad \text{and} \quad E_{i+1} = K(\zeta, \alpha_1, \alpha_2, \ldots, \alpha_i) \quad \text{for } 1 \le i \le n.$$

Note that $E_{n+1} = F$. Also, it is readily seen that E_{i+1} is abelian over E_i for $0 \le i \le n$: This is clear from 3.5.5 when $i = 0$, because E_1 is a cyclotomic field of order k over E_0; and it follows from 3.9.6 when $1 \le i \le n$, for in this case we have $E_{i+1} = E_i(\alpha_i)$ and $\alpha_i^{k} \in E_i$.

Next we put $\Gamma_i = \mathrm{Gal}(F/E_i)$ for $0 \le i \le n+1$. Then $(\Gamma_i)_{0 \le i \le n+1}$ is a decreasing sequence of subgroups of $\mathrm{Gal}(F/K)$ such that $\Gamma_0 = \mathrm{Gal}(F/K)$ and $\Gamma_{n+1} = \{i_F\}$. And as we have just seen, E_{i+1} is abelian over E_i for $0 \le i \le n$; according to 3.3.9 and 3.2.21, this implies that Γ_{i+1} is normal in Γ_i, and that the quotient group Γ_i / Γ_{i+1} is isomorphic to the abelian group $\mathrm{Gal}(E_{i+1}/E_i)$. In conclusion, $(\Gamma_i)_{0 \le i \le n+1}$ is a solvable series for $\mathrm{Gal}(F/K)$, and hence $\mathrm{Gal}(F/K)$ is solvable. The solvability of the Galois group of f over K can now be deduced without difficulty: Indeed, since f splits in $F[X]$, there exists a subfield E of F that is a splitting field of f over K; by 2.3.18, E is normal over K, and 3.2.21 then shows that $\mathrm{Gal}(E/K)$ is a homomorphic image of $\mathrm{Gal}(F/K)$, which implies that $\mathrm{Gal}(E/K)$ is solvable.

Conversely, suppose now that the Galois group of f over K is solvable. Let L be a splitting field of f over K; our assumption means that $\mathrm{Gal}(L/K)$ is solvable. Put $n = \mathrm{Card}(\mathrm{Gal}(L/K))$, and let F be a cyclotomic field of order n over L. Since $\mathrm{Char}(K) = 0$, it follows from 2.3.18 that L is Galois over K; and using 3.5.5, we see that $F = L(\zeta)$, where ζ is a primitive nth root of unity in F.

We contend that $\mathrm{Gal}(F/K(\zeta))$ is solvable. To prove this, note first that L is normal over $L \cap K(\zeta)$, because $K \subseteq L \cap K(\zeta) \subseteq L$. Furthermore, we have $F = L(\zeta) = L \vee K(\zeta)$. It then follows from 3.4.7 that $\mathrm{Gal}(F/K(\zeta))$ is isomorphic to the subgroup $\mathrm{Gal}(L/L \cap K(\zeta))$ of $\mathrm{Gal}(L/K)$. Therefore $\mathrm{Gal}(F/K(\zeta))$ is embeddable in $\mathrm{Gal}(L/K)$, and as such it is solvable.

Thus, a solvable series $(\Gamma_i)_{0 \le i \le k}$ for $\mathrm{Gal}(F/K(\zeta))$ can be chosen in such a way that Γ_{i-1}/Γ_i is cyclic for $0 < i \le k$. Letting $E_i = \mathrm{Inv}(\Gamma_i)$ for $0 \le i \le k$, we obtain an increasing sequence $(E_i)_{0 \le i \le k}$ of intermediate fields between $K(\zeta)$ and F such that $E_0 = K(\zeta)$ and $E_k = F$; and by virtue of 3.4.1, we have $\mathrm{Gal}(F/E_i) = \Gamma_i$ for $0 \le i \le k$. Applying 3.3.9 and 3.2.21, we now see that if $0 < i \le k$, then E_i is cyclic over E_{i-1} and $\mathrm{Gal}(E_i/E_{i-1})$ is isomorphic to Γ_{i-1}/Γ_i; writing $n_i = \mathrm{Card}(\Gamma_{i-1}/\Gamma_i)$, it follows from 3.4.2

that

$$[E_i : E_{i-1}] = \text{Card}(\text{Gal}(E_i/E_{i-1})) = n_i.$$

If $0 < i \leq k$, then $\zeta \in E_i$ and $n_i | n$, and 3.5.2 shows that E_i contains a primitive n_ith root of unity; therefore by 3.9.5, there exists a $\rho_i \in E_i$ such that

$$E_i = E_{i-1}(\rho_i) \quad \text{and} \quad \rho_i^{n_i} \in E_{i-1}.$$

It then follows that $(\rho_i)_{1 \leq i \leq k}$ is a radical sequence for F over $K(\zeta)$; and since $\zeta^n = 1 \in K$, we see that $(\zeta, \rho_1, \rho_2, \ldots, \rho_k)$ is a radical sequence for F over K. Thus, F is a radical extension of K; and since f splits in $F[X]$, we conclude that the equation $f(X) = 0$ is solvable by radicals over K. □

Here we have chosen to present Galois's criterion in its original form: In characteristic 0, the solvability of the Galois group is a necessary and sufficient condition for the solvability of the polynomial equation by radicals. In reality, this restriction on the characteristic can be relaxed. The necessity of the condition can be established in the general case; and its sufficiency can be proved for a polynomial of degree n provided that $n!$ is not divisible by the characteristic. The proofs of these more general assertions, in our view, would merely require few additional technicalities of little conceptual significance. For this reason, we have preferred to leave them as an exercise for the interested reader (problem 4).

With the help of Galois's theorem, it is possible to exhibit polynomial equations that are not solvable by radicals. We shall close the present section with a brief discussion of some of the classical illustrations.

3.10.5. Example. Let K be a field, and let n be a positive integer. The polynomial p_n in $K(X_1, X_2, \ldots, X_n)[Y]$ defined by

$$p_n(Y) = \sum_{i=0}^{n-1} (-1)^{n-i} X_{n-i} Y^i + Y^n$$

is said to be the **general polynomial of degree n over K**; and the equation $p_n(Y) = 0$ is said to be the **general polynomial equation of degree n over K**. The origin of this terminology lies in the fact that every monic polynomial of degree n in $K[Y]$ can be obtained by a suitable "specialization of the indeterminates $X_1, X_2, \ldots, X_n$", that is, every polynomial in $K[Y]$ of the form $\sum_{i=0}^{n-1} \alpha_i Y^i + Y^n$ is the value at $p_n(Y)$ of the K-homomorphism from $K[X_1, X_2, \ldots, X_n][Y]$ to $K[Y]$ such that

$$Y \to Y \quad \text{and} \quad X_i \to (-1)^i \alpha_{n-i} \quad \text{for } 1 \leq i \leq n.$$

We shall apply properties of the symmetric polynomials in order to prove that the Galois group of p_n over $K(X_1, X_2, \ldots, X_n)$ is the symmetric group $\text{Sym}(n)$.

Our discussion will be a continuation of that in 3.4.22, with the symbols L, E, f, and $e_1, e_2, \ldots, e_n$ retaining the same meaning. As we have seen, L is a splitting field of f over E, and

$$\text{Card}(\text{Gal}(L/E)) = [L : E] = n!.$$

Let M be a splitting field of p_n over L. We have to show that $\text{Gal}(M/L)$, viewed as a subgroup of $\text{Sym}(n)$, is identical with the whole group $\text{Sym}(n)$. To do this, it will suffice to prove that there exists an isomorphism v from M to L such that $v(L) = E$; this would then define the isomorphism $s \to v \circ s \circ v^{-1}$ from $\text{Gal}(M/L)$ to $\text{Gal}(L/E)$, and so

$$\text{Card}(\text{Gal}(M/L)) = \text{Card}(\text{Gal}(L/E)) = n!,$$

which implies that $\text{Gal}(M/L)$ and $\text{Sym}(n)$ are equipotent.

To obtain our isomorphism, note first that by 0.0.7, it is meaningful to speak of the K-isomorphism from $K[X_1, X_2, \ldots, X_n]$ to $K[e_1, e_2, \ldots, e_n]$ such that $X_i \to e_i(X_1, X_2, \ldots, X_n)$ for $1 \leq i \leq n$. Since L and E are, respectively, fields of fractions of $K[X_1, X_2, \ldots, X_n]$ and $K[e_1, e_2, \ldots, e_n]$, this is extendible to a K-isomorphism u from L to E. Note next that

$$up_n(Y) = \sum_{i=0}^{n-1} (-1)^{n-i} u(X_{n-i}) Y^i + Y^n$$

$$= \sum_{i=0}^{n-1} (-1)^{n-i} e_{n-i}(X_1, X_2, \ldots, X_n) Y^i + Y^n = f(Y).$$

Since M and L are, respectively, splitting fields of p_n over L and of f over E, and since u is an isomorphism from L to E such that $up_n = f$, it now follows from 2.3.20 that there exists an isomorphism v from M to L extending u. In particular, we have $v(L) = u(L) = E$, and our argument is complete. $\quad\square$

If n is a positive integer, the symmetric group $\text{Sym}(n)$ is solvable if and only if $n \leq 4$. Combining Galois's theorem with the result in the example just discussed, we obtain the **Ruffini–Abel theorem**: *If K is a field of characteristic 0, and if n is a positive integer such that $n > 4$, then the general polynomial equation of degree n over K is not solvable by radicals over $K(X_1, X_2, \ldots, X_n)$.*

The unsolvability by radicals expressed by the Ruffini–Abel theorem may be interpreted as follows. Given a field K of characteristic 0 and an integer $n > 4$, no "general formula" involving exclusively rational operations and extraction of roots can be found that, when applied to elements of K, yields the roots of every polynomial equation of degree n with coefficients in K.

Note that this does not mean that every polynomial equation of degree greater than 4 is unsolvable by radicals. As an extreme example of the opposite, consider the case of the field $\mathbf{R}$: The roots of every polynomial equation with coefficients in $\mathbf{R}$ belong to $\mathbf{C}$, which is a radical extension of

R; therefore, every polynomial equation with coefficients in **R** is solvable by radicals over **R**.

Although the Ruffini-Abel theorem proves the existence of polynomial equations unsolvable by radicals, it does not tell us how to find such equations with coefficients in a given field. In the next and final example, we shall consider a special type of polynomial equation with coefficients in **Q**.

3.10.6. Example.   Let p be a prime, and let f be an irreducible polynomial in **Q**$[X]$ of degree p with exactly $p-2$ zeros in **R**. Then the Galois group of f over **Q** is Sym(p); and hence the equation $f(X)=0$ is solvable by radicals over **Q** if and only if $p=2$ or $p=3$.

Let G denote the Galois group of f over **Q**. We shall show that G contains an element of order p and a transposition. To do this, let us denote by $\alpha_1, \alpha_2, \ldots, \alpha_p$ the complex zeros of f, with $\alpha_1, \alpha_2 \notin$ **R**; then put $F =$ **Q**$(\alpha_1, \alpha_2, \ldots, \alpha_p)$, so that F is a splitting field of f over **Q**, with Gal($F/$**Q**) identified with G in the usual way.

Note first that

$$\mathbf{Q} \subseteq \mathbf{Q}(\alpha_1) \subseteq F \quad \text{and} \quad [\mathbf{Q}(\alpha_1):\mathbf{Q}] = \deg(f) = p;$$

therefore $p \mid [F:\mathbf{Q}]$, which means that $p \mid$ Card(G). Applying Sylow's theorem, we see that G contains a subgroup of order p, and hence an element of order p.

Next, let us now denote complex conjugation by s. Since F is normal over **Q**, we have $s(F) = F$, and so $s_F \in$ Gal($F/$**Q**). For $1 \leq i \leq p$, we have $s(\alpha_i) = \alpha_i$ if and only if $\alpha_i \in$ **R**. Consequently

$$s_F(\alpha_1) = \alpha_2, \quad s_F(\alpha_2) = \alpha_1, \quad \text{and} \quad s_F(\alpha_i) = \alpha_i \quad \text{for} \quad 2 < i \leq p,$$

which shows that G contains the transposition (12).

To conclude, let c be an element of order p in G. Then c is a p-cycle, and so there exists a positive integer k for which $c^k = (12\ldots)$. Since $c^k \in G$ and $(12) \in G$, a suitable change of notation allows us to assume that $(123\ldots p) \in G$ and $(12) \in G$; and since

$$(123\ldots p)(12)(123\ldots p)^{-1} = (23)$$

$$(123\ldots p)(23)(123\ldots p)^{-1} = (34)$$

$$\vdots \quad \vdots \quad \vdots \qquad \quad \vdots$$

$$(123\ldots p)(p-2 \ \ p-1)(123\ldots p)^{-1} = (p-1 \ \ p).$$

we see that $(12),(23),\ldots,(p-1 \ \ p) \in G$, whence $G =$ Sym(p). $\square$

It is not difficult to give examples of polynomials in **Q**$[X]$ satisfying the conditions of this example. In fact,

$$X^5 - 4X + 2 \quad \text{and} \quad 2X^5 - 5X^4 + 5$$

are two such polynomials: Their irreducibility in $Q[X]$ follows immediately from Eisenstein's criterion; and with a little help from calculus, it is readily seen that each of them admits exactly three real zeros.

More generally, an explicit construction can be given to obtain this type of polynomial in $Q[X]$ of every possible prime degree (problem 5).

PROBLEMS

1. Let K be a field, and let L be a radical separable extension of K. Show that if $(\alpha_i)_{1 \le i \le n}$ is a radical sequence for L over K, then there exists a sequence $(k_i)_{1 \le i \le n}$ of positive integers such that:
   a. $\alpha_1^{k_1} \in K$.
   b. $\alpha_i^{k_i} \in K(\alpha_1, \alpha_2, \ldots, \alpha_{i-1})$ for $1 < i \le n$.
   c. $\mathrm{Char}(K) \nmid k_i$ for $1 \le i \le n$.

2. Let K be a field, let L be an extension field of K, and let M be an extension field of L that is normal over K. Prove that if $\mathrm{Gal}(M/K)$ is solvable, then $\mathrm{Gal}(L/K)$ is solvable.

3. Let K be a field, and let L be a finite extension of K. Prove that if L admits an extension field that is a radical extension of K, then $\mathrm{Gal}(L/K)$ is solvable. Prove also that, conversely, if $\mathrm{Gal}(L/K)$ is solvable, if L is Galois over K, and if $\mathrm{Char}(K) \nmid [L:K]$, then L admits an extension field that is a radical extension of K.

4. Let K be a field, and let f be a nonconstant polynomial in $K[X]$. Show that if the equation $f(X) = 0$ is solvable by radicals over K, then the Galois group of f over K is solvable. Show also that, conversely, if the Galois group of f over K is solvable and $\mathrm{Char}(K) \nmid \deg(f)!$, then the equation $f(X) = 0$ is solvable by radicals over K.

   (Thus, Galois's criterion is valid in situations more general than that described in 3.10.4. And accordingly, the Ruffini-Abel theorem holds under more general conditions than those stated in the text.)

5. Let p be a prime such that $p \ge 5$, and let $(k_i)_{1 \le i \le p-2}$ be a strictly increasing sequence of even positive integers. Prove that if m is an even integer such that $m > \sum_{i=1}^{p-2} k_i^2$, then the polynomial

$$(X^2 + m)\left(\prod_{i=1}^{p-2} (X - k_i) \right)$$

in $Q[X]$ satisfies the conditions described in 3.10.6 (i.e., it is irreducible in $Q[X]$ and has exactly $p - 2$ real zeros), and hence its Galois group is $\mathrm{Sym}(p)$. (This is due to Brauer.)

3.11. FINITE FIELDS

We now are in a position to derive in rapid succession the basic general properties of finite fields. These fields were discovered and investigated by Galois, and are often referred to as **Galois fields**.

Finite fields play an important role in several areas of pure mathematics. In recent years, moreover, interesting applications of finite fields have been discovered. A substantial effort would indeed be required to give an adequate treatment of these developments. In this book, we shall not attempt such an undertaking; instead, we shall content ourselves with the discussion of those results on finite fields bearing directly on our presentation of the general theory.

3.11.1. Proposition. *Let K be a finite field of cardinality q. Then $\alpha^{q-1} = 1$ for every $\alpha \in K^*$, and $\alpha^q = \alpha$ for every $\alpha \in K$. Moreover, the equalities*

$$X^{q-1} - 1 = \prod_{\alpha \in K^*} (X - \alpha) \quad and \quad X^q - X = \prod_{\alpha \in K} (X - \alpha)$$

hold in $K[X]$.

Proof. Since K^* is a group of order $q - 1$, we see that $\alpha^{q-1} = 1$ for every $\alpha \in K^*$; and this implies that $\alpha^q = \alpha$ for every $\alpha \in K$.

Thus, every element of K^* is a zero of $X^{q-1} - 1$, and every element of K is a zero of $X^q - X$. Consequently, $\prod_{\alpha \in K^*}(X - \alpha)$ and $\prod_{\alpha \in K}(X - \alpha)$ divide, respectively, $X^{q-1} - 1$ and $X^q - X$ in $K[X]$. Since these four polynomials are monic, and since

$$\deg\left(\prod_{\alpha \in K^*} (X - \alpha) \right) = q - 1 = \deg(X^{q-1} - 1)$$

and

$$\deg\left(\prod_{\alpha \in K} (X - \alpha) \right) = q = \deg(X^q - X),$$

the two stated polynomial equalities in $K[X]$ now follow at once. □

3.11.2. Proposition. *Let K be a field. If p is a prime and n is a positive integer, and if P denotes the prime subfield of K, then the following conditions are equivalent:*
  (a) $\mathrm{Card}(K) = p^n$.
  (b) $\mathrm{Char}(K) = p$ and $[K : P] = n$.
  (c) $\mathrm{Char}(K) = p$ and K is a splitting field of $X^{p^n} - X$ over P.

Proof. If K is finite, then K is finite over P and $\mathrm{Char}(K) = \mathrm{Card}(P)$, whence

$$\mathrm{Card}(K) = \mathrm{Card}(P)^{[K:P]} = \mathrm{Char}(K)^{[K:P]};$$

and since $\mathrm{Char}(K)$ is a prime, the equivalence of (a) and (b) is evident.

To show that (a) implies (c), assume that $\mathrm{Card}(K) = p^n$. Since (a) and (b) are equivalent, we have $\mathrm{Char}(K) = p$. Furthermore, according to 3.11.1, the equality $X^{p^n} - X = \prod_{\alpha \in K}(X - \alpha)$ holds in $K[X]$, so that $X^{p^n} - X$ splits in $K[X]$ and K is the set of all zeros of $X^{p^n} - X$ in K. This clearly implies that K is a splitting field of $X^{p^n} - X$ over P.

Finally, to verify that (c) implies (a), assume that $\mathrm{Char}(K) = p$ and that K is a splitting field of $X^{p^n} - X$ over P. Let D denote the set of all zeros of $X^{p^n} - X$ in K. Since $(X^{p^n} - X)' = -1$, we see that $X^{p^n} - X$ is separable and splits in $K[X]$, whence $\mathrm{Card}(D) = p^n$. Moreover, since D consists of the fixed points of the endomorphism $\alpha \to \alpha^{p^n}$ of K, it is a subfield of K. But then $P \subseteq D$, so that $K = P(D) = D$, and hence $\mathrm{Card}(K) = p^n$. $\square$

The following result is an immediate consequence of the preceding two propositions.

3.11.3. Proposition. *Let K be a finite field of cardinality q, and let L be an extension field of K. If k is a nonnegative integer, then the mapping $\alpha \to \alpha^{q^k}$ from L to L is a K-endomorphism; and it is a K-automorphism if L is perfect.*

3.11.4. Examples

Let K be a finite field of cardinality q, and let $K(\xi)$ be a simple transcendental extension of K. We shall now combine 3.2.7 and 3.4.1 with the preceding propositions in order to determine the fields of invariants of certain subgroups of $\mathrm{Gal}(K(\xi)/K)$.

a. The mapping from K^* to $\mathrm{Gal}(K(\xi)/K)$ assigning to each $\alpha \in K^*$ the K-automorphism of $K(\xi)$ such that $\xi \to \alpha\xi$ is a monomorphism, and the field of invariants of its image is $K(\xi^{q-1})$.

The mapping described in our assertion obviously is a monomorphism. Let Γ denote its image, so that Γ is a subgroup of order $q - 1$ of $\mathrm{Gal}(K(\xi)/K)$. Applying 3.4.1, we obtain $[K(\xi):\mathrm{Inv}(\Gamma)] = q - 1$. Also, note that ξ is a zero of the polynomial $X^{q-1} - \xi^{q-1}$ in $K(\xi^{q-1})[X]$, whence $[K(\xi): K(\xi^{q-1})] \leq q - 1$.

For every $\alpha \in K^*$, we know that $\alpha^{q-1} = 1$, and hence

$$(\alpha\xi)^{q-1} = \alpha^{q-1}\xi^{q-1} = \xi^{q-1}.$$

Consequently, $\xi^{q-1} \in \mathrm{Inv}(\Gamma)$, which implies that $K(\xi^{q-1}) \subseteq \mathrm{Inv}(\Gamma)$. Taking into account the equality and the inequality in the preceding paragraph, we see that the desired equality $K(\xi^{q-1}) = \mathrm{Inv}(\Gamma)$ follows at once.

b. The mapping from K^+ to $\mathrm{Gal}(K(\xi)/K)$ assigning to each $\alpha \in K^+$ the K-automorphism of $K(\xi)$ such that $\xi \to \xi + \alpha$ is a monomorphism, and the field of invariants of its image is $K(\xi^q - \xi)$.

As in example a, it is clear that the mapping in question is a monomorphism. If Δ denotes its image, then Δ is a subgroup of order q of $\mathrm{Gal}(K(\xi)/K)$. According to 3.4.1, we have $[K(\xi):\mathrm{Inv}(\Delta)] = q$. Moreover, since ξ is a zero of the polynomial $X^q - X - (\xi^q - \xi)$ in $K(\xi^q - \xi)[X]$, we see that $[K(\xi):K(\xi^q - \xi)] \leq q$.

Our conclusion will be established, exactly as in example a, if we verify the inclusion $K(\xi^q - \xi) \subseteq \mathrm{Inv}(\Delta)$. To do this, note that for each $\alpha \in K^+$ we have $\alpha^q = \alpha$, whence

$$(\xi + \alpha)^q - (\xi + \alpha) = (\xi^q + \alpha^q) - (\xi + \alpha)$$
$$= (\xi^q - \xi) + (\alpha^q - \alpha) = \xi^q - \xi.$$

But this means that $\xi^q - \xi \in \mathrm{Inv}(\Delta)$; therefore $K(\xi^q - \xi) \subseteq \mathrm{Inv}(\Delta)$, which is what we wanted.

c. $\mathrm{Inv}(\mathrm{Gal}(K(\xi)/K)) = K((\xi^{q^2} - \xi)^{q+1}/(\xi^q - \xi)^{q^2+1})$.

To verify this, we shall argue as in examples a and b. This time, however, some tedious computations will be needed.

We saw in 3.2.7 that $\mathrm{Gal}(K(\xi)/K)$ is of order $q^3 - q$. By virtue of 3.4.1, it then follows that

$$[K(\xi):\mathrm{Inv}(\mathrm{Gal}(K(\xi)/K))] = q^3 - q.$$

Now put

$$\sigma = (\xi^{q^2} - \xi)^{q+1}/(\xi^q - \xi)^{q^2+1}.$$

Note that

$$\sigma = \left((\xi^{q^2} - \xi)^{q+1}/(\xi^q - \xi)^{q+1}\right)/(\xi^q - \xi)^{q^2-q}$$

$$= \left(((\xi^{q^2} - \xi^q) + (\xi^q - \xi))/(\xi^q - \xi)\right)^{q+1}/(\xi^q - \xi)^{q^2-q}$$

$$= \left(((\xi^q - \xi)^q + (\xi^q - \xi))/(\xi^q - \xi)\right)^{q+1}/(\xi^q - \xi)^{q^2-q}$$

$$= \left((\xi^q - \xi)^{q-1} + 1\right)^{q+1}/(\xi^q - \xi)^{q^2-q},$$

and hence

$$\left((\xi^q - \xi)^{q-1} + 1\right)^{q+1} - \sigma(\xi^q - \xi)^{q^2-q} = 0.$$

Thus, ξ is a zero of the polynomial $((X^q - X)^{q-1} + 1)^{q+1} - \sigma(X^q - X)^{q^2-q}$ in $K(\sigma)[X]$, which implies that $[K(\xi):K(\sigma)] \leq q^3 - q$.

To conclude, it remains for us to verify that $K(\sigma) \subseteq \mathrm{Inv}(\mathrm{Gal}(K(\xi)/K))$. To this end, we first recall that $\alpha^q = \alpha = \alpha^{q^2}$ for every

$\alpha \in K$. If $\alpha \in K^*$, then

$$\frac{\left((\alpha\xi)^{q^2}-(\alpha\xi)\right)^{q+1}}{\left((\alpha\xi)^{q}-(\alpha\xi)\right)^{q^2+1}}=\frac{\left(\alpha\xi^{q^2}-\alpha\xi\right)^{q+1}}{\left(\alpha\xi^{q}-\alpha\xi\right)^{q^2+1}}$$

$$=\frac{\alpha^{q+1}\left(\xi^{q^2}-\xi\right)^{q+1}}{\alpha^{q^2+1}\left(\xi^{q}-\xi\right)^{q^2+1}}$$

$$=\frac{\alpha^2\left(\xi^{q^2}-\xi\right)^{q+1}}{\alpha^2\left(\xi^{q}-\xi\right)^{q^2+1}}=\sigma$$

Similarly, if $\alpha \in K^+$, then

$$\frac{\left((\xi+\alpha)^{q^2}-(\xi+\alpha)\right)^{q+1}}{\left((\xi+\alpha)^{q}-(\xi+\alpha)\right)^{q^2+1}}=\frac{\left((\xi^{q^2}+\alpha^{q^2})-(\xi+\alpha)\right)^{q+1}}{\left((\xi^{q}+\alpha^{q})-(\xi+\alpha)\right)^{q^2+1}}$$

$$=\frac{\left(\xi^{q^2}-\xi\right)^{q+1}}{\left(\xi^{q}-\xi\right)^{q^2+1}}=\sigma.$$

Finally, note that

$$\frac{\left((1/\xi)^{q^2}-(1/\xi)\right)^{q+1}}{\left((1/\xi)^{q}-(1/\xi)\right)^{q^2+1}}=\frac{\left((\xi-\xi^{q^2})/\xi^{q^2+1}\right)^{q+1}}{\left((\xi-\xi^{q})/\xi^{q+1}\right)^{q^2+1}}$$

$$=\frac{\left(\xi^{q^2}-\xi\right)^{q+1}}{\left(\xi^{q}-\xi\right)^{q^2+1}}=\sigma.$$

If Γ and Δ are as in examples a and b, what has just been shown means that σ is a fixed point of every element of Γ, of every element of Δ, and of the K-automorphism of $K(\xi)$ such that $\xi \to 1/\xi$. On the other hand, it has been proved in 3.2.7 that every K-automorphism of $K(\xi)$ sends ξ to an element of $K(\xi)$ of the form $(\alpha\xi+\beta)/(\gamma\xi+\delta)$, where $\begin{bmatrix} \alpha & \beta \\ \gamma & \delta \end{bmatrix} \in GL_2(K)$; therefore, it is expressible as the product of a finite sequence of K-automorphisms of $K(\xi)$, each of which is of one of the three aforementioned types. It then follows that $\sigma \in \mathrm{Inv}(\mathrm{Gal}(K(\xi)/K))$, and hence $K(\sigma) \subseteq \mathrm{Inv}(\mathrm{Gal}(K(\xi)/K))$. $\square$

The examples discussed in 3.2.7 and 3.11.4 provide a good illustration of the contrasting behavior that finite and infinite fields may exhibit in certain situations.

As we have seen, the cardinality of a finite field is a prime power. The following remarkable result states the existence and essential uniqueness of a field with a prescribed prime power as its cardinality.

3.11.5. Theorem (Moore). *If p is a prime and n is a positive integer, then there exists a field of cardinality p^n, and every two such fields are isomorphic.*

Proof. According to 3.11.2, every splitting field of $X^{p^n} - X$ over $\mathbf{Z}/p$ has cardinality p^n. The first assertion then follows immediately from 2.3.21.

To prove the second, suppose that K and $\bar{K}$ are fields of cardinality p^n, and let P and $\bar{P}$ denote, respectively, their prime subfields. By 3.11.2, we have $\mathrm{Char}(K) = p = \mathrm{Char}(\bar{K})$, so that P and $\bar{P}$ are isomorphic to $\mathbf{Z}/p$. Consequently, P and $\bar{P}$ are isomorphic; and furthermore, we know from 3.11.2 that K and $\bar{K}$ are, respectively, splitting fields of $X^{p^n} - X$ over P and $\bar{P}$. Applying 2.3.20, we now conclude that K and $\bar{K}$ are isomorphic.   □

It follows from the foregoing results that if two finite fields are equipotent, then they are isomorphic. This is of course false for infinite fields: **Q** and **A** are equipotent, but not isomorphic; and the same holds for **R** and **C**.

3.11.6. Proposition. *Let p be a prime, and let m and n be positive integers.*

(i) *If K and L are, respectively, fields of cardinalities p^m and p^n, and if K is a subfield of L, then $m \mid n$ and $[L : K] = n/m$.*

(ii) *If $m \mid n$, and if L is a field of cardinality p^n, then L possesses a unique subfield of cardinality p^m; in fact, the mapping $\alpha \to \alpha^{p^m}$ from L to L is an automorphism, and the elements of this subfield are its fixed points.*

Proof. The verification of (i) is easy: If K and L are as indicated in its statement, and if P denotes the prime subfield of K, then 3.11.2 shows that

$$n = [L : P] = [L : K][K : P] = [L : K]m,$$

which implies the desired conclusion.

To prove (ii), assume that $m \mid n$ and that L is a field of cardinality p^n. Let P denote the prime subfield of L. It then follows from 3.11.2 that $\mathrm{Char}(L) = p$ and that L is a splitting field of $X^{p^n} - X$ over P. Since L is perfect, the mapping $\alpha \to \alpha^{p^m}$ from L to L is an automorphism. Let K denote the subfield of L consisting of the fixed points of this automorphism. We shall show that K is the only subfield of L having cardinality p^m.

First, let $d = n/m$; then

$$p^n - 1 = p^{md} - 1 = (p^m)^d - 1,$$

which implies that $(p^m - 1) \mid (p^n - 1)$. Next let $e = (p^n - 1)/(p^m - 1)$; in $P[X]$ we then have

$$X^{p^n - 1} - 1 = X^{(p^m - 1)e} - 1 = \left(X^{p^m - 1} \right)^e - 1,$$

so that $X^{p^{m-1}} - 1$ divides $X^{p^{n-1}} - 1$ in $P[X]$, and hence $X^{p^m} - X$ divides $X^{p^n} - X$ in $P[X]$. Since $X^{p^n} - X$ splits in $L[X]$, it now follows that $X^{p^m} - X$ splits in $L[X]$; and since K consists of the zeros of $X^{p^m} - X$ in L, we conclude that K is a splitting field of $X^{p^m} - X$ over P. By 3.11.2, this implies that $\mathrm{Card}(K) = p^m$.

Finally, suppose that E is a subfield of L of cardinaltiy p^m. By 3.11.1, we then have $\alpha^{p^m} = \alpha$ for every $\alpha \in E$, which shows that $E \subseteq K$. Since E and K are finite and equipotent, we now conclude that $E = K$. $\qquad\square$

3.11.7. Proposition. *If K is a finite field and d is a positive integer, then K admits an extension field of linear degree d over K, and every two such extension fields of K are K-isomorphic.*

Proof. This can be obtained as a consequence of 3.11.2, 3.11.5, and 3.11.6. Write $\mathrm{Card}(K) = p^m$, where $p = \mathrm{Char}(K)$ and m is a positive integer.

There exists a field of cardinality p^{md}; and every such field contains a subfield of cardinality p^m, hence a subfield isomorphic to K. Therefore K is embeddable in a field of cardinality p^{md}. According to 0.0.1, this implies the existence of an extension field of K of cardinality p^{md}; and for every such extension field F of K, we have $[F : K] = md / m = d$.

To conclude, suppose that L and M are extension fields of K such that $[L : K] = d = [M : K]$. We then have $\mathrm{Card}(L) = p^{md} = \mathrm{Card}(M)$. Consequently, L and M are splitting fields of $X^{p^{md}} - X$ over the prime subfield of K, and hence also over K. By 2.3.21, it now follows that L and M are K-isomorphic. $\qquad\square$

It is easy to see that the proposition just proved does not remain valid for infinite fields (problem 11).

As shown by the next proposition and its corollary, the Galois theory of finite fields can be described completely in very simple terms.

3.11.8. Proposition. *Let K be a finite field, and let L be a finite extension of K. Then L is cyclic over K; in fact, if $q = \mathrm{Card}(K)$, then $\mathrm{Gal}(L/K)$ admits the K-automorphism $\alpha \to \alpha^q$ of L as a generator.*

Proof. Let $p = \mathrm{Char}(K)$. Then $q = \mathrm{Card}(K) = p^m$ and $\mathrm{Card}(L) = p^n$, where m and n are positive integers; and if we let $d = [L : K]$, it follows from 3.11.6 that $n = md$.

Let s denote the K-automorphism $\alpha \to \alpha^{p^m}$ of L. According to 3.11.6, we have $K = \mathrm{Inv}(\{s\})$. Consequently, L is a finite Galois extension of K, and

$$\mathrm{Card}(\mathrm{Gal}(L/K)) = [L : K] = d.$$

To conclude, it now suffices to show that s is an element of order d in $\mathrm{Gal}(L/K)$. To do this, note that for every nonnegative integer k, the element s^k of $\mathrm{Gal}(L/K)$ is the K-automorphism $\alpha \to \alpha^{p^{mk}}$ of L.

If $\alpha \in L$, we see from 3.11.1 that

$$s^d(\alpha) = \alpha^{p^{md}} = \alpha^{p^n} = \alpha.$$

Therefore $s^d = i_L$. Finally, assume that $1 \le k < d$. Then $mk < md = n$, so that $p^{mk} < p^n$. If $s^k = i_L$, we would have $\alpha^{p^{mk}} = s^k(\alpha) = \alpha$ for every $\alpha \in L$, and every element of L would be a zero of $X^{p^{mk}} - X$; but this is impossible, because

$$\text{Card}(L) = p^n > p^{mk} = \deg(X^{p^{mk}} - X).$$

It then follows that $s^k \ne i_L$, which is what was needed.   □

3.11.9. Corollary. *If K is a finite field then* $\text{Aut}(K)$ *is cyclic, and admits the Frobenius mapping of K as a generator.*

Proof. Let P denote the prime subfield of K. Then $\text{Aut}(K) = \text{Gal}(K/P)$ and $\text{Char}(K) = \text{Card}(P)$, and the conclusion follows at once from the proposition.   □

In order to prove the next result, we shall use the fact that the multiplicative group of every finite field is cyclic. This property, which is an obvious consequence of 3.5.1, actually characterizes finite fields (problem 10).

3.11.10. Proposition. *If K is a finite field and d is a positive integer, then there exists an irreducible polynomial in $K[X]$ of degree d.*

Proof. By virtue of 3.11.7, there exists an extension field L of K such that $[L:K] = d$. Since L is finite, the group L^* is cyclic. If α is a generator of L^*, it is clear that $L = K(\alpha)$; and if f denotes the minimal polynomial of α over K, then f is irreducible in $K[X]$, and $\deg(f) = [K(\alpha):K] = d$.   □

The hypothesis of finiteness is essential for the validity of the preceding proposition. We have seen, for example, that if K is an algebraically closed field, then the irreducible polynomials in $K[X]$ are those of degree 1.

We shall now conclude this section with a comment on notation. It has been shown that if q is a prime power and $q > 1$, then there exists a field of cardinality q, and that every two such fields are isomorphic. This justifies the common use of the symbols $\mathbf{F}_q$ and $GF(q)$ to denote "the" field of cardinality q; $\mathbf{F}_q$ stands for "finite field with q elements", and $GF(q)$ for "Galois field with q elements".

PROBLEMS

1. Let p be a prime. Use 3.11.1 to verify the following assertions:
    a. If n is an integer such that $p \nmid n$, then $n^{p-1} \equiv 1 \pmod{p}$.
    b. If n is an integer, then $n^p \equiv n \pmod{p}$.

c. $(p-1)! \equiv -1 \pmod{p}$.
(Assertions a and b are known as **Fermat's little theorem**, and assertion c as **Wilson's theorem**.)

2. Prove that if K is a finite field, and if L is an algebraic extension of K, then every element of L^* is a root of unity in L.

3. Use Lagrange's interpolation formula in order to prove that if K is a finite field, and if n is a positive integer, then every mapping from $K^{(n)}$ to K is a polynomial mapping.

4. Let p be a prime. Use 3.11.6 to verify that if m and n are positive integers such that $(p^m - 1)|(p^n - 1)$, then $m|n$.

5. Let n be an odd positive integer, and let K be a field of cardinality 2^n. Prove that if $\alpha, \beta \in K$ and $\alpha^2 + \alpha\beta + \beta^2 = 0$, then $\alpha = 0 = \beta$.

6. Let p be a prime. Use 3.11.8 to verify that $n|\varphi(p^n - 1)$ for every positive integer n.

7. Let K be a finite field of cardinality q. Verify the following assertions:
   a. If n is a positive integer such that $(q-1)|n$, then $\sum_{\alpha \in K} \alpha^n = -1$.
   b. If n is a positive integer such that $(q-1) \nmid n$, then $\sum_{\alpha \in K} \alpha^n = 0$.

8. Let K be a finite field of prime characteristic p, and let n be a positive integer. Show that if $(f_i)_{i \in I}$ is a nonempty finite family of nonzero polynomials in $K[X_1, X_2, \ldots, X_n]$ such that $\sum_{i \in I} \deg(f_i) < n$, and if V denotes the set of all $(\alpha_1, \alpha_2, \ldots, \alpha_n) \in K^{(n)}$ such that $f_i(\alpha_1, \alpha_2, \ldots, \alpha_n) = 0$ for every $i \in I$, then $p | \mathrm{Card}(V)$. (This result is known as **Warning's theorem**.)

9. Let K be a finite field, and let n be a positive integer. Prove that if $(f_i)_{i \in I}$ is a nonempty finite family of nonzero polynomials without constant term in $K[X_1, X_2, \ldots, X_n]$ such that $\sum_{i \in I} \deg(f_i) < n$, then there exists an $(\alpha_1, \alpha_2, \ldots, \alpha_n) \in K^{(n)}$ such that $(\alpha_1, \alpha_2, \ldots, \alpha_n) \neq (0, 0, \ldots, 0)$ and $f_i(\alpha_1, \alpha_2, \ldots, \alpha_n) = 0$ for every $i \in I$. (The particular case of this result in which the index set I consists of a single element is known as **Chevalley's theorem**.)

10. Show that if K is an infinite field, then the group K^* is not cyclic.

11. Prove that there exist infinite fields K of every possible characteristic admitting two finite extensions that are not K-isomorphic, but have the same linear degree over K.

12. Let F be an algebraically closed field of prime characteristic p. Prove that for every positive integer n, the mapping $\alpha \to \alpha^{p^n}$ from F to F is an automorphism, and its fixed points make up the only subfield of F of cardinality p^n.

13. Let K be a finite field of cardinality q, and let L be a finite extension of K. Show that if k is a positive integer, and if r is the highest common factor of $[L:K]$ and k, then the fixed points of the K-automorphism $\alpha \to \alpha^{q^k}$ of L make up the subfield of L of cardinality q^r.

14. Let F be an algebraically closed field. Prove the following assertions:
   a. If F has characteristic 0, then there exists no nonzero polynomial in $F[X]$ whose zeros in F form a subfield of F.

b. If F has prime characteristic p, and if f is a monic separable polynomial in $F[X]$ whose zeros in F form a subfield of F, then $f(X) = X^{p^n} - X$ for some positive integer n.

15. Let K be a finite field of cardinality q, let f be an irreducible polynomial in $K[X]$, and let n be a positive integer. Show that f divides $X^{q^n} - X$ in $K[X]$ if and only if $\deg(f)|n$.

16. Let F be an algebraically closed field of prime characteristic. Show that there exists an infinite subfield of F that is algebraic over the prime subfield of F, but is not algebraically closed.

17. Let K be a finite field of cardinality q, let L be an algebraic closure of K, and let $s \in \mathrm{Gal}(L/K)$. Prove that if E is a finite intermediate field between K and L, then there exists a positive integer k such that $s(\alpha) = \alpha^{q^k}$ for every $\alpha \in E$.

18. Let K be a finite field, and let L be an algebraic closure of K. Prove that L is abelian over K, and that i_L is the only element of finite order in $\mathrm{Gal}(L/K)$.

19. Show that if p and q are primes, and if C_p and C_q are, respectively, algebraic closures of prime fields of characteristic p and q, then $\mathrm{Aut}(C_p)$ and $\mathrm{Aut}(C_q)$ are isomorphic.

20. Prove that if K is a finite field, and if L is a finite extension of K, then $N_{L/K}$ and $T_{L/K}$ are surjective.

3.12. INFINITE GALOIS THEORY

As we have seen, every finite group of automorphisms of a field is a Galois group. Also, in the Galois correspondence defined by a finite Galois extension, every intermediate field is related to a subgroup of the Galois group, and every subgroup of the Galois group is related to an intermediate field.

It was discovered by Dedekind that when the above finiteness restrictions are omitted, the assertions no longer remain valid. In other words, an infinite group of automorphisms of a field need not be a Galois group. And, although in the Galois correspondence defined by an infinite Galois extension every intermediate field is related to a subgroup of the Galois group, not every subgroup of the Galois group need be related to an intermediate field.

3.12.1. Example. Let P be a prime field of prime characteristic, and let F be an algebraic closure of P. Since P is finite, it follows that F is an infinite Galois extension of P. Let Γ denote the subgroup of $\mathrm{Aut}(F)$ generated by the Frobenius mapping of F. We shall now show that Γ is not a Galois group on F; this will show, in particular, that Γ is a subgroup of $\mathrm{Gal}(F/P)$ that, in the Galois correspondence, is not related to an intermediate field between P and F.

The discussion will require most of the properties of finite fields proved previously, which we shall apply without explicit comment. First, let us write $p = \text{Char}(P)$, so that P and $\mathbf{Z}/p$ are isomorphic. It is readily seen that if n is a positive integer, then F contains a unique subfield of cardinality p^n: Indeed, the polynomial $X^{p^n} - X$ in $P[X]$ splits in $F[X]$; if D denotes the set of all its zeros in F, then $P(D)$ obviously is the only subfield of F that is a splitting field of $X^{p^n} - X$ over P, and this amounts to saying that $P(D)$ is the only subfield of F of cardinality p^n.

For every positive integer n, let E_n denote the subfield of F of cardinality p^{2^n}. Since $2^n | 2^{n+1}$ for every $n \geq 1$, it follows that $(E_n)_{n \geq 1}$ is a strictly increasing sequence of subfields of F. Now write $E = \cup_{n=1}^{\infty} E_n$, so that E is an infinite subfield of F. We cannot have $E = F$, since this equality would imply that the subfield of F of cardinality p^3 is contained in E_n for some $n \geq 1$, and this would lead to the contradictory conclusion that $3|2^n$. Therefore $E \subset F$.

Since F is Galois over P, it also is Galois over E, whence $E = \text{Inv}(\text{Gal}(F/E))$. In view of the strict inclusion $E \subset F$, this equality implies that $\text{Gal}(F/E)$ is nontrivial. If s is an E-automorphism of F such that $s \neq i_F$, then there exists no positive integer n such that $s(\alpha) = \alpha^{p^n}$ for every $\alpha \in F$: otherwise every element of E, being a fixed point of s, would be a zero of $X^{p^n} - X$, which is incompatible with the infiniteness of E.

It now follows that $\text{Gal}(F/E) \not\subseteq \Gamma$, and so $\Gamma \neq \text{Aut}(F)$. Since P consists of the fixed points of the Frobenius mapping of F, we now have

$$\text{Inv}(\Gamma) = P = \text{Inv}(\text{Aut}(F)).$$

Thus, Γ and $\text{Aut}(F)$ are two distinct groups of automorphisms of F with the same field of invariants; and $\text{Aut}(F)$ is a Galois group on F. We conclude that Γ is not a Galois group on F, which is what we wanted.   □

In a certain sense, the situation illustrated by the preceding example is typical. It can be shown in effect that, in the Galois correspondence defined by an infinite Galois extension, there always exist subgroups of the Galois group that are not related to intermediate fields (problem 5).

Several decades after Dedekind's negative discovery, Krull defined a topology for the Galois groups of infinite Galois extensions, and proved that the subgroups of the Galois groups related to intermediate fields in the Galois correspondence are precisely the *closed* subgroups.

We shall now proceed to develop the topological notions required for our presentation of the Galois theory of infinite extensions. Instead of following Krull's original approach, we shall begin by defining a topology for every group of field automorphisms.

Let F be a field, and let s be an automorphism of F. For every finite subset A of F, we denote by $\Omega_s(A)$ the set of all automorphisms of F that agree with s on A, and say that $\Omega_s(A)$ is the **basic set at s defined by A.** If n

is a positive integer and $\alpha_1, \alpha_2, \ldots, \alpha_n \in F$, we write $\Omega_s(\alpha_1, \alpha_2, \ldots, \alpha_n)$ instead of $\Omega_s(\{\alpha_1, \alpha_2, \ldots, \alpha_n\})$.

The following assertions are evident:

(i) $\Omega_s(\phi) = \text{Aut}(F)$.

(ii) If A is a finite subset of F, then $s \in \Omega_s(A)$.

(iii) If A and B are finite subsets of F, then

$$\Omega_s(A) \cap \Omega_s(B) = \Omega_s(A \cup B).$$

Let F be a field. It follows from the foregoing remarks that for every $s \in \text{Aut}(F)$, the basic sets at s form a filter base, and s belongs to each basic set at s. Consequently, there exists a unique topology on $\text{Aut}(F)$ with respect to which, for every $s \in \text{Aut}(F)$, the basic sets at s form a fundamental system of neighborhoods of s. This topology is called the **finite topology on** $\text{Aut}(F)$.

The usefulness of the finite topology in the study of groups of field automorphisms lies in the fact that it is suitably related to the group structure. This is stated in precise terms in the following proposition.

3.12.2. Proposition. *If F is a field, then the group structure and the finite topology on* $\text{Aut}(F)$ *are compatible.*

Proof. It suffices to prove that the mapping $(s, t) \to st^{-1}$ from $\text{Aut}(F) \times \text{Aut}(F)$ to $\text{Aut}(F)$ is continuous. Let $(u, v) \in \text{Aut}(F) \times \text{Aut}(F)$; we shall show that our mapping is continuous at (u, v). To do this, it is sufficient to verify that for every finite subset A of F, the image of $\Omega_u(v^{-1}(A)) \times \Omega_v(v^{-1}(A))$ by the mapping in question is contained in $\Omega_{uv^{-1}}(A)$. Thus, let $(s, t) \in \Omega_u(v^{-1}(A)) \times \Omega_v(v^{-1}(A))$. If $\alpha \in A$, then $v^{-1}(\alpha) \in v^{-1}(A)$; since t and v agree on $v^{-1}(A)$, we have $t(v^{-1}(\alpha)) = v(v^{-1}(\alpha)) = \alpha$, whence $v^{-1}(\alpha) = t^{-1}(\alpha)$; and since s and u agree on $v^{-1}(A)$, we see that $s(t^{-1}(\alpha)) = s(v^{-1}(\alpha)) = u(v^{-1}(\alpha))$. This shows that st^{-1} and uv^{-1} agree on A, whence $st^{-1} \in \Omega_{uv^{-1}}(A)$. $\qquad\square$

Let F be a field. According to the preceding proposition, a topological group is defined by providing $\text{Aut}(F)$ with its finite topology. From now on, whenever a group Γ of automorphisms of F is considered as a topological group, it will be tacitly understood that Γ has been provided with the topology induced by the finite topology on $\text{Aut}(F)$; and, for every subset Ω of $\text{Aut}(F)$, the symbol $\overline{\Omega}$ will be used to denote the closure of Ω in $\text{Aut}(F)$.

3.12.3. Proposition. *Let F be a field. If Γ is a group of automorphisms of F, then* $\text{Inv}(\Gamma) = \text{Inv}(\overline{\Gamma})$.

Proof. It is clear that $\text{Inv}(\Gamma) \supseteq \text{Inf}(\overline{\Gamma})$, because $\Gamma \subseteq \overline{\Gamma}$. To verify that $\text{Inv}(\Gamma) \subseteq \text{Inv}(\overline{\Gamma})$, let $\alpha \in \text{Inv}(\Gamma)$. If $s \in \overline{\Gamma}$, then there exists a $t \in \Gamma \cap \Omega_s(\alpha)$; and, since the conditions $t \in \Omega_s(\alpha)$ and $t \in \Gamma$ imply that $t(\alpha) = s(\alpha)$ and $t(\alpha) = \alpha$, we see that $s(\alpha) = \alpha$. Therefore $\alpha \in \text{Inv}(\overline{\Gamma})$, as required. $\qquad\square$

3.12.4. Proposition. *Let K be a field, and let L be an extension field of K. Then* $\mathrm{Gal}(L/K)$ *is closed in* $\mathrm{Aut}(L)$.

Proof. Let $s \in \overline{\mathrm{Gal}(L/K)}$. We have to show that $s \in \mathrm{Gal}(L/K)$, that is, that $s(\alpha) = \alpha$ for every $\alpha \in K$. But this is evident: If $\alpha \in K$, then there exists a $t \in \mathrm{Gal}(L/K) \cap \Omega_s(\alpha)$, and so $s(\alpha) = t(\alpha) = \alpha$. □

3.12.5. Example. Let us continue the discussion of the example given in 3.12.1, with P a prime field of prime characteristic, F an algebraic closure of P, and Γ the subgroup of $\mathrm{Aut}(F)$ generated by the Frobenius mapping of F. It will now be shown that Γ is dense in $\mathrm{Aut}(F)$.

We have to show that every $s \in \mathrm{Aut}(F)$ is an adherent point of Γ in $\mathrm{Aut}(F)$. To do this, let A be a finite subset of F. Then write $E = P(A)$, so that E is a finite subfield of F, and hence $\mathrm{Aut}(E)$ is cyclic and admits the Frobenius mapping of E as a generator; also, since E is normal over P, it is meaningful to speak of the automorphism s_E of E. It follows that s_E is an integer power of the Frobenius mapping of E, and hence s agrees on A with an integer power of the Frobenius mapping of F, that is, with an element of Γ. This shows that $\Gamma \cap \Omega_s(A) \neq \varnothing$, which is what was needed. □

The topology originally introduced by Krull for the Galois group of an algebraic extension, which is commonly known as the **Krull topology**, was defined by specifying a filter base consisting of certain subgroups of the Galois group as a fundamental system of neighborhoods of the neutral element. This filter base will be explicitly described in the next result, which shows, in particular, that the Krull topology on the Galois group of an algebraic extension coincides with the topology defined previously.

3.12.6. Proposition. *Let K be a field, and let L be an algebraic extension of K. Then the Galois groups of L over its subfields that are finite extensions of K make up a fundamental system of neighborhoods of the neutral element in* $\mathrm{Gal}(L/K)$.

Proof. Taking into account that the subfields of L that are finite extensions of K are those obtained by adjoining finite subsets of L to K, our conclusion is an immediate consequence of the fact that

$$\mathrm{Gal}(L/K(A)) = \mathrm{Gal}(L/K) \cap \Omega_{i_L}(A)$$

for every finite subset A of L. □

We show next that, in the case of a normal extension, not all the subgroups in the fundamental system of neighborhoods described in the preceding proposition are essential.

3.12.7. Proposition. *Let K be a field, and let L be a normal extension of K. Then the Galois groups of L over its subfields that are finite normal* *extensions of K make up a fundamental system of neighborhoods of the neutral element in* Gal(L/K).

Proof. This follows from 2.3.11, 2.3.13, and 3.12.6. Indeed, if E is a subfield of L that is a finite extension of K, and if F denotes the normal closure of E over K in L, then F is normal and finite over K, and Gal(L/F) $\subseteq$ Gal(L/E).   □

As the next proposition shows, the topological concepts with which we are dealing in this section will not yield interesting results for finite extensions.

3.12.8. Proposition. *Let K be a field, and let L be an extension field of K. If L is finite over K, then* Gal(L/K) *is discrete. Conversely, if* Gal(L/K) *is discrete and L is Galois over K, then L is finite over K.*

Proof. The first assertion follows immediately from 3.12.6. For if L is finite over K, then Gal(L/L) is a neighborhood of i_L in Gal(L/K); and since Gal(L/L) $= \{i_L\}$, this means that Gal(L/K) is discrete.

Next suppose, conversely, that Gal(L/K) is discrete and L is Galois over K. Then the trivial subgroup $\{i_L\}$ of Gal(L/K) is a neighborhood of i_L in Gal(L/K); therefore, by 3.12.7, we have Gal(L/E) $= \{i_L\}$ for some subfield E of L that is a finite extension of K. Since L is Galois over E, we then have

$$E = \text{Inv}(\text{Gal}(L/E)) = \text{Inv}(\{i_L\}) = L,$$

whence L is finite over K.   □

We now turn to the study of compactness for groups of field automorphisms. We shall begin with the following auxiliary result on the finite topology.

3.12.9. Proposition. *Let F be a field. For every $\alpha \in F$, let D_α denote the discrete topological space having F as its set of points. Then* Aut(F) $\subseteq$ $\times_{\alpha \in F} D_\alpha$, *and the finite topology on* Aut(F) *coincides with the topology on* Aut(F) *induced by that on the product space* $\times_{\alpha \in F} D_\alpha$.

Proof. Since the points of the product space $\times_{\alpha \in F} D_\alpha$ are just the mappings from F to F, we see that Aut(F) $\subseteq \times_{\alpha \in F} D_\alpha$. The final assertion follows from the definitions of the finite and product topologies, by observing that if $s \in$ Aut(F) and A is a finite subset of F, and if we let $U_\alpha = \{s(\alpha)\}$ for $\alpha \in A$ and $U_\alpha = F$ for $\alpha \in F - A$, then

$$\Omega_s(A) = \text{Aut}(F) \cap \left(\underset{\alpha \in F}{\times} U_\alpha \right).$$
□

3.12.10. Corollary. *Let F be a field.*

(i) *Every group of automorphisms of F is separated and totally disconnected.*

(ii) *If* $\alpha \in F$, *then the mapping* $s \to s(\alpha)$ *from* Aut(F) *to the discrete topological space having* F *as its set of points is continuous.*

We can now establish the main topological property of the Galois group of an algebraic extension.

3.12.11. Theorem. *Let* K *be a field, and let* L *be an algebraic extension of* K. *Then* Gal(L/K) *is compact.*

Proof. We shall apply 3.12.9 to the field L. Let $\Pi = \times_{\lambda \in L} D_\lambda$, where D_λ denotes the discrete topological space with L as its set of points for every $\lambda \in L$. Then Gal(L/K) $\subseteq \Pi$, and the topology of Gal(L/K) is that induced by the topology of the product space Π. Since this space is separated, the compactness of Gal(L/K) will be established if it is verified that Gal(L/K) is closed and relatively compact in Π.

To prove that Gal(L/K) is closed in Π, we have to verify that every adherent point of Gal(L/K) in Π belongs to Gal(L/K). Thus, let u be such an adherent point; then u is a mapping from L to L. Let $\alpha, \beta \in L$ and $\gamma \in K$; next let $A = \{\alpha, \beta, \alpha + \beta, \alpha\beta, \gamma\}$, and write $U_\lambda = \{u(\lambda)\}$ for $\lambda \in A$ and $U_\lambda = L$ for $\lambda \in L - A$. Our assumption on u implies the existence of an $s \in$ Gal(L/K)$\cap(\times_{\lambda \in L} U_\lambda)$; such an s is a K-automorphism of L such that $s(\lambda) = u(\lambda)$ for every $\lambda \in A$, and so

$$u(\alpha + \beta) = s(\alpha + \beta) = s(\alpha) + s(\beta) = u(\alpha) + u(\beta),$$

$$u(\alpha\beta) = s(\alpha\beta) = s(\alpha)s(\beta) = u(\alpha)u(\beta),$$

$$u(\gamma) = s(\gamma) = \gamma.$$

This shows that u is a K-endomorphism of L. Since L is algebraic over K, it now follows from 2.1.21 that u is a K-automorphism of L, which means that $u \in$ Gal(L/K).

To conclude, we shall now prove that Gal(L/K) is relatively compact in Π. For each $\lambda \in L$, let C_λ denote the set of all zeros in L of the minimal polynomial of λ over K. If $\lambda \in L$, then C_λ is finite, and hence it is a compact subset of D_λ. Therefore, by Tihonov's theorem, $\times_{\lambda \in L} C_\lambda$ is a compact subset of Π. On the other hand, it is clear that if $s \in$ Gal(L/K), then $s(\lambda) \in C_\lambda$ for every $\lambda \in L$, whence $s \in \times_{\lambda \in L} C_\lambda$. Therefore Gal($L/K$) $\subseteq \times_{\lambda \in L} C_\lambda$, and Gal($L/K$) is indeed relatively compact in Π. □

The compactness of the Galois group does not characterize algebraic extensions. In fact, we saw in 3.2.7 that there exist transcendental extensions with a finite Galois group. It can be shown, however, that if a field extension has a compact Galois group and its top field is algebraically closed, then it is an algebraic extension (section 4.2, problem 10).

3.12.12. Example. Let us show that if K is a field, then Gal($K(X_n)_{n \geq 0}/K$) is not compact.

Write $L = K(X_n)_{n \geq 0}$. For every $n \geq 0$, let s_n denote the K-automorphism of L such that

$$X_0 \to X_n, \quad X_n \to X_0, \quad \text{and} \quad X_k \to X_k \quad \text{for } k \neq 0, n.$$

We then have $s_m \neq s_n$ whenever $m \neq n$; if we now write $\Omega = \{s_0, s_1, s_2, \ldots\}$, then Ω is an infinite subset of $\mathrm{Gal}(L/K)$. If $\mathrm{Gal}(L/K)$ were compact, then Ω would possess a limit point s in $\mathrm{Gal}(L/K)$; but this is impossible, since it would imply that $s_n \in \Omega_s(X_0)$ for infinitely many $n \geq 0$, that is, that $X_n = s_n(X_0) = s(X_0)$ for infinitely many $n \geq 0$.   □

The preceding results allow us to deduce the following characterization of compact groups of field automorphisms.

3.12.13. Proposition. *Let F be a field. Then a group of automorphisms of F is compact if and only if it is closed in $\mathrm{Aut}(F)$ and has finite orbits.*

Proof. Let Γ be a group of automorphisms of F.

The direct implication is readily seen to follow from 3.12.10. Indeed, suppose that Γ is compact. Since $\mathrm{Aut}(F)$ is separated, it is seen that Γ is closed in $\mathrm{Aut}(F)$. And if $\alpha \in F$, then $O_\Gamma(\alpha)$ is the image of Γ by the mapping $s \to s(\alpha)$ from $\mathrm{Aut}(F)$ to the discrete topological space with F as its set of points; since this mapping is continuous, it follows that $O_\Gamma(\alpha)$ is a compact subset of a discrete topological space, whence $O_\Gamma(\alpha)$ is finite.

To prove the opposite implication, suppose now that Γ is closed in $\mathrm{Aut}(F)$ and has finite orbits. Since $\Gamma \subseteq \mathrm{Gal}(F/\mathrm{Inv}(\Gamma))$, we see that Γ is closed in $\mathrm{Gal}(F/\mathrm{Inv}(\Gamma))$. Moreover, according to 3.2.22, F is algebraic over $\mathrm{Inv}(\Gamma)$; and this, by 3.12.11, implies that $\mathrm{Gal}(F/\mathrm{Inv}(\Gamma))$ is compact. Therefore Γ is compact, as required.   □

We now are in a position to state and prove the central result of the section. It can be said to be the analog of the Dedekind–Artin theorem in the infinite Galois theory. In fact, we shall presently see that in setting up the Galois correspondences, its role in the infinite theory is the same as that of the Dedekind–Artin theorem in the finite theory.

3.12.14. Theorem. *If F is a field, then every compact group of automorphisms of F is a Galois group on F.*

Proof. The proof will be based on the following auxiliary technicality.

Lemma. *Let F be a field, and let Γ be a group of automorphisms of F. If P is a subfield of F that is a finite normal extension of $\mathrm{Inv}(\Gamma)$, then $\mathrm{Gal}(P/\mathrm{Inv}(\Gamma))$ is the image of Γ by the homomorphism $s \to s_P$ from $\mathrm{Gal}(F/\mathrm{Inv}(\Gamma))$ to $\mathrm{Gal}(P/\mathrm{Inv}(\Gamma))$.*

Proof. The normality of P over $\mathrm{Inv}(\Gamma)$ makes it meaningful to speak of the homomorphism $s \to s_P$ from $\mathrm{Gal}(F/\mathrm{Inv}(\Gamma))$ to $\mathrm{Gal}(P/\mathrm{Inv}(\Gamma))$. Let Δ denote the image of Γ by this homomorphism. Then Δ is a subgroup of $\mathrm{Gal}(P/\mathrm{Inv}(\Gamma))$; we have to show that $\Delta = \mathrm{Gal}(P/\mathrm{Inv}(\Gamma))$.

Assume, on the contrary, that $\Delta \subset \mathrm{Gal}(P/\mathrm{Inv}(\Gamma))$. Since P is finite over $\mathrm{Inv}(\Gamma)$, the inequalities

$$\mathrm{Card}(\Delta) < \mathrm{Card}(\mathrm{Gal}(P/\mathrm{Inv}(\Gamma))) \le [P:\mathrm{Inv}(\Gamma)]$$

obtain. In particular, Δ is finite, and the Dedekind–Artin theorem shows that

$$[P:\mathrm{Inv}(\Delta)] = \mathrm{Card}(\Delta) < [P:\mathrm{Inv}(\Gamma)].$$

Since $\mathrm{Inv}(\Gamma) \subseteq \mathrm{Inv}(\Delta)$, this now implies that $\mathrm{Inv}(\Gamma) \subset \mathrm{Inv}(\Delta)$.

To conclude, choose an $\alpha \in \mathrm{Inv}(\Delta)$ so that $\alpha \notin \mathrm{Inv}(\Gamma)$. Then $s(\alpha) \ne \alpha$ for some $s \in \Gamma$; but since $s_P \in \Delta$, we also have $s_P(\alpha) = \alpha$. Since $s_P(\alpha) = s(\alpha)$, this is a contradiction. The lemma is proved.

Let Γ be a compact group of automorphisms of F. By 3.12.13, 3.2.22, and 3.3.8, we know that Γ is closed in $\mathrm{Aut}(F)$ and that F is Galois over $\mathrm{Inv}(\Gamma)$. Since $\Gamma \subseteq \mathrm{Gal}(F/\mathrm{Inv}(\Gamma))$, the desired conclusion will be established if it is shown that every $s \in \mathrm{Gal}(F/\mathrm{Inv}(\Gamma))$ is an adherent point of Γ in $\mathrm{Aut}(F)$. This is seen to be a consequence of 3.12.7: Indeed, let P be a subfield of F that is a finite normal extension of $\mathrm{Inv}(\Gamma)$. Then $s_P \in \mathrm{Gal}(P/\mathrm{Inv}(\Gamma))$, and the lemma implies the existence of a $t \in \Gamma$ with $s_P = t_P$. Since this means that s and t agree on P, we have $s\,\mathrm{Gal}(F/P) = t\,\mathrm{Gal}(F/P)$; therefore $t \in \Gamma \cap s\,\mathrm{Gal}(F/P)$, so that $\Gamma \cap s\,\mathrm{Gal}(F/P) \ne \varnothing$. $\square$

3.12.15.   Corollary. *Let F be a field, and let Γ be a group of automorphisms of F. If Γ is relatively compact in $\mathrm{Aut}(F)$, then $\overline{\Gamma} = \mathrm{Gal}(F/\mathrm{Inv}(\Gamma))$.*

Proof. According to 3.12.3, we have $\mathrm{Inv}(\Gamma) = \mathrm{Inv}(\overline{\Gamma})$. Moreover, the hypothesis implies that $\overline{\Gamma}$ is compact, and the theorem shows that $\overline{\Gamma} = \mathrm{Gal}(F/\mathrm{Inv}(\overline{\Gamma}))$. The conclusion now follows at once. $\square$

3.12.16.   Corollary. *Let K be a field, and let L be an algebraic extension of K. Then a subgroup Γ of $\mathrm{Gal}(L/K)$ is dense in $\mathrm{Gal}(L/K)$ if and only if $\mathrm{Inv}(\Gamma) = \mathrm{Inv}(\mathrm{Gal}(L/K))$.*

Proof. By 3.12.11, $\mathrm{Gal}(L/K)$ is compact; therefore Γ is relatively compact in $\mathrm{Aut}(L)$, and the preceding corollary shows that $\overline{\Gamma} = \mathrm{Gal}(L/\mathrm{Inv}(\Gamma))$. Thus, to say that Γ is dense in $\mathrm{Gal}(L/K)$ means that

$$\mathrm{Gal}(L/\mathrm{Inv}(\Gamma)) = \mathrm{Gal}(L/K);$$

and since the two groups appearing in this equality are Galois groups on L, we conclude that Γ is dense in $\mathrm{Gal}(L/K)$ if and only if

$$\mathrm{Inv}(\mathrm{Gal}(L/\mathrm{Inv}(\Gamma))) = \mathrm{Inv}(\mathrm{Gal}(L/K)),$$

and hence if and only if $\mathrm{Inv}(\Gamma) = \mathrm{Inv}(\mathrm{Gal}(L/K))$. $\square$

3.12.17. **Corollary.** *Let K be a field, and let L be a Galois extension of K. Then a subgroup Γ of $\mathrm{Gal}(L/K)$ is dense in $\mathrm{Gal}(L/K)$ if and only if $\mathrm{Inv}(\Gamma) = K$.*

Proof. Here we have $\mathrm{Inv}(\mathrm{Gal}(L/K)) = K$, so that this is merely a particular case of the preceding corollary.   □

It should be noted that what was shown in 3.12.5 is an evident consequence of corollary 3.12.17.

The results in the foregoing discussion now lead us to the natural formulation of the **fundamental theorems of infinite Galois theory**. The real work having been done already, all that remains is the formality of assembling the various components into a coherent unit.

3.12.18. **Theorem.** *Let F be a field.*

(i) *If P is a field admitting F as a Galois extension, then $\mathrm{Gal}(F/P)$ is a compact group of automorphisms of F.*

(ii) *If Γ is a compact group of automorphisms of F, then $\mathrm{Inv}(\Gamma)$ is a field admitting F as a Galois extension.*

(iii) *The mapping $P \to \mathrm{Gal}(F/P)$ from the set of all fields admitting F as a Galois extension to the set of all compact groups of automorphisms of F and the mapping $\Gamma \to \mathrm{Inv}(\Gamma)$ from the set of all compact groups of automorphisms of F to the set of all fields admitting F as a Galois extension are mutually inverse inclusion-reversing bijections.*

Proof. Assertion (i) is evident from 3.12.11, and assertion (ii) from 3.12.13, 3.2.22, and 3.3.8.

To conclude, note that according to 3.12.14 and 3.3.8, every compact group of automorphisms of F is a Galois group on F, and every field admitting F as a Galois extension is invariant in F. Consequently, (iii) follows from (i), (ii), and 3.2.2.   □

3.12.19. **Theorem** (Krull). *Let K be a field, and let L be a Galois extension of K.*

(i) *If E is an intermediate field between K and L, then $\mathrm{Gal}(L/E)$ is a closed subgroup of $\mathrm{Gal}(L/K)$.*

(ii) *If Γ is a closed subgroup of $\mathrm{Gal}(L/K)$, then $\mathrm{Inv}(\Gamma)$ is an intermediate field between K and L.*

(iii) *The mapping $E \to \mathrm{Gal}(L/E)$ from the set of all intermediate fields between K and L to the set of all closed subgroups of $\mathrm{Gal}(L/K)$ and the mapping $\Gamma \to \mathrm{Inv}(\Gamma)$ from the set of all closed subgroups of $\mathrm{Gal}(L/K)$ to the set of all intermediate fields between K and L are mutually inverse inclusion-reversing bijections.*

Proof. This follows easily from the preceding theorem by observing that, by 3.12.10 and 3.12.11, the closed subgroups of $\mathrm{Gal}(L/K)$ are precisely the compact groups of automorphisms of L contained in

Gal(L/K); and that, by 3.3.1, every intermediate field between K and L admits L as a Galois extension.   □

For a given Galois extension, we can regard the set of all intermediate fields and the set of all closed subgroups of the Galois group as ordered by the inclusion relation. The inclusion-reversing property of the bijections described in Krull's theorem immediately yields the following result.

3.12.20.   Corollary.   *Let K be a field, and let L be a Galois extension of K.*

   (i) *If $(\Gamma_i)_{i \in I}$ is a nonempty family of closed subgroups of Gal(L/K), then*

$$\text{Inv}\left(\bigcap_{i \in I} \Gamma_i \right) = \bigvee_{i \in I} \text{Inv}(\Gamma_i).$$

   (ii) *If $(E_i)_{i \in I}$ is a nonempty family of intermediate fields between K and L, then*

$$\text{Gal}\left(L \Big/ \bigcap_{i \in I} E_i \right) = \overline{\bigvee_{i \in I} \text{Gal}(L/E_i)}.$$

In view of 3.12.8, it is clear that, in the case of a finite Galois extension, Krull's theorem 3.12.19 reduces to Galois's theorem 3.4.5; and similarly, 3.12.20 reduces to 3.4.6.

We now are in a position to give the complete answers to the two questions on the Galois correspondences raised in the comments following 3.3.8.

First, given a field F, we have proved that the Galois groups on F related to the fields admitting F as a Galois extension are the *compact* groups of automorphisms of F. And, given a field K and a Galois extension L of K, we have seen that the Galois groups on L related to the intermediate fields between K and L are the *closed* subgroups of Gal(L/K).

On the negative side, we note that the Galois correspondence defined by an *infinite* Galois extension definitely fails to relate every subgroup of the Galois group to an intermediate field, because the Galois group always contains subgroups that are not closed (problem 5).

We proceed next to discuss how the homomorphisms of Galois groups considered in 3.2.20 and 3.2.21 behave with respect to the Krull topology.

3.12.21.   Proposition.   *Let K be a field, let L be a normal extension of K, and let M be an algebraic extension of L. Then the homomorphism $s \to s_L$ from Gal(M/K) to Gal(L/K) is continuous.*

Proof.   This follows at once from 3.12.6 by observing that, for every intermediate field E between K and L that is finite over K, the image of Gal(M/E) by the homomorphism in question is contained in Gal(L/E).   □

3.12.22. Proposition. *Let K be a field, let L be a normal extension of K, and let M be an extension field of L that is normal over K. Then the homomorphism from* $\mathrm{Gal}(M/K)/\mathrm{Gal}(M/L)$ *to* $\mathrm{Gal}(L/K)$ *such that* $s\,\mathrm{Gal}(M/L) \to s_L$ *for every* $s \in \mathrm{Gal}(M/K)$ *is a topological isomorphism.*

Proof. We already know that our homomorphism is an isomorphism. Moreover, its continuity follows from the preceding proposition. To conclude, we now need only note that $\mathrm{Gal}(M/K)/\mathrm{Gal}(M/L)$ is compact and $\mathrm{Gal}(L/K)$ is separated. $\square$

The following two propositions are the natural generalizations of 3.4.7 and 3.4.10 to the case of arbitrary normal extensions.

3.12.23. Proposition. *Let F be a field, and let P and Q be subfields of F such that P is normal over* $P \cap Q$. *Then the homomorphism* $s \to s_p$ *from* $\mathrm{Gal}(P \vee Q/Q)$ *to* $\mathrm{Gal}(P/P \cap Q)$ *is a topological isomorphism.*

Proof. It will suffice to adapt the proof of 3.4.7. Let the notation be as in that proof. Using the first two paragraphs and 3.12.21, we see that our homomorphism is continuous and injective. The finiteness assumption in 3.4.7 was used only in order to prove that Γ is finite, and hence a Galois group on P; in the present situation, this conclusion can be deduced from the compactness of Γ, which is a consequence of the fact that Γ is a continuous image of the compact group $\mathrm{Gal}(P \vee Q/Q)$. The surjectivity of our homomorphism can now be established exactly as before. Thus, our homomorphism is a continuous isomorphism, and the argument is completed by observing that $\mathrm{Gal}(P \vee Q/Q)$ is compact and $\mathrm{Gal}(P/P \cap Q)$ is separated. $\square$

3.12.24. Proposition. *Let K be a field, let N be an extension field of K, and let L and M be intermediate fields between K and N that are normal over K. Then the mapping* $s \to (s_L, s_M)$ *from* $\mathrm{Gal}(L \vee M/K)$ *to* $\mathrm{Gal}(L/K) \times \mathrm{Gal}(M/K)$ *is a continuous injective homomorphism; and if* $L \cap M = K$, *then it is a topological isomorphism.*

Proof. The finiteness assumption in 3.4.10 allowed us to invoke 3.4.7 in its proof. Consequently, the same proof can be used here, invoking 3.12.23 instead of 3.4.7. We conclude, therefore, that the mapping in question is an injective homomorphism; and that it is an isomorphism if $L \cap M = K$. Its continuity follows directly from 3.12.21; and since $\mathrm{Gal}(L \vee M/K)$ is compact and $\mathrm{Gal}(L/K) \times \mathrm{Gal}(M/K)$ is separated, we see that it is a topological isomorphism if $L \cap M = K$. $\square$

To close the section, we now give a result that, in the case of a finite Galois extension, reduces to 3.4.11.

3.12.25. Proposition. *Let K be a field, let L be a Galois extension of K, and let Γ and Δ be closed subgroups of $\mathrm{Gal}(L/K)$ such that*

$$\mathrm{Gal}(L/K) = \Gamma \vee \Delta \quad and \quad \Gamma \cap \Delta = \{i_L\}.$$

Then

$$L = \mathrm{Inv}(\Gamma) \vee \mathrm{Inv}(\Delta) \quad and \quad \mathrm{Inv}(\Gamma) \cap \mathrm{Inv}(\Delta) = K.$$

Proof. Argue as in the proof of 3.4.11, using 3.12.20 instead of 3.4.6. □

PROBLEMS

1. Let K be a field, and let $L = K(X_n)_{n \geq 0}$. For every $\lambda \in L$, let D_λ denote the discrete topological space with L as its set of points. Prove that $\mathrm{Gal}(L/K)$ is not closed in the product space $\times_{\lambda \in L} D_\lambda$.

2. Show that if K is an infinite field and $K(\xi)$ is a simple transcendental extension of K, then $\mathrm{Gal}(K(\xi)/K)$ is not compact.

3. Let K be a field, let L be a normal extension of K, and let Γ be a subgroup of $\mathrm{Gal}(L/K)$. Verify the following assertions:
   a. If Γ is open in $\mathrm{Gal}(L/K)$, then Γ has finite index in $\mathrm{Gal}(L/K)$.
   b. If Γ is closed in $\mathrm{Gal}(L/K)$, then

   $$[\mathrm{Gal}(L/K) : \Gamma] = [\mathrm{Inv}(\Gamma) : K].$$

4. Let K be a field, let L be a normal extension of K, and let $\Omega \subseteq \mathrm{Gal}(L/K)$. Prove that if Ω is dense in $\mathrm{Gal}(L/K)$ and E is an intermediate field between K and L that is finite over K, then every K-automorphism of E is defined by restriction of an element of Ω. Prove also that, conversely, if for every intermediate field E between K and L that is normal and finite over K, every K-automorphism of E is defined by restriction of an element of Ω, then Ω is dense in $\mathrm{Gal}(L/K)$.

5. Let K be a field, and let L be an infinite Galois extension of K. Show that $\mathrm{Gal}(L/K)$ contains denumerable subgroups, and that no such subgroup is closed in $\mathrm{Gal}(L/K)$.

6. Let I be an infinite set of primes, and put $F = \vee_{p \in I} \mathbf{Q}(\sqrt{p})$. Prove the following assertions:
   a. F is Galois and infinite over $\mathbf{Q}$.
   b. The set of all subfields of F that are quadratic extensions of $\mathbf{Q}$ has cardinality $\aleph_0$.
   c. The set of all closed subgroups of index 2 of $\mathrm{Gal}(F/\mathbf{Q})$ has cardinality $\aleph_0$.
   d. Every $\mathbf{Q}$-automorphism of F different from i_F is an element of order 2 in $\mathrm{Gal}(F/\mathbf{Q})$; and hence $\mathrm{Gal}(F/\mathbf{Q})$ admits a unique $\mathbf{Z}/2$-space structure.

e. The dimension of $\text{Gal}(F/\mathbf{Q})$ as a $\mathbf{Z}/2$-space is $2^{\aleph_0}$.

f. The set of all subgroups of index 2 of $\text{Gal}(F/\mathbf{Q})$ and that are not closed in $\text{Gal}(F/\mathbf{Q})$ has cardinality $2^{\aleph_0}$.

(These results show that the Galois group of an infinite Galois extension may contain "relatively few" closed subgroups of a given finite index.)

7. Let K be a field, and let L be a Galois extension of K. Show that the field of invariants of the commutator subgroup of $\text{Gal}(L/K)$ is the largest element of the set of all intermediate fields between K and L that are abelian over K.

NOTES

Section 3.1. The Dedekind independence theorem (3.1.2), stated as the nonvanishing of a certain determinant, was first proved for algebraic number fields by Dedekind [4: §161]; the general formulation and proof presented here are due to Artin [3: sec. 2.F].

A discussion of the Jacobson–Bourbaki correspondence (problems 7–13) is given in Jacobson [4: sec. 1.2]. This correspondence was developed by Jacobson [2] in order to treat various extensions of Galois theory in a unified manner.

Section 3.2. Artin's theorem (problem 12) and its application to the study of separable extensions are discussed in Bourbaki [3: sec. 5.7.1, 5.7.2].

Section 3.4. The Dedekind–Artin theorem (3.4.1) was first proved, for algebraic number fields, by Dedekind [4: §166]; and later, in full generality, by Artin [3: sec. 2.H]. The first suggested alternative proof of this theorem (problem 3) is given in Jacobson [4: sec. 1.4]; and the second (problem 4) in Bourbaki [3: sec. 5.10.5]. For the generalization of 3.4.3 (problem 2), see Kaplansky [3: sec. 1.3].

Detailed discussions on Galois groups of polynomials can be found in Hungerford [1: sec. 5.4] and Jacobson [5: I, sec. 4.8]. For Galois's original presentation, we refer to Galois [1: pp. 25–61].

Section 3.5. The study of cyclotomy was undertaken by Gauss [2: sec. VII] at the early stage of his career. His important results led him to the proof of the constructibility of the regular n-sided polygon when

$$n = 2^j p_1 p_2 \cdots p_k,$$

where $p_1, p_2, \ldots, p_k$ are distinct odd primes of the form $2^{2^i} + 1$. For a detailed discussion of the constructibility of regular polygons, consult Stewart [1: ch. 11]; and, especially, Hadlock [1: sec. 2.5–2.7], where Gauss's original argument is presented.

Gauss also investigated the irreducibility of cyclotomic polynomials, and succeeded in proving that the pth cyclotomic polynomial is irreducible for every prime p. The first general proof was given by Kronecker, and many others can be found in the literature. The particularly simple proof given here is due to Dedekind [3].

The theorem stating that every finite abelian extension of $\mathbf{Q}$ is a subfield of a cyclotomic field over $\mathbf{Q}$ was enunciated by Kronecker, but it was Weber who gave the first complete proof (see Weber [2: II, §208] and Hilbert [1: I, p. 53]). This result can be derived from the main theorems of class field theory, as is done in Lang [3: sec. 10.3]; a more elementary, but longer, proof is given in Ribenboim [1: ch. 12].

The result on the absolute values of the coefficients of cyclotomic polynomials stated in the remarks made after 3.5.6 was given by Schur [1: I, pp. 460–461] in a letter to Frobenius. The properties of algebraically closed fields of characteristic 0 stated in problem 7 are discussed in Miller and Guralnick [1]. The Vahlen–Capelli criterion (problem 13) was first proved by Vahlen [1] for the rational field, and it was extended by Capelli [1; 2] to arbitrary fields of characteristic 0; the general result was proved by Rédei [1: I, §172]. The sketch provided in problems 10–13 is taken from Kaplansky [3: sec. 1.12].

Section 3.6. The characterization of finite simple extensions (3.6.1) and the theorem on primitive elements (3.6.3) are due to Steinitz [1: §§13, 11, 15]. Galois [2] asserted the validity of the theorem on primitive elements for extensions of finite fields, but did not give a proof. Alternative proofs of 3.6.4 have been given by Isaacs [1] and Gilmer [1].

Section 3.8. Euler's theorem (problem 11) is treated in the standard texts on algebraic number theory, for example, Weiss [1: sec. 3.7] and Lang [3: sec. 3.1]. It is useful in the study of the different and the discriminant.

For infinite fields, the normal base theorem (problem 18) was first proved by Noether [1]; a surprisingly simple proof in this case has been given recently by Waterhouse [1]. The general case was first treated in Deuring [1]. The proof usually found in the literature is that outlined in problems 16–18, and is given in Bourbaki [3: sec. 5.9.8; 2: sec. 7.5.7] and Jacobson [4: sec. 1.12, 1.13; 5: I, sec. 4.14]. A proof based on a special case of the Jacobson–Bourbaki theorem has been given by Berger and Reiner [1].

Section 3.9. The first part of Hilbert's theorem (3.9.3) appears in Hilbert [3: §54] as "Satz 90".

For fields of prime characteristic p, the characterization of cyclic extensions of linear degree p (3.9.7) was given in Artin and Schreier [1]; in this work, cyclic extensions of linear degree p^2 are also studied. The related results on cyclic extensions stated in problems 8–14 are due to Albert [1].

The results stated in problems 17 and 18 were proved in Artin and Schreier [1; 2]. These are discussed in Jacobson [4: sec. 6.11], Lang [4: sec. 8.9], and Nagata [2: sec. 6.1]. The outline given in problems 15–18 follows Kaplansky [3: sec. 1.12].

Section 3.10. Most texts on abstract algebra cover, with varying degree of generality, the topic of solvability of polynomial equations by radicals. More detailed discussions than that given here can be found in Jacobson [5: I, sec. 4.9] and van der Waerden [2: I, §§57–59]. The texts of Čebotarev [1], Dehn [1], and Weber [2: I, II] contain thorough presentations of the classical Galois theory of polynomial equations. For a very readable introductory account on rational polynomials with symmetric groups, we refer to Hadlock [1: ch. 4].

Gauss's proof of the constructibility of the regular n-sided polygon, for the values of n of the form described in the notes to section 3.5, can be simplified considerably by using Galois's theorem (3.10.4). For this, we refer to Hadlock [1: sec. 3.7] and Jacobson [5: I, sec. 4.11]. The reader interested in pursuing the study of geometric constructions may consult the encyclopedic treatise by Bieberbach [1].

Section 3.11. As stated in the introduction, our treatment of finite fields is limited by the restrictions imposed by our objectives. For a lucid and comprehensive exposition on finite fields and their applications to geometry, combinatorics, and coding theory, we refer the reader to the book by Lidl and Niederreiter [1].

The study of finite fields was initiated by Galois [2] in his work on higher congruences modulo a prime. The uniqueness part of theorem 3.11.5 is due to Moore [1]. The theorems stated in problems 8 and 9 were given by Warning [1] and Chevalley [1], respectively.

Section 3.12. The Galois theory for infinite Galois extensions is due to Krull [1]. The use of the finite topology in the study of rings and fields first appeared in Jacobson [3: sec. 2.3, 7.6]. Brief expositions of the Galois theory for infinite Galois extensions can be found in Bourbaki [3: app. II], Jacobson [4: sec. 4.2; 5: II, sec. 8.6], and Nagata [2: sec. 7.1, 7.2]. The main theorem (3.12.4) is stated in Jacobson [4: sec. 4.2] as an exercise.

Suggestions for further reading

1. Two types of abelian extensions that can be studied with the help of relatively elementary tools are the Kummer n-extensions and the abelian p-extensions. The Kummer n-extensions of a field containing a primitive nth root of unity can be described by using basic facts on characters of finite abelian groups. A method for analyzing the abelian p-extensions of a field of prime characteristic p has been given by Witt [1]; this is based on the properties of a special ring—the ring of Witt vectors—that can be associated with every ring of prime characteristic. The results of Artin and Schreier (3.9.7) and Albert (section 3.9, problem 10) can also be derived from Witt's method. Detailed expositions on Kummer n-extensions, Witt vectors, and abelian p-extensions are given in Jacobson [4: ch. 3; 5: II, sec. 8.9–8.11]; alternatively, the reader may consult Lang [4: ch. 8, sec. 8 and exer. 21–26]. Witt vectors have other interesting applica-

tions (see Greenberg [1: ch. 6], Jacobson [4: sec. 5.5], Serre [1: sec. 2.6], and the original article of Witt [1]).

 2. The theory of formally real fields was initiated in Artin and Schreier [1; 2], and was successfully applied by Artin [1] to the solution of a problem of Hilbert on the representability of a positive-definite rational function with rational coefficients as a sum of squares of rational functions with rational coefficients. Although formally real fields have not been defined here, they have made two appearances in disguise: the first in the proof of the "fundamental theorem of algebra" given in 3.4.14, which depends on special properties of the real field as a formally real field; and the second in the outline on the theorem of Artin and Schreier given in section 3.9, problems 15–18. Introductory discussions on formally real fields can be found in Bourbaki [2: ch. 6], Lang [4: sec. 11.1, 11.2], and Nagata [2: sec. 6.1–6.3]; more detailed presentations—which include connections with mathematical logic—are given in Jacobson [4: ch. 6; 5: I, ch. 5 and II, sec. 11.1–11.4].

 3. The Galois theory of fields has been extended in several directions. The pertinent literature is very voluminous, and we shall limit ourselves to brief remarks on the more significant contributions. There is a Galois theory of commutative rings with no idempotents other than 0 and 1; it was given in Chase, Harrison, and Rosenberg [1], and includes the Galois theory of fields. This general theory has been used recently by Saltman [1] to develop the concept of a generic Galois extension, with the help of which he has been able to obtain interesting results (for example, simplifications in the construction of examples related to the problem of Noether stated at the end of section 4.2). There is a Galois theory of division rings, which was discovered independently by Cartan [1] and Jacobson [2], and which also includes the Galois theory of fields; for a thorough presentation of this Galois theory, the reader may consult Jacobson [3: ch. 7]. A further generalization in the noncommutative case is the Galois theory of simple rings, which is due to Tominaga and Nagahara [1]. Finally, we would like to mention the Galois theory of differential equations, which in some sense can be said to be parallel to that of polynomial equations; the standard reference for this Galois theory is the book by Kaplansky [1].

 4. The principal application of Galois theory are in modern number theory. With modest additional background, the reader who has studied Galois theory may proceed to the books of Artin [4; 5], Borevič and Šafarevič [1], Long [1], Ribenboim [1; 2], Samuel [1], and Weiss [1]; and to the book of O'Meara [1], in which the arithmetic theory of quadratic forms over fields and domains is developed. More advanced expositions on number theory, which will require more extensive prerequisites, are given in the works of Artin and Tate [1], Cassels and Fröhlich [1], Iyanaga [1], Koch [1], Serre [1; 2], and Shatz [1], in which the cohomological point of view is adopted; and in the books of Fröhlich [1], Goldstein [1], Janusz [1], Lang [3], and Weil [1], in which analytic methods play a dominant role.

Chapter 4

Transcendental Extensions

4.1. DIMENSIONAL OPERATORS

This section is set-theoretical in its entirety. Our objective is to present an axiomatic treatment of dimension theory, essentially due to Steinitz, that can be suitably applied in a number of situations. To the reader familiar with the dimension theory for vector spaces, it will become apparent that here we are merely selecting some simple facts on vector spaces as axioms, and then deriving their consequences formally.

It should be mentioned, incidentally, that the dimension theory for vector spaces is itself one of the instances covered by the axiomatic theory discussed here.

Let E be a set. Recall that the symbol $\mathscr{P}(E)$ denotes the power set of E, that is, the set consisting of all subsets of E. By a **dimensional operator on** E we understand a mapping d from $\mathscr{P}(E)$ to $\mathscr{P}(E)$ having the following properties:

(i) If $S \subseteq E$, then $S \subseteq d(S)$.

(ii) If $S \subseteq E$, then $d(S) = d(d(S))$.

(iii) If $S \subseteq T \subseteq E$, then $d(S) \subseteq d(T)$.

(iv) If $S \subseteq E$, and if Ω denotes the set of all finite subsets of S, then
$d(S) = \cup_{A \in \Omega} d(A)$.

(v) If $S \subseteq E$, $x \in E$, and $y \in d(S \cup \{x\}) - d(S)$, then
$x \in d(S \cup \{y\})$.

The last property in the preceding definition is known as the **exchange axiom**.

4.1.1. Examples

It is easy to give examples of dimensional operators:

a. If E is a set, the identity mapping $i_{\mathscr{P}(E)}$ is a dimensional operator on E.

b. If E is a set, the mapping $S \to E$ from $\mathscr{P}(E)$ to $\mathscr{P}(E)$ is a dimensional operator on E.

c. Let F be a field, and let E be an F-space. Then the mapping $S \to FS$ from $\mathscr{P}(E)$ to $\mathscr{P}(E)$ is a dimensional operator on E.

Conditions (i)–(iv) in the definition of dimensional operator are trivially satisfied; and condition (v) is the **Steinitz exchange theorem** in linear algebra. $\qquad\qquad\square$

Let E be a set, and let d be a dimensional operator on E. A subset S of E is said to be d-**free** when $x \notin d(S - \{x\})$ for every $x \in S$, and is said to be d-**dense** when $d(S) = E$. By a d-**base** we understand a subset of E that is both d-free and d-dense.

4.1.2. Examples

Let us go back to the examples discussed previously.

a. For the dimensional operator defined in 4.1.1a, every subset of E is free, and E is the only dense subset of E; therefore E is the only base.

b. For the dimensional operator defined in 4.1.1b, $\varnothing$ is the only free subset of E, and every subset of E is dense; therefore $\varnothing$ is the only base.

c. In the case of the dimensional operator defined in 4.1.1c, we find familiar notions from linear algebra: A subset of E is free if and only if it is linearly independent, and it is dense if and only if it is a generating system. Consequently, a base in the present sense is the same thing as a base in the usual sense of linear algebra. $\qquad\qquad\square$

The example introduced in 4.1.1c should be kept in mind throughout the general discussion that follows. Each result, when interpreted in this context, will become a familiar fact about vector spaces.

We begin our discussion with some elementary results on free and dense sets. These will be consequences of the first four conditions defining a dimensional operator.

4.1.3. *Proposition.* *Let E be a set, and let d be a dimensional operator on E.*

(i) $\varnothing$ *is d-free.*

(ii) E *is d-dense.*

(iii) *If $S \subseteq T \subseteq E$ and T is d-free, then S is d-free.*
(iv) *If $S \subseteq T \subseteq E$ and S is d-dense, then T is d-dense.*
(v) *If $S \subseteq E$, then S is d-dense if and only if $d(S)$ is d-dense.*

4.1.4. Proposition. *Let E be a set, and let d be a dimensional operator on E. If $(S_i)_{i \in I}$ is a nonempty filtered family of d-free subsets of E, then $\cup_{i \in I} S_i$ is d-free.*

Proof. Assume that $\cup_{i \in I} S_i$ is not d-free, and choose an $x \in \cup_{i \in I} S_i$ so that $x \in d((\cup_{i \in I} S_i) - \{x\})$. We then have $x \in d(A)$ for some finite subset A of $(\cup_{i \in I} S_i) - \{x\}$. Thus, $A \cup \{x\}$ is a finite subset of $\cup_{i \in I} S_i$, and our hypothesis implies that $A \cup \{x\} \subseteq S_j$ for some $j \in I$. But then $A \subseteq S_j - \{x\}$, and hence $d(A) \subseteq d(S_j - \{x\})$. Thus $x \in S_j$ and $x \in d(S_j - \{x\})$, which contradicts the assumption that S_j is d-free. $\square$

4.1.5. Corollary. *Let E be a set, and let d be a dimensional operator on E. If $S \subseteq E$ and every finite subset of S is d-free, then S is d-free.*

Proof. This is immediate from the proposition and from the fact that the set of all finite subsets of S is filtered. $\square$

Up to now, the only condition in the definition of a dimensional operator that has not been applied is the exchange axiom. We shall now use it in the proof of the next proposition.

4.1.6. Proposition. *Let E be a set, and let d be a dimensional operator on E. If S is a d-free subset of E and $x \in E - d(S)$, then $S \cup \{x\}$ is d-free.*

Proof. Assume, on the contrary, that $S \cup \{x\}$ is not d-free. Then choose a $y \in S \cup \{x\}$ so that $y \in d((S \cup \{x\}) - \{y\})$. Since $x \notin d(S)$, it is clear that $x \notin S$. If $y = x$, then $x \in d((S \cup \{x\}) - \{x\}) = d(S)$, which contradicts the hypothesis. Thus $y \neq x$, and hence $y \in S$. Since S is d-free, it now follows that $y \notin d(S - \{y\})$. In conclusion, since

$$(S \cup \{x\}) - \{y\} = (S - \{y\}) \cup \{x\},$$

we have

$$y \in d((S - \{y\}) \cup \{x\}) - d(S - \{y\}).$$

But this implies that $x \in d((S - \{y\}) \cup \{y\}) = d(S)$, which also contradicts the hypothesis. $\square$

Before proceeding to the main results, we shall give two useful characterizations of bases.

4.1.7. Proposition. *Let E be a set, let d be a dimensional operator on E, and let $B \subseteq E$. Then the following conditions are equivalent:*
  (a) *B is a d-base.*
  (b) *B is a minimal element of the set of all d-dense subsets of E.*
  (c) *B is a maximal element of the set of all d-free subsets of E.*

Proof. We first show that (a) implies (b). Suppose that B is a d-base. If $S \subset B$, there exists an $x \in B$ such that $S \subseteq B - \{x\}$; we then have $x \notin d(B - \{x\})$ and $d(S) \subseteq d(B - \{x\})$, whence $x \notin d(S)$. Therefore, no proper subset of B is d-dense.

Next, we show that (b) implies (c). Suppose that B is a minimal element of the set of all d-dense subsets of E. It is readily seen that B is d-free: Indeed, if $x \in B$ and $x \in d(B - \{x\})$, then

$$B = (B - \{x\}) \cup \{x\} \subseteq d(B - \{x\}),$$

which implies that $d(B - \{x\})$ is d-dense, and hence that $B - \{x\}$ is d-dense; but this is impossible, because $B - \{x\} \subset B$. To conclude, suppose that $B \subset S \subseteq E$, and choose an $x \in S$ so that $B \subseteq S - \{x\}$; then $S - \{x\}$ is d-dense, which implies that $x \in d(S - \{x\})$. Consequently, no subset of E strictly containing B is d-free.

Finally, we verify that (c) implies (a). Suppose that B is a maximal element of the set of all d-free subsets of E. If $d(B) \subset E$, we can choose an $x \in E - d(B)$; but then $B \cup \{x\}$ is d-free and $B \subset B \cup \{x\}$, which contradicts our assumption. Therefore $d(B) = E$, and B is d-dense. □

We shall now deal with the fundamental questions of dimension theory. We shall begin with the following important theorem, which will be proved with the help of Zorn's lemma.

4.1.8. Theorem. *Let E be a set, and let d be a dimensional operator on E. If S and T are, respectively, a d-free and a d-dense subset of E, then there exists a subset C of T such that $S \cup C$ is a d-base and $S \cap C = \emptyset$.*

Proof. Let Ω denote the set of all subsets A of T such that $S \cup A$ is d-free and $S \cap A = \emptyset$. Then order Ω by the inclusion relation. Clearly $\emptyset \in \Omega$, so that Ω is nonempty. Furthermore, it is seen from 4.1.4 that Ω is inductive. According to Zorn's lemma, there exists a maximal element C in Ω.

To prove the theorem, it remains to verify that $S \cup C$ is d-dense. Assume that this is false. Then T is d-dense but $d(S \cup C)$ is not, which implies that $T \not\subseteq d(S \cup C)$. Thus, an $x \in T$ can be chosen so that $x \notin d(S \cup C)$. Note, in particular, that $x \notin S \cup C$.

Since $S \cup C$ is d-free and $x \notin d(S \cup C)$, it follows from 4.1.6 that $(S \cup C) \cup \{x\}$ is d-free, which means that $S \cup (C \cup \{x\})$ is d-free. We also have $C \subseteq T$ and $x \in T$, so that $C \cup \{x\} \subseteq T$. Finally, since $S \cap C = \emptyset$ and $x \notin S$, we have $S \cap (C \cup \{x\}) = \emptyset$. This shows that $C \cup \{x\} \in \Omega$; and since

$x \notin C$, we see that $C \subset C \cup \{x\}$. But this is incompatible with the maximality of C in Ω.   □

4.1.9. Corollary. *Let E be a set, and let d be a dimensional operator on E. If S and T are, respectively, a d-free and a d-dense subset of E such that $S \subseteq T$, then there exists a d-base B such that $S \subseteq B \subseteq T$.*

4.1.10. Corollary. *Let E be a set, and let d be a dimensional operator on E.*

(i) *Every d-free subset of E is contained in a d-base.*
(ii) *Every d-dense subset of E contains a d-base.*
(iii) *There exists a d-base.*

In the next and final result of the section, elementary properties of cardinal numbers will be used to establish an important property of bases.

4.1.11. Theorem. *Let E be a set, and let d be a dimensional operator on E. Then every two d-bases are equipotent.*

Proof. We shall consider two cases.

Case 1. There exists a finite d-base.

We proceed by induction. For every nonnegative integer n, consider the following statement: If B and C are d-bases such that C is finite and $\mathrm{Card}(C - B) = n$, then B and C are equipotent.

It will suffice to prove that this statement is true for every $n \geq 0$. It is trivially true for $n = 0$: If B and C are d-bases such that $\mathrm{Card}(C - B) = 0$, then $C - B = \varnothing$, which means that $C \subseteq B$; but this, according to 4.1.7, implies that $C = B$.

Now assume that n is a nonnegative integer for which our statement is true. And then consider d-bases B and C such that C is finite and $\mathrm{Card}(C - B) = n + 1$.

Since $C - B \neq \varnothing$, we can choose an $x \in C$ so that $x \notin B$. Then $C - \{x\} \subset C$, and 4.1.7 shows that $C - \{x\}$ is not d-dense; therefore $d(C - \{x\})$ is not d-dense, which implies that $B \not\subseteq d(C - \{x\})$. Thus, we can choose a $y \in B$ so that $y \notin d(C - \{x\})$. Since $C - \{x\}$ is d-free, it now follows from 4.1.6 that $(C - (x)) \cup \{y\}$ is d-free. Furthermore, since $y \in d(C)$, we have

$$y \in d((C - \{x\}) \cup \{x\}) - d(C - \{x\}),$$

whence $x \in d((C - \{x\}) \cup \{y\})$; but then

$$C = (C - \{x\}) \cup \{x\} \subseteq d((C - \{x\}) \cup \{y\}),$$

which shows that $d((C - \{x\}) \cup \{y\})$ is d-dense, so that $(C - \{x\}) \cup \{y\}$ also is d-dense. Consequently $(C - \{x\}) \cup \{y\}$ is a d-base; let it be denoted by $\overline{C}$. Since $x \in C$ and $y \notin C - \{x\}$, we see that C and $\overline{C}$ are equipotent.

In particular, $\overline{C}$ is finite. Also, note that $y \in B$ and $x \in C - B$, whence

$$\overline{C} - B = ((C - \{x\}) \cup \{y\}) - B = (C - B) - \{x\}$$

and

$$\text{Card}(\overline{C} - B) = \text{Card}(C - B) - 1 = n.$$

It then follows from the induction assumption that B and $\overline{C}$ are equipotent. Therefore, B and C are equipotent, as was to be shown.

Case 2. Every d-base is infinite.

It suffices to prove that $\text{Card}(B) \le \text{Card}(C)$ for every two d-bases B and C.

To this end, we first note that $C \subseteq d(B)$ because B is d-dense. For each $x \in C$, we can next choose a finite subset B_x of B for which $x \in d(B_x)$. Let $\overline{B} = \cup_{x \in C} B_x$. It is evident that $C \subseteq d(\overline{B})$: Indeed, if $x \in C$, then $x \in d(B_x)$ and $B_x \subseteq \overline{B}$, which implies that $x \in d(\overline{B})$.

It follows that $d(\overline{B})$ is d-dense, so that $\overline{B}$ is a d-dense subset of B; according to 4.1.7, this shows that $\overline{B} = B$. Finally, the infiniteness of C implies that

$$\text{Card}(B) = \text{Card}(\overline{B}) = \text{Card}\left(\bigcup_{x \in C} B_x\right)$$

$$\le \sum_{x \in C} \text{Card}(B_x) \le \aleph_0 \text{Card}(C) = \text{Card}(C),$$

as required. $\qquad\square$

To conclude the discussion, it only remains to make the obvious definition. We have proved that a base always exists, and that every two bases are equipotent.

Given a set E and a dimensional operator d on E, by the d-**dimension of** E we shall understand the common cardinality of the d-bases.

PROBLEMS

1. Let E be a set, and let d be a dimensional operator on E. Prove the following assertions:
   a. If S is a d-free subset of E and $x \in d(S)$, then the set of all subsets A of S for which $x \in d(A)$ has a smallest element.
   b. If $V \subseteq E$, then the set of all d-free subsets of V admits a maximal element, and $d(S) = d(V)$ for every such maximal element S.
2. Let E be a set, let d be a dimensional operator on E, and let $V \subseteq E$. The mapping $S \to d(V \cup S)$ from $\mathscr{P}(E)$ to $\mathscr{P}(E)$ will be denoted by

d_V. Show that d_V is a dimensional operator on E, and that every d-dense subset of E is d_V-dense.

3. Let E be a set, let d be a dimensional operator on E, and let $V, W \subseteq E$. Prove that the following conditions are equivalent:
   a. If S and T are, respectively, d-free subsets of V and W, then $S \cap T = \varnothing$ and $S \cup T$ is d-free.
   b. Every d-free subset of V is d_W-free.
   c. Every d-free subset of W is d_V-free.
       When these conditions are satisfied, V and W are said to be **d-disjoint**.

4. Let K be a field of prime characteristic p, and let L be an extension field of K. Prove that the mapping $S \to K \vee L^p(S)$ from $\mathscr{P}(L)$ to $\mathscr{P}(L)$ is a dimensional operator on L.
       A subset of L will be said to be **p-independent in L over K**, **p-dense in L over K**, or a **p-base of L over K** according as it is free, dense, or a base relative to the dimensional operator just defined. The dimension of L relative to the latter will be called the **degree of imperfection of L over K**.

5. Show that if K is a field of prime characteristic, and if L is a finitely generated extension of K, then the degree of imperfection of L over K is finite.

6. Let K be a field of prime characteristic p, and let $K(\alpha)$ be a simple extension of K such that $\alpha^p \notin K$ and $\alpha^{p^2} \in K$. Verify the following assertions:
   a. $\{\alpha\}$ is a p-base of $K(\alpha)$ over K.
   b. $\{\alpha^p\}$ is neither p-independent nor p-dense in $K(\alpha)$ over K.
   c. $\{\alpha^p\}$ is a p-base of $K(\alpha^p)$ over K.

7. Let K be a field of prime characteristic p, let L be an extension field of K, and let $S \subseteq L$. Prove the following assertions:
   a. If S is p-dense in L over K, then $L = K \vee L^{p^n}(S)$ for every nonnegative integer n.
   b. S is p-independent in L over K if and only if $K \vee L^p(A) \subset K \vee L^p(S)$ whenever $A \subset S$.
   c. If S is finite, then S is p-independent in L over K if and only if $[K \vee L^p(S) : K \vee L^p] = p^{\mathrm{Card}(S)}$.
   d. If S is nonempty and finite, and if $n = \mathrm{Card}(S)$ and $S = \{\alpha_1, \alpha_2, \ldots, \alpha_n\}$, then S is p-independent in L over K if and only if the family $(\prod_{i=1}^{n} \alpha_i^{k_i})_{0 \leq k_1, k_2, \ldots, k_n \leq p-1}$ is linearly independent over $K \vee L^p$; and S is a p-base of L over K if and only if this family is a linear base of L over $K \vee L^p$.

8. Let K be a field of prime characteristic p, let L be an extension field of K, and let n be a nonnegative integer. Show that the degree of imperfection of L over K is n if and only if $[L : K \vee L^p] = p^n$.

9. Let K be a field of prime characteristic p, let L be an extension field of K having finite degree of imperfection over K, and suppose that $L^{p^n} \subseteq K$ for some positive integer n. Verify the following assertions:

   a. L is finite over K.

   b. The degree of imperfection of L over K is the smallest element of the set of all nonnegative integers k such that there exists a subset of L having cardinality k and generating L over K.

   c. If m_k denotes the degree of imperfection of $K \vee L^{p^k}$ over K for $0 \leq k \leq n$, and if $m = \sum_{k=0}^{n} m_k$, then $[L : K] = p^m$.

10. Let K be a field of prime characteristic, and let L be a finite extension of K. Prove that L is a simple extension of K if and only if its degree of imperfection over K is 0 or 1.

11. Let K be a field of prime characteristic, and let L be a finite extension of K. Show that if the degree of imperfection of L over K is positive, then it is the smallest element of the set of all positive integers k such that there exists a subset of L having cardinality k and generating L over K.

12. Let k be a field of prime characteristic p. Show that every finite extension of K is simple if and only if the degree of imperfection of K over K^p is 0 or 1.

13. Let K be a field of prime characteristic p, let L be an extension field of K, and let M be an extension field of L. Prove that if $S \subseteq L$ and S is p-dense in L over K, and if $T \subseteq M$ and T is p-dense in M over L, then $S \cup T$ is p-dense in M over K.

4.2. TRANSCENDENCE BASES AND TRANSCENDENCE DEGREE

It will now be shown that a certain dimensional operator can be associated in a natural way with every field extension. The corresponding bases and dimension will be seen to play a fundamental role in the study of transcendental extensions.

The dimensional operators in question are defined with the help of the notion of relative algebraic closure. The precise description is contained in the following basic result.

4.2.1. Proposition. *Let K be a field, and let L be an extension field of K. Then the mapping $S \to A(L/K(S))$ from $\mathscr{P}(L)$ to $\mathscr{P}(L)$ is a dimensional operator on L.*

Proof. Let d denote the mapping in question. Of the five conditions defining a dimensional operator, the first and third obviously are satisfied by d. Moreover, the same is true of the second: We need only take into account that for every subset S of L, the field $A(L/K(S))$ is algebraically closed in L.

To show that d satisfies the fourth condition, let $S \subseteq L$ and $\alpha \in d(S)$. Then $\alpha \in L$ and α is algebraic over $K(S)$, so that we have an equality of the form $\Sigma_{i=0}^{n} \gamma_i \alpha^k = 0$, where n is a nonnegative integer, $\gamma_0, \gamma_1, \ldots, \gamma_n \in K(S)$, and $\gamma_n \neq 0$. If Ω denotes the set of all finite subsets of S, then $(K(A))_{A \in \Omega}$ is filtered, and so we have $K(S) = \cup_{A \in \Omega} K(A)$. Therefore, we can choose an $A \in \Omega$ for which $\gamma_0, \gamma_1, \ldots, \gamma_n \in A$; the same equality $\Sigma_{i=0}^{n} \gamma_i \alpha^i = 0$ then shows that α is algebraic over $K(A)$, and hence $\alpha \in d(A)$.

To conclude, we now have to prove that d satisfies the exchange axiom. To do this, suppose that $S \subseteq L, \alpha \in L$, and $\beta \in d(S \cup \{\alpha\}) - d(S)$. Write $P = K(S)$; then the condition $\beta \in d(S \cup \{\alpha\}) - d(S)$ implies that β is algebraic over $P(\alpha)$ and transcendental over P.

Since β is algebraic over $P(\alpha)$, we have an equality of the form $\Sigma_{i=0}^{n} \gamma_i \beta^i = 0$, where n is a nonnegative integer, $\gamma_0, \gamma_1, \ldots, \gamma_n \in P(\alpha)$, and $\gamma_n \neq 0$. Moreover, it is seen that we may take the coefficients $\gamma_0, \gamma_1, \ldots, \gamma_n$ in $P[\alpha]$: Indeed, $P(\alpha)$ is the field of fractions of $P[\alpha]$ in L, and hence $\gamma_0, \gamma_1, \ldots, \gamma_n$ can be expressed as fractions with numerators and denominators in $P[\alpha]$; multiplying our equality by the product of these denominators, we get an equality of the same type, but with coefficients in $P[\alpha]$.

For $0 \leq i \leq n$, we now can write $\gamma_i = f_i(\alpha)$ with $f_i \in P[X]$. Defining a polynomial h in $P[X, Y]$ by $h(X, Y) = \Sigma_{i=0}^{n} f_i(X) Y^i$, we see that $h \neq 0$ and

$$h(\alpha, \beta) = \sum_{i=0}^{n} f_i(\alpha) \beta^i = \sum_{i=0}^{n} \gamma_i \beta^i = 0.$$

Next, write $h(X, Y) = \Sigma_{j=0}^{m} g_j(Y) X^j$, where m is a nonnegative integer, $g_0, g_1, \ldots, g_m \in P[Y]$, and $g_m \neq 0$. Since β is transcendental over P, it follows that $g_m(\beta) \neq 0$. Finally, defining a polynomial g in $P(\beta)[X]$ by $g(X) = \Sigma_{j=0}^{m} g_j(\beta) X^j$, we have $g \neq 0$ and

$$g(\alpha) = \sum_{j=0}^{m} g_j(\beta) \alpha^j = h(\alpha, \beta) = 0,$$

whence α is algebraic over $P(\beta)$. But this means that $\alpha \in d(S \cup \{\beta\})$, as required. $\square$

Before proceeding to the main definitions, we shall show that the free and dense sets associated with our dimensional operators can be described in familiar terms.

4.2.2. Proposition. *Let K be a field, and let L be an extension field of K. Then, with respect to the dimensional operator $S \to A(L/K(S))$ on L, we have:*

(i) A subset S of L is dense if and only if L is algebraic over $K(S)$.

(ii) A subset S of L is free if and only if S is algebraically independent over K.

Proof. Let d denote the dimensional operator under consideration.

Assertion (i) is evident. For if $S \subseteq L$, to say that S is d-dense amounts to saying that $L = A(L/K(S))$, which means that L is algebraic over $K(S)$.

To prove (ii), suppose first that $S \subseteq L$ and S is algebraically dependent over K. Then there exist finite subsets of S that are algebraically dependent over K. From among these, select one with the smallest possible cardinality, and denote it by A. Clearly $A \neq \varnothing$; write $n = \text{Card}(A)$ and $A = \{\alpha_1, \alpha_2, \ldots, \alpha_n\}$, and choose a polynomial f in $K[X_1, X_2, \ldots, X_n]$ such that $f \neq 0$ and $f(\alpha_1, \alpha_2, \ldots, \alpha_n) = 0$. We want to show that S is not d-free; to this end, it will suffice to verify that $\alpha_n \in d(A - \{\alpha_n\})$, since this will imply that A is not d-free, and hence that S is not d-free.

If $n = 1$, then α_n is algebraic over K and $A - \{\alpha_n\} = \varnothing$, so that $\alpha_n \in d(A - \{\alpha_n\})$. Assume next that $n > 1$, and write

$$f(X_1, X_2, \ldots, X_n) = \sum_{i=0}^{m} f_i(X_1, X_2, \ldots, X_{n-1}) X_n^i,$$

where m is a nonnegative integer, $f_0, f_1, \ldots, f_m \in K[X_1, X_2, \ldots, X_{n-1}]$, and $f_m \neq 0$. The choice of A implies then that $f_m(\alpha_1, \alpha_2, \ldots, \alpha_{n-1}) \neq 0$; defining a polynomial g in $K(A - \{\alpha_n\})[Y]$ by

$$g(Y) = \sum_{i=0}^{m} f_i(\alpha_1, \alpha_2, \ldots, \alpha_{n-1}) Y^i,$$

we see that $g \neq 0$ and

$$g(\alpha_n) = \sum_{i=0}^{m} f_i(\alpha_1, \alpha_2, \ldots, \alpha_{n-1}) \alpha_n^i$$

$$= f(\alpha_1, \alpha_2, \ldots, \alpha_n) = 0.$$

Therefore α_n is algebraic over $K(A - \{\alpha_n\})$, which means that $\alpha_n \in d(A - \{\alpha_n\})$.

To conclude the proof of (ii), suppose now that $S \subseteq L$ and S is not d-free. Then there exists a finite subset A of S that is not d-free. Choose an $\alpha \in A$ so that $\alpha \in d(A - \{\alpha\})$. Thus, α is algebraic over $K(A - \{\alpha\})$, and we have a relation of the form $\sum_{i=0}^{m} \gamma_i \alpha^i = 0$, where m is a nonnegative integer, $\gamma_0, \gamma_1, \ldots, \gamma_m \in K(A - \{\alpha\})$, and $\gamma_m \neq 0$. Furthermore, exactly as in the proof of 4.2.1, we can take the coefficients $\gamma_0, \gamma_1, \ldots, \gamma_m$ in $K[A - \{\alpha\}]$. We have to show that S is algebraically dependent over K; to do this, it will suffice to verify that A is algebraically dependent over K.

If $A = \{\alpha\}$, then $A - \{\alpha\} = \varnothing$, so that α is algebraic over K, and hence A is algebraically dependent over K. Now assume that $A \neq \{\alpha\}$; then write $n = \text{Card}(A)$ and $A = \{\alpha_1, \alpha_2, \ldots, \alpha_n\}$, with $\alpha_n = \alpha$. For $0 \leq i \leq m$, we also have $\gamma_i = f_i(\alpha_1, \alpha_2, \ldots, \alpha_{n-1})$, where $f_i \in K[X_1, X_2, \ldots, X_{n-1}]$. Since $\gamma_m \neq 0$, we see that $f_m \neq 0$. Finally, let a polynomial f in $K[X_1, X_2, \ldots, X_n]$ be defined by

$$f(X_1, X_2, \ldots, X_n) = \sum_{i=0}^{m} f_i(X_1, X_2, \ldots, X_{n-1}) X_n^i.$$

It then follows that $f \neq 0$ and

$$f(\alpha_1, \alpha_2, \ldots, \alpha_n) = \sum_{i=0}^{m} f_i(\alpha_1, \alpha_2, \ldots, \alpha_{n-1}) \alpha_n^i$$

$$= \sum_{i=0}^{m} \gamma_i \alpha^i = 0,$$

which implies that A is algebraically dependent over K.   □

Let K be a field, and let L be an extension field of K. By a **transcendence base of L over K** we understand a base with respect to the dimensional operator $S \rightarrow A(L/K(S))$ on L. The dimension of L with respect to this dimensional operator is called the **transcendence degree of L over K**, and is denoted by tr.deg.(L/K).

It is clear that algebraic extensions can be described as the field extensions having transcendence degree 0.

For each of the basic notions defined by a dimensional operator of the type under consideration in this section, we now have a field-theoretical description or term. From now on, the results from the section on dimensional operators, suitably restated in this field-theoretical vocabulary, will be applied freely without explicit comment. Since the reader undoubtedly is familiar with the corresponding results from linear algebra, this should cause no difficulty.

Let K be a field, let L be an extension field of K, and let B and C be two transcendence bases of L over K. Then B and C are equipotent; and L is algebraic over each of $K(B)$ and $K(C)$, so that L has transcendence degree 0 over each of them. However, we cannot assert that L has the same linear degree over $K(B)$ as over $K(C)$, even when L is finite over each of these subfields (problem 5).

4.2.3. Proposition. *Let K be a field, and let L be a finitely generated extension of K. Then L has finite transcendence degree over K; and if B is a transcendence base of L over K, then L is finite over $K(B)$.*

Proof. By hypothesis, $L = K(S)$ for some finite subset S of L. Then L is algebraic over $K(S)$, and hence S contains a transcendence base of L over K; and since S is finite, such a base a finite, whence tr.deg.(L/K) is finite.

Now let B be a transcendence base of L over K. Then L is algebraic over $K(B)$; and since L is finitely generated over $K(B)$, we conclude that L is finite over $K(B)$.   □

4.2.4. Corollary. *Let K be a field, and let L be a finitely generated extension of K. If B is a transcendence base of L over K, and if $T \subseteq L$ and $L = K(T)$, then there exists a finite subset S of T such that $B \cap S = \varnothing$ and $L = K(B \cup S)$.*

Proof. By the proposition, L is finite over $K(B)$. Consequently, $[L : K(B)]$ is a positive integer such that

$$[K(B)(S) : K(B)] \leq [L : K(B)]$$

for every finite subset S of $T - B$. From among the finite subsets of $T - B$, let S be selected so that $[K(B)(S) : K(B)]$ is as large as possible. We claim that $L = K(B)(S)$: Indeed, assume that $K(B)(S) \subset L$; then $T - B \not\subseteq K(B)(S)$, for otherwise

$$L = K(T) = K(B)(T - B) \subseteq K(B)(S) \subset L;$$

choosing an $\alpha \in T - B$ so that $\alpha \notin K(B)(S)$, we see that $S \cup \{\alpha\}$ is a finite subset of $T - B$ and

$$[K(B)(S \cup \{\alpha\}) : K(B)] = [K(B)(S)(\alpha) : K(B)]$$
$$= [K(B)(S)(\alpha) : K(B)(S)][K(B)(S) : K(B)]$$
$$> [K(B)(S) : K(B)],$$

which is incompatible with the choice of S. Therefore $L = K(B)(S)$, and we conclude that S is a finite subset of T for which $B \cap S = \varnothing$ and $L = K(B \cup S)$. $\qquad\square$

It is customary to refer to a finitely generated transcendental extension of a field K as an **algebraic function field over** K. If L is such a field and $n = \mathrm{tr.deg.}(L/K)$, the preceding proposition shows that n is a positive integer, and L is said to be an **algebraic function field in** n **variables over** K.

4.2.5. Proposition. *Let K be a field, let L be an extension field of K, and let M be an extension field of L.*

(i) *If B and C are, respectively, transcendence bases of L over K and of M over L, then $B \cap C = \varnothing$ and $B \cup C$ is a transcendence base of M over K.*

(ii) $\mathrm{tr.deg.}(M/K) = \mathrm{tr.deg.}(M/L) + \mathrm{tr.deg.}(L/K)$.

Proof. We need only concern ourselves with (i). Let B and C be as described in its statement.

Since B and C are, respectively, algebraically independent over K and L, it clearly follows that $B \cup C$ is algebraically independent over K and $B \cap C = \varnothing$.

It remains, therefore, to verify that M is algebraic over $K(B \cup C)$. To do this, note first that since L is algebraic over $K(B)$, every element of L is algebraic over $K(B \cup C)$; since $L(C) = K(B \cup C)(L)$, we see that $L(C)$ is algebraic over $K(B \cup C)$. Also, M is algebraic over $L(C)$. The required conclusion now follows at once. $\qquad\square$

4.2.6. Proposition.   *Let K be a field, let N be an extension field of K, and let L and M be intermediate fields between K and N.*

(i) $\text{tr.deg.}(L \vee M/M) \leq \text{tr.deg.}(L/K)$.

(ii) $\text{tr.deg.}(L \vee M/K) \leq \text{tr.deg.}(L/K) + \text{tr.deg.}(M/K)$.

Proof.   It is evident that (ii) follows from (i) and 4.2.5.

To prove (i), let C be a transcendence base of L over K. Then L is algebraic over $K(C)$, and so every element of L is algebraic over $M(C)$. Since $L \vee M = M(C)(L)$, it now follows that $L \vee M$ is algebraic over $M(C)$. Consequently, C contains a transcendence base B of $L \vee M$ over M, and hence

$$\text{tr.deg.}(L \vee M/M) = \text{Card}(B) \leq \text{Card}(C) = \text{tr.deg.}(L/K),$$

which is what we wanted.   □

It is easy to give an example showing that, even when the transcendence degrees involved are finite, the inequalities in the preceding proposition may be strict (problem 6).

To continue the discussion, we now proceed to derive some results on the extendibility of field monomorphisms from the properties of transcendence bases and degrees established in the foregoing discussion.

4.2.7. Proposition.   *Let K and $\overline{K}$ be fields, and let L and $\overline{L}$ be, respectively, algebraically closed extension fields of K and $\overline{K}$ such that*

$$\text{tr.deg.}(L/K) = \text{tr.deg.}(\overline{L}/\overline{K}).$$

Then every isomorphism from K to $\overline{K}$ is extendible to an isomorphism from L to $\overline{L}$.

Proof.   Let u be an isomorphism form K to $\overline{K}$; and choose transcendence bases B and $\overline{B}$ of L over K and of $\overline{L}$ over $\overline{K}$, respectively. By hypothesis, B and $\overline{B}$ are equipotent, and so we can choose a bijection b from B to $\overline{B}$. Since B and $\overline{B}$ are, respectively, algebraically independent over K and $\overline{K}$, there exists an isomorphism from $K[\mathrm{B}]$ to $\overline{K}[\overline{B}]$ extending u and b. Moreover, since $K(B)$ and $\overline{K}(\overline{B})$ are, respectively, fields of fractions of $K[B]$ and $\overline{K}[\overline{B}]$, this isomorphism can be extended to an isomorphism from $K(B)$ to $\overline{K}(\overline{B})$. Finally, since L and $\overline{L}$ are, respectively, algebraic closures of $K(B)$ and $\overline{K}(\overline{B})$, it follows from 2.2.13 that the latter is extendible to an isomorphism from L to $\overline{L}$; and it is evident that such an isomorphism extends u.   □

We state next two evident consequences of the result just proved.

4.2.8. Corollary.   *If K is a field, then every two algebraically closed extension fields of K having the same transcendence degree over K are K-isomorphic.*

4.2.9. Corollary. *Let K be a field, and let L be an algebraically closed extension field of K. Then every automorphism of K is extendible to an automorphism of L.*

Note that 4.2.7 is a generalization of Steinitz's theorem 2.2.13; and that 4.2.8 is a generalization of the part concerning essential uniqueness in Steinitz's theorem 2.2.14.

4.2.10. Proposition. *Let K be a field, let L be an extension field of K having finite transcendence degree over K, and let M be an algebraically closed extension field of L. Then every K-monomorphism from L to M is extendible to a K-automorphism of M.*

Proof. Let u be a K-monomorphism from L to M. Then

$$\mathrm{tr.deg.}(L/K) = \mathrm{tr.deg.}(u(L)/K),$$

so that these transcendence degrees are equal and finite; but according to 4.2.5, we have

$$\mathrm{tr.deg.}(M/L) + \mathrm{tr.deg.}(L/K) = \mathrm{tr.deg.}(M/K)$$

$$= \mathrm{tr.deg.}(M/u(L)) + \mathrm{tr.deg.}(u(L)/K),$$

and hence

$$\mathrm{tr.deg.}(M/L) = \mathrm{tr.deg.}(M/u(L)).$$

The extendibility of u to a K-automorphism of M now follows at once from 4.2.7. □

4.2.11. Corollary. *Let K be a field, and let L be an algebraically closed extension field of K having finite transcendence degree over K. Then every K-endomorphism of L is a K-automorphism.*

It can be shown that the last proposition and its corollary do not remain valid if the finiteness hypotheses are omitted (problem 11). We shall now end the present section with a brief discussion on the simplest type of transcendental extension. Let K be a field, and let L be a transcendental extension of K. By a **pure base of L over K** we understand a transcendence base of L over K that generates L over K. In other words, a pure base of L over K is a subset of L that generates L over K and is algebraically independent over K. We say that L is **purely transcendental over K**, or that L is a **purely transcendental extension of K**, when L admits a pure base over K.

There exist transcendental extensions that are not purely transcendental (problem 16; and the example given in 4.4.11). Also, it can be shown that every purely transcendental extension admits transcendence bases that are not pure (problems 4 and 5).

Let K be a field. It can be readily seen that the fields of rational functions over K essentially describe the purely transcendental extensions of K.

Indeed, if I is a nonempty set, then $K(X_i)_{i \in I}$ is purely transcendental over K, and

$$\text{tr.deg.}(K(X_i)_{i \in I}/K) = \text{Card}(I).$$

Conversely, consider a purely transcendental extension L of K. If S is a pure base of L over K, we have the K-isomorphism from $K[X_\alpha]_{\alpha \in S}$ to $K[S]$ such that $X_\alpha \to \alpha$ for every $\alpha \in S$; and since $K(X_\alpha)_{\alpha \in S}$ and L are, respectively, fields of fractions of $K[X_\alpha]_{\alpha \in S}$ and $K[S]$, this is extendible to a K-isomorphism from $K(X_\alpha)_{\alpha \in S}$ to L.

Consequently, we can say that given a nonzero cardinal number $\mathfrak{m}$, there exist purely transcendental extensions of K having transcendence degree $\mathfrak{m}$ over K; and if I is a set such that $\text{Card}(I) = \mathfrak{m}$, an extension field of K is purely transcendental and has transcendence degree $\mathfrak{m}$ over K if and only if it is K-isomorphic to $K(X_i)_{i \in I}$.

Finally, note that if K is a field and L is a transcendental extension of K, to say that there exists a transcendence base of L over K amounts to saying that there exists an intermediate field between K and L that is purely transcendental over K and admits L as an algebraic extension; every such intermediate field, of course, has the same transcendence degree over K as L.

The preceding remarks should not induce the reader to think that the study of purely transcendental extensions is devoid of interest. On the contrary, certain problems on purely transcendental extensions have been the subject of extensive investigations.

It is natural, for instance, to pose the following question: Given a field K and a purely transcendental extension L of K, is every intermediate field properly between K and L purely transcendental over K?

This was answered affirmatively when $\text{tr.deg.}(L/K) = 1$ by Lüroth (problem 17); more generally, it was shown by Weber and Igusa that every intermediate field of transcendence degree 1 over K is purely transcendental over K. When K is algebraically closed, it was shown by Castelnuovo and Zariski that every intermediate field of transcendence degree 2 over K is purely transcendental over K. On the other hand, it has been shown that when K is not algebraically closed, an intermediate field of transcendence degree 2 over K need not be purely transcendental over K; and that, even when K is algebraically closed, intermediate fields of transcendence degree greater than 2 need not be purely transcendental over K.

To conclude, let us mention a classical problem formulated by Noether. Suppose that $L = K(X_1, X_2, \ldots, X_n)$, where n is a positive integer. We have seen in 3.4.22 that $\text{Sym}(n)$ is naturally embeddable in $\text{Gal}(L/K)$; in this way, every subgroup of $\text{Sym}(n)$ can be "identified" with a subgroup of $\mathrm{Gal}(L/K)$. Noether's problem is whether the field of invariants of every subgroup of $\mathrm{Sym}(n)$ is purely transcendental over K. As shown in 3.4.22, this is true for the whole group $\mathrm{Sym}(n)$; however, it was found by Swan that when $n = 47$, there exists a transitive cyclic subgroup whose field of invariants is not purely transcendental over K.

PROBLEMS

1. Let K be a field, let L be an extension field of K, and let $S, T \subseteq L$. Show that the following conditions are equivalent.
   a. $S \cup T$ is algebraically independent over K, and $S \cap T = \varnothing$.
   b. S is algebraically independent over K, and T is algebraically independent over $K(S)$.
   c. T is algebraically independent over K, and S is algebraically independent over $K(T)$.
2. Let K be a field, let L be an algebraic extension of K, and let M be an extension field of L. Show that, if a subset of M is algebraically independent over K, then it is algebraically independent over L.
3. Let K be a field, and let L be an extension field of M such that $\mathrm{tr.deg.}(L/K) > 1$. Prove that L is not a simple extension of K.
4. Let K be a field, let L be an extension field of K, and let B be a transcendence base of L over K. Let $(k_\alpha)_{\alpha \in B}$ be a family of positive integers indexed by B, and let C be the set consisting of all elements of L of the form α^{k_α} with $\alpha \in B$. Prove that C is a transcendence base of L over K.
5. Let K be a field, and let $K(\alpha)$ be a simple transcendental extension of K. Show that if n is a positive integer, then $\{\alpha^n\}$ is a transcendence base of $K(\alpha)$ over K, and $[K(\alpha): K(\alpha^n)] = n$.
6. Let K be a field, and let $K(\alpha)$ be a simple transcendental extension of K. Prove that there exist intermediate fields L and M between K and $K(\alpha)$ for which the inequalities appearing in 4.2.6 are strict.
7. Let K be a field, and let L be a transcendental extension of K. Prove that
$$\mathrm{Card}(L) = \aleph_0 \mathrm{Card}(K)\mathrm{tr.deg.}(L/K).$$
8. Prove that each of $\mathbf{R}$ and $\mathbf{C}$ has transcendence degree $2^{\aleph_0}$ over $\mathbf{Q}$.
9. Prove that $\mathrm{Card}(\mathrm{Aut}(\mathbf{C})) = 2^{2^{\aleph_0}}$.
10. Let K be a field, and let L be an algebraically closed extension field of K. Prove the following assertions:
    a. If K has characteristic 0, then K is invariant in L.
    b. If K has prime characteristic, then $P(L/K) = \mathrm{Inv}(\mathrm{Gal}(L/K))$.
    c. If $\mathrm{Gal}(L/K)$ is compact, then L is algebraic over K.
11. Let K be a field, and let L be an algebraically closed extension field of K having infinite transcendence degree over K. Verify the following

assertions:

   a. If $\mathfrak{m}$ is a cardinal number such that $\mathfrak{m} \leq \text{tr.deg.}(L/K)$, then the set of all K-endomorphisms u of L such that $\text{tr.deg.}(L/u(L)) = \mathfrak{m}$ is infinite.

   b. The set of all K-endomorphisms of L that are not K-automorphisms is infinite.

12. Let K be a field, let L be a transcendental extension of K, and let M be an algebraically closed extension field of L. Prove that the set of all K-monomorphisms from L to M is infinite.

13. Let K be a field, let L be an extension field of K having infinite transcendence degree over K, and let M be an algebraically closed extension field of L such that

$$\text{tr.deg.}(L/K) < \text{tr.deg.}(M/K).$$

Show that every K-monomorphism from L to M is extendible to a K-automorphism of M.

14. Let K be a field of prime characteristic, and let L be a finitely generated transcendental extension of K. Show that the degree of imperfection of L over K is positive.

15. Show that if K is a field and L is a purely transcendental extension of K, then K is algebraically closed in L.

16. Let K be a field such that $\text{Char}(K) \neq 3$, let $K(\alpha)$ be a simple transcendental extension of K, and form a simple extension $K(\alpha, \beta)$ of $K(\alpha)$ by adjunction of a cube root β of $1 - \alpha^3$ in an extension field of $K(\alpha)$. Show that L is not purely transcendental over K.

17. The following statement is known as **Lüroth's theorem**: If K is a field, and if $K(\alpha)$ is a simple transcendental extension of K, then every intermediate field properly between K and $K(\alpha)$ is a simple transcendental extension of K.

     Prove this result by verifying the assertions in the sketch that follows.

Let E be a subfield of $K(\alpha)$ such that $K \subset E$. Recall that, as shown in 3.2.7, $K(\alpha)$ is finite over E. Denote by f the minimal polynomial of α over E, and write $f(X) = \sum_{i=0}^{n-1} \gamma_i X^i + X^n$, where n is a positive integer and $\gamma_0, \gamma_1, \ldots, \gamma_{n-1} \in E$. Note that $\{\gamma_0, \gamma_1, \ldots, \gamma_{n-1}\} \not\subseteq K$, for otherwise α would be algebraic over K; thus, choose a $\gamma \in \{\gamma_0, \gamma_1, \ldots, \gamma_{n-1}\}$ so that $\gamma \notin K$, and write $\gamma = u(\alpha)/v(\alpha)$, where u and v are relatively prime polynomials in $K[X]$. Then:

   a. $\gamma v(X) - u(X)$ is divisible by $f(X)$ in $E[X]$.

   b. There exist polynomials $c_0, c_1, \ldots, c_n$ in $K[X]$ such that $\gamma_i = c_i(\alpha)/c_n(\alpha)$ for $0 \leq i \leq n-1$ and such that $\sum_{i=0}^{n} c_i(X) Y^i$ is primitive in $K[X][Y]$.

c. If $c_0, c_1, \ldots, c_n$ are as described in (ii), then

$$\deg_X \left(\sum_{i=0}^{n} c_i(X) Y^i \right) \geq [K(\alpha) : K(\gamma)],$$

and there exists a $\lambda \in K^*$ such that the equality

$$u(X) v(Y) - u(Y) v(X) = \lambda \left(\sum_{i=0}^{n} c_i(X) Y^i \right)$$

holds in $K[X, Y]$.
   d. $E = K(\gamma)$.

4.3. SPECIALIZATIONS AND PLACES OF FIELDS

Given a homomorphism from a subdomain of a field K to a field F, it is natural to investigate its extendibility to a homomorphism from a larger subdomain of K to F. We shall devote this section to the study of this problem. We shall see that in an important particular case, a complete description can be given of those homomorphisms that are not so extendible. This will then be used to derive the famous Nullstellensatz—the zeros theorem—of Hilbert.

In order to deal with the extendibility problem just formulated, we shall have to introduce the notions of local ring and valuation domain. Although these are most interesting and of fundamental importance in commutative algebra, their detailed study cannot be undertaken in this book. We shall therefore confine ourselves to the most elementary properties in the particular case needed here.

A ring is said to be **local** when it possesses a unique maximal ideal. If A is a local ring, its maximal ideal will be denoted by $m(A)$.

It is clear that a ring is a field if and only if it is local and its maximal ideal is null. Also, as noted in 0.0.4, an element of a ring is noninvertible if and only if it belongs to a maximal ideal; therefore a ring is local if and only if its noninvertible elements make up an ideal, in which case this ideal is necessarily the maximal ideal.

We proceed next to describe what in commutative algebra is known as the process of localization of a ring at a prime ideal. Although this can be done for arbitrary rings, we shall concern ourselves exclusively with domains, since this is what is needed here. It should be mentioned, however, that the process of localization for general rings requires slightly more involved technicalities.

Consider a field K, a subdomain A of K, and a prime ideal $\mathscr{P}$ of A. It is readily verified that the elements of K expressible as fractions with numerators in A and denominators in $A - \mathscr{P}$ make up an intermediate domain between A and the field of fractions of A in K. This domain will be denoted by $A_{\mathscr{P}}$, and will be called the **localization of A at $\mathscr{P}$ in K**.

The notion just defined is essentially independent of the field containing the given domain. For suppose that A is a domain and $\mathscr{P}$ is a prime ideal of A, and that K and L are fields containing A as a common subdomain. Then the localizations of A at $\mathscr{P}$ in K and L correspond to each other by the A-isomorphism between the fields of fractions of A in K and L. In other words, the localizations of A at $\mathscr{P}$ are "identified" when the fields of fractions of A are "identified" in the usual manner.

4.3.1. Examples

a. If A is a domain, then the localizations of A at its null ideal are the fields of fractions of A.

b. If A is a local domain, then $A - \mathrm{m}(A) = A^*$, and hence $A_{\mathrm{m}(A)} = A$.

c. Let A be a factorial domain, and let π be an irreducible element in A. If K is a field containing A as a subdomain, then the subdomain $A_{\pi A}$ of K consists of the elements of K expressible as fractions α/β with $\alpha, \beta \in A$ and β not divisible by π in A.   □

As stated in the following proposition, the preceding construction associates a local domain with every prime ideal of a given domain.

4.3.2. Proposition.   *Let K be a field, let A be a subdomain of K, and let $\mathscr{P}$ be a prime ideal of A.*

(i) *$\mathscr{P}A_{\mathscr{P}}$ consists of all fractions in K of the form α/β with $\alpha \in \mathscr{P}$ and $\beta \in A - \mathscr{P}$.*

(ii) *$A_{\mathscr{P}}^*$ consists of all fractions in K of the form α/β with $\alpha, \beta \in A - \mathscr{P}$.*

(iii) *$A_{\mathscr{P}}$ is local and $\mathrm{m}(A_{\mathscr{P}}) = \mathscr{P}A_{\mathscr{P}}$.*

(iv) *$\mathscr{P} = A \cap \mathscr{P}A_{\mathscr{P}}$.*

We can now turn to the main topic of the section. Prime ideals occur naturally in the discussion, because the kernels of the homomorphisms with which we shall be concerned are prime ideals.

Let K and F be fields. By a **specialization of K in F** we shall understand a homomorphism from a subdomain of K to F. If u is a specialization of K in F, the subdomain of K on which u is defined will be called the **domain of definition of u**.

As usual, if u and v are specializations of K in F, to say that v extends u means that the domain of definition of u is contained in that of v, and that u and v agree on the domain of definition of u; and to say that v properly extends u means that v extends u and $u \neq v$.

We shall show first that a specialization can always be uniquely extended to the localization of its domain of definition at its kernel.

4.3.3. Proposition. *Let K and F be fields, and let u be a specialization of K in F. If A denotes the domain of definition of u, and if $\mathscr{P} = \text{Ker}(u)$, then there exists a unique specialization of K in F extending u and having $A_{\mathscr{P}}$ as its domain of definition; and its kernel is $\mathscr{P}A_{\mathscr{P}}$.*

Proof. Since $u(\beta) \neq 0$ for every $\beta \in A - \mathscr{P}$, it is seen that there exists a unique homomorphism v from $A_{\mathscr{P}}$ to F such that $\alpha/\beta \to u(\alpha)/u(\beta)$ whenever $\alpha \in A$ and $\beta \in A - \mathscr{P}$. Clearly, v is the only homomorphism from $A_{\mathscr{P}}$ to F extending u. As to the final assertion, note that if $\alpha \in A$ and $\beta \in A - \mathscr{P}$, then $v(\alpha/\beta) = 0$ if and only if $u(\alpha) = 0$, so that $\alpha/\beta \in \text{Ker}(v)$ if and only if $\alpha \in \mathscr{P}$; but this, according to 4.3.2, means that $\text{Ker}(v) = \mathscr{P}A_{\mathscr{P}}$. $\square$

4.3.4. Corollary. *Let K and F be fields, and let u be a specialization of K in F that is not properly extendible to a specialization of K in F. Then the domain of definition of u is local and has $\text{Ker}(u)$ as its maximal ideal; and $\text{Im}(u)$ is a subfield of F.*

Proof. The first assertion is clear from the proposition. And the second follows from the first by taking into account that if A denotes the domain of definition of u, the rings $A/\text{Ker}(u)$ and $\text{Im}(u)$ are isomorphic. $\square$

Let K and F be fields, and let u be a specialization of K in F. Let A denote the domain of definition of u, and write $\mathscr{P} = \text{Ker}(u)$. Note that $\mathscr{P}A_{\mathscr{P}}$ is null if and only if $\mathscr{P}$ is null, that is, if and only if u is monomorphism. The nonzero elements of $\mathscr{P}A_{\mathscr{P}}$ are the fractions α/β with $\alpha \in \mathscr{P} - \{0\}$ and $\beta \in A - \mathscr{P}$, and the inverse in K of such a fraction is the fraction β/α; and the latter cannot belong to the domain of definition of a specialization v of K in F extending u, for otherwise

$$u(\beta) = v(\beta) = v(\alpha(\beta/\alpha)) = v(\alpha)v(\beta/\alpha) = u(\alpha)v(\beta/\alpha),$$

which is impossible because $u(\alpha) = 0$ and $u(\beta) \neq 0$.

In conclusion, u can be extended to a specialization of K in F having $A_{\mathscr{P}}$ and $\mathscr{P}A_{\mathscr{P}}$ as its domain of definition and kernel, respectively; and a condition that will ensure no further extendibility of u to a specialization of K in F is that the inverse in K of every element of $K - A_{\mathscr{P}}$ should belong to $\mathscr{P}A_{\mathscr{P}}$.

Thus, for given fields K and F, we are led to the consideration of those specializations u of K in F having the property that the inverse in K of every element of K not in the domain of definition of u belongs to $\text{Ker}(u)$. Before studying these specializations, however, it will be convenient to look at this property as a condition on their domains of definition.

Let K be a field. By a **valuation domain** in K we shall understand a subdomain A of K such that $1/\alpha \in A$ for every $\alpha \in K - A$.

The following three remarks are evident from the definition just given. First, if K is a field, then K is the only valuation domain in K that is a field. Next, if K is a field and A is a valuation domain in K, then K is a field of fractions of A, and every intermediate domain between A and K is a valuation domain in K. Finally, if K is a field and L is an extension field of K, and if A is a valuation domain in L, then $A \cap K$ is a valuation domain in K.

4.3.5. Example. Let A be a factorial domain, and let K be a field of fractions of A. Then, for every irreducible element π in A, the localization $A_{\pi A}$ of A at πA is a valuation domain in K.

Indeed, let $\alpha \in K - A_{\pi A}$; and write $\alpha = \sigma/\tau$, where σ and τ are relatively prime elements in A. Since $\alpha \notin A_{\pi A}$, we have $\tau \in \pi A$, and hence $\sigma \notin \pi A$; therefore $1/\alpha = \tau/\sigma \in A_{\pi A}$, as required.   □

4.3.6. Proposition. *Let A be a domain, and let K be a field of fractions of A. Then the following conditions are equivalent:*
   (a) *A is a valuation domain in K.*
   (b) *The set of all principal ideals of A is a chain.*
   (c) *The set of all ideals of A is a chain.*

Proof. To show that (a) implies (b), suppose that A is a valuation domain in K, and let $\alpha, \beta \in A - \{0\}$. If $\alpha/\beta \in A$, then $\alpha \in \beta A$, and hence $\alpha A \subseteq \beta A$. On the other hand, if $\alpha/\beta \notin A$, then $\beta/\alpha = 1/(\alpha/\beta) \in A$; but then $\beta \in \alpha A$, so that $\beta A \subseteq \alpha A$. We conclude that either $\alpha A \subseteq \beta A$ or $\beta A \subseteq \alpha A$.

To prove that (b) implies (c), suppose that the principal ideals of A form a chain, and let $\mathscr{A}$ and $\mathscr{B}$ be ideals of A. Assume that $\mathscr{A} \not\subseteq \mathscr{B}$; we have to show that $\mathscr{B} \subseteq \mathscr{A}$. To this end, choose an $\alpha \in \mathscr{A}$ so that $\alpha \notin \mathscr{B}$. For every $\beta \in \mathscr{B}$, we then have $\alpha \notin \beta A$, so that $\alpha A \not\subseteq \beta A$; therefore $\beta A \subseteq \alpha A$, and hence $\beta \in \mathscr{A}$. Thus $\mathscr{B} \subseteq \mathscr{A}$.

Finally, we have to prove that (c) implies (a). Suppose that the ideals of A form a chain, and let $\alpha \in K - A$. Write $\alpha = \sigma/\tau$, where $\sigma, \tau \in A - \{0\}$. We cannot have $\sigma A \subseteq \tau A$, since this would imply that $\alpha = \sigma/\tau \in A$. Consequently $\tau A \subseteq \sigma A$, so that $1/\alpha = \tau/\sigma \in A$.   □

4.3.7. Corollary. *Let K be a field, and let A be a valuation domain in K.*
   (i) *A is local.*
   (ii) *If $\alpha \in K - A$, then $1/\alpha \in \mathfrak{m}(A)$.*

Proof. To prove (i), we first recall from 0.0.4 that A possesses a maximal ideal; and then note that, by the proposition, A cannot possess two distinct maximal ideals.

To verify (ii), let $\alpha \in K - A$. Then $1/\alpha \in A$; and since $1/(1/\alpha) = \alpha \notin A$, we see that $1/\alpha \notin A^*$. Therefore $1/\alpha \in A - A^* = \mathfrak{m}(A)$.   □

We are now prepared to define and study the main concept of the present section. Let K and F be fields. By a **place of** K **in** F we shall understand a specialization u of K in F such that $1/\alpha \in \operatorname{Ker}(u)$ for every $\alpha \in K$ not belonging to the domain of definition of u.

4.3.8. Example. Let K be a field. If A is a valuation domain in K, then the natural projection from A to $A/\mathrm{m}(A)$ is a place of K in $A/\mathrm{m}(A)$. $\square$

Let A be a factorial domain, and let π be an irreducible element in A. We saw in 4.3.5 that if K is a field of fractions of K, then $A_{\pi A}$ is a valuation domain in K; therefore, according to 4.3.8, the natural projection from $A_{\pi A}$ to $A_{\pi A}/\pi A_{\pi A}$ is a place of K in $A_{\pi A}/\pi A_{\pi A}$. This place is said to be the π-**adic place of K**.

4.3.9. Proposition. *Let K and F be fields. Then a specialization of K in F is a place if and only if it is not properly extendible to a specialization of K in F and its domain of definition is a valuation domain in K.*

Proof. Let u be a specialization of K in F, and denote its domain of definition by A.

Suppose first that u is a place. Since $\operatorname{Ker}(u) \subset A$, we clearly have $1/\alpha \in A$ for every $\alpha \in K - A$, and so A is a valuation domain in K. Assume next that there exists a specialization v of K in F properly extending u, and denote its domain of definition by B. Then $A \subset B$, and we can choose an $\alpha \in B$ so that $\alpha \notin A$. Consequently $1/\alpha \in \operatorname{Ker}(u)$, which means that $1/\alpha \in A$ and $u(1/\alpha) = 0$; also, we have $\alpha, 1/\alpha \in B$, so that $1/\alpha \in B^*$ and $v(1/\alpha) \neq 0$. But this is a contradiction, because $u(1/\alpha) = v(1/\alpha)$. Therefore, such v cannot exist.

Now suppose, conversely, that A is a valuation domain in K and that u is not properly extendible to a specialization of K in F. By 4.3.7, it then follows that A is local, and that $1/\alpha \in \mathrm{m}(A)$ for every $\alpha \in K - A$; moreover, according to 4.3.4, we have $\operatorname{Ker}(u) = \mathrm{m}(A)$. Thus, $1/\alpha \in \operatorname{Ker}(u)$ for every $\alpha \in K - A$, which means that u is a place. $\square$

It can now be seen that, in a certain sense, the example given in 4.3.8 describes all possible places.

More precisely, let K and F be fields. If u is a place of K in F, and if A denotes its domain of definition, it follows from 4.3.9 and 4.3.4 that A is a valuation domain in K and $\mathrm{m}(A) = \operatorname{Ker}(u)$; therefore, there exists an isomorphism from $A/\mathrm{m}(A)$ to $\operatorname{Im}(u)$ such that $\alpha + \mathrm{m}(A) \to u(\alpha)$ for every $\alpha \in A$.

Consequently, if u and v are places of K in F with the same domain of definition, then there exists an isomorphism from $\operatorname{Im}(u)$ to $\operatorname{Im}(v)$ such that $u(\alpha) \to v(\alpha)$ for every α in the common domain of definition. Two such places, therefore, can be considered as essentially identical.

So far in our discussion on specializations and places of a field K in a field F we have been dealing with two arbitrary fields K and F. The role of the field F in the discussion, especially in the light of the preceding remarks, does not appear to be of major significance.

We shall now show that under an additional assumption ensuring that F is "sufficiently large", the places of K in F can be characterized by a weaker condition than that given in the preceding proposition.

4.3.10. Theorem. *Let K be a field, and let F be an algebraically closed field. Then every specialization of K in F that is not properly extendible to a specialization of K in F is a place.*

Proof. Let u be a specialization of K in F with the indicated property, and denote by A and $\mathscr{P}$, respectively, its domain of definition and its kernel. According to the preceding proposition, in order to prove that u is a place, it suffices to verify that A is a valuation domain in K. To do this, we shall require the following lemma.

Lemma. *Let P be a field, let C be a subdomain of P, and let $\mathscr{J}$ be a proper ideal of C. If $\pi \in P^*$, then either $\mathscr{J}C[\pi] \subset C[\pi]$ or $\mathscr{J}C[1/\pi] \subset C[1/\pi]$.*

Proof. Assume, on the contrary, that $\mathscr{J}C[\pi] = C[\pi]$ and $\mathscr{J}C[1/\pi] = C[1/\pi]$. Then there obtain equalities of the form

$$\sum_{i=0}^{m} \gamma_i \pi^i = 1 \quad \text{and} \quad \sum_{j=0}^{n} \frac{\delta_j}{\pi^j} = 1,$$

where m and n are nonnegative integers and $\gamma_0, \gamma_1, \ldots, \gamma_m, \delta_0, \delta_1, \ldots, \delta_n \in \mathscr{J}$. Since $\mathscr{J} \subset C$, it is seen that m and n are positive. We may assume, furthermore, that m and n are the smallest positive integers for which equalities of the type above hold. Finally, by the symmetry of the situation, we may assume that $m \leq n$.

Dividing the first of our equalities by π^n and multiplying the second by $1 - \gamma_0$, we get

$$\sum_{i=0}^{m} \frac{\gamma_i}{\pi^{n-i}} = \frac{1}{\pi^n} \quad \text{and} \quad \sum_{j=0}^{n} \frac{(1-\gamma_0)\delta_j}{\pi^j} = 1 - \gamma_0,$$

whence

$$\frac{1-\gamma_0}{\pi^n} = \sum_{i=1}^{m} \frac{\gamma_i}{\pi^{n-i}}$$

and

$$1 - \gamma_0 = \sum_{j=0}^{n-1} \frac{(1-\gamma_0)\delta_j}{\pi^j} + \sum_{i=1}^{m} \frac{\gamma_i \delta_n}{\pi^{n-i}}.$$

Thus

$$\gamma_0 + \sum_{j=0}^{n-1} \frac{(1-\gamma_0)\delta_j}{\pi^j} + \sum_{i=1}^{m} \frac{\gamma_i \delta_n}{\pi^{n-i}} = 1,$$

which shows that 1 can be written as a polynomial expression in $1/\pi$ with coefficients in $\mathcal{I}$ and in which the exponents are less than n. This, however, is incompatible with the minimal choice of n. The lemma is proved.

We can now proceed to prove that A is a valuation domain in K. By the lemma, it will suffice to show that the conditions $\pi \in K^*$ and $\mathcal{P}A[\pi] \subset A[\pi]$ imply $\pi \in A$. To this end, let $E = \mathrm{Im}(u)$; then denote by $\bar{u}$ the homomorphism from A to E defined by restriction of u, and by $\bar{\bar{u}}$ the homomorphism from $A[X]$ to $E[X]$ defined by $\bar{u}$.

Let $\mathcal{I}$ denote the kernel of the homomorphism $f(X) \rightarrow f(\pi)$ from $A[X]$ to $A[\pi]$. Since $\bar{u}$ is surjective, so is $\bar{\bar{u}}$; therefore $\bar{\bar{u}}(\mathcal{I})$ is an ideal of $E[X]$. We now claim that $\bar{\bar{u}}(\mathcal{I}) \subset E[X]$. Indeed, the equality $\bar{\bar{u}}(\mathcal{I}) = E[X]$ implies the existence of a polynomial f in $A[X]$ such that $f(\pi) = 0$ and $\bar{u}f = 1$; writing $f(X) = \sum_{i=0}^{n} \gamma_i X^i$ with a nonnegative integer n and $\gamma_0, \gamma_1, \dots, \gamma_n \in A$, we see that

$$\sum_{i=0}^{n} \gamma_i \pi^i = 0, u(\gamma_0) = 1 \quad \text{and} \quad u(\gamma_i) = 0 \quad \text{for } 1 \le i \le n;$$

therefore $1 - \gamma_0, \gamma_1, \gamma_2, \dots, \gamma_n \in \mathcal{P}$, and the equalities

$$(1 - \gamma_0) - \sum_{i=1}^{n} \gamma_i \pi^i = 1 - \sum_{i=0}^{n} \gamma_i \pi^i = 1$$

lead to the conclusion that $1 \in \mathcal{P}A[\pi]$, which contradicts the assumed strict inclusion $\mathcal{P}A[\pi] \subset A[\pi]$.

Thus, $\bar{\bar{u}}(\mathcal{I})$ is a proper ideal of $E[X]$. Moreover, according to 4.3.9 and 4.3.4, E is a field, so that $E[X]$ is a principal ideal domain. We can then write $\bar{\bar{u}}(\mathcal{I}) = hE[X]$, where h is either the zero polynomial or a nonconstant polynomial in $E[X]$. In either case, since F is algebraically closed, we can choose a $\rho \in F$ such that $h(\rho) = 0$.

Consequently, for every $f \in \mathcal{I}$ we have $\bar{u}f = \bar{\bar{u}}(f) \in \bar{\bar{u}}(\mathcal{I})$, so that $\bar{u}f \in hE[X]$ and $\bar{u}f(\rho) = 0$. This means that the kernel of the surjective homomorphism $f(X) \rightarrow f(\pi)$ from $A[X]$ to $A[\pi]$ is contained in that of the homomorphism $f(X) \rightarrow \bar{u}f(\rho)$ from $A[X]$ to F. It then follows that there exists a homomorphism from $A[\pi]$ to F such that $f(\pi) \rightarrow \bar{u}f(\rho)$ for every $f \in A[X]$. Specializing the latter condition to the case of constant polynomials, we see that this homomorphism extends u, and hence it is a specialization of K in F extending u. Our hypothesis on u now implies that $A = A[\pi]$, whence $\pi \in A$.   □

The preceding theorem is basic in the study of places. It further exemplifies the remarkable properties of algebraically closed fields. It will now be used to give a quick proof of the following important extension theorem.

4.3.11.   Theorem (van der Waerden).   *Let K be a field, and let F be an algebraically closed field. Then every specialization of K in F is extendible to a place of K in F.*

Proof. An obvious application of Zorn's lemma shows that if u is a specialization of K in F, then there exists a specialization v of K in F extending u and that cannot be properly extended to a specialization of K in F; and according to the preceding theorem, such a specialization v is a place.   □

4.3.12. Corollary (Chevalley).   *Let K be a field. If A is a subdomain of K and $\mathscr{P}$ is a prime ideal of A, then there exists a valuation domain B in K such that $A \subseteq B$ and $\mathscr{P} = A \cap \mathrm{m}(B)$.*

Proof. Let F be an algebraic closure of a field of fractions of the domain $A/\mathscr{P}$, and let u denote the homomorphism $\alpha \to \alpha + \mathscr{P}$ from A to F. Then u is a specialization of K in F with A as its domain of definition and such that $\mathrm{Ker}(u) = \mathscr{P}$. By the theorem, u can be extended to a place v of K in F. If B denotes the domain of definition of v, then 4.3.9 and 4.3.4 show that B is a valuation domain in K and $\mathrm{Ker}(v) = \mathrm{m}(B)$; and since v extends u, we have

$$A \subseteq B \quad \text{and} \quad \mathscr{P} = \mathrm{Ker}(u) = A \cap \mathrm{Ker}(v) = A \cap \mathrm{m}(B),$$

which is what was needed.   □

We shall not pursue the general study of places beyond the extension theorem just proved (except for some additional results given as problems). Instead, we shall conclude this section by deriving the celebrated result of Hilbert commonly known as the Nullstellensatz.

We shall follow an argument due to Zariski. The extension theorem will be applied in the proof of the following result (to which reference was made in the remarks on 2.1.11 immediately after 2.1.12).

4.3.13. Theorem (Zariski).   *Let K be a field, let L be an extension field of K, and let D be a finite subset of L. If $K[D] = K(D)$, then $D \subseteq A(L/K)$.*

Proof. Assume, on the contrary, that $K[D] = K(D)$ but $D \nsubseteq A(L/K)$. Put $n = \mathrm{Card}(D)$ and $m = \mathrm{tr.deg.}(K(D)/K)$. It is then seen that $1 \le m < n$: Indeed, since $K(D)$ is transcendental over K, we have $1 \le m \le n$; and the equality $m = n$ cannot hold, for otherwise D would be algebraically independent over K, and hence $K[D]$ would be K-isomorphic to $K[X_1, X_2, \ldots, X_n]$, which is not a field.

We can now write $D = \{\delta_1, \delta_2, \ldots, \delta_n\}$, where $\{\delta_1, \delta_2, \ldots, \delta_m\}$ is a transcendence base of $K(D)$ over K. Then the elements $\delta_{m+1}, \delta_{m+2}, \ldots, \delta_n$ of $K(D)$ are algebraic over $K(\delta_1, \delta_2, \ldots, \delta_m)$. Since $K(\delta_1, \delta_2, \ldots, \delta_m)$ is a field of fractions of $K[\delta_1, \delta_2, \ldots, \delta_m]$, for $m + 1 \le i \le n$ we have an equality of the form

$$\sum_{j=0}^{d_i} h_{i,j}(\delta_1, \delta_2, \ldots, \delta_m) \delta_i^j = 0,$$

where d_i is a positive integer, $h_{i,0}, h_{i,1}, \ldots, h_{i,d_i} \in K[X_1, X_2, \ldots, X_m]$, and $h_{i,d_i} \neq 0$. Write $h = \prod_{i=m+1}^n h_{i,d_i}$; then $h \in K[X_1, X_2, \ldots, X_m]$ and $h \neq 0$.

Let F be an algebraic closure of K. Since F is infinite, we can choose a $(\rho_1, \rho_2, \ldots, \rho_m) \in F^{(m)}$ so that $h(\rho_1, \rho_2, \ldots, \rho_m) \neq 0$. The algebraic independence of $(\delta_1, \delta_2, \ldots, \delta_m)$ over K now implies the existence of a K-homomorphism from $K[\delta_1, \delta_2, \ldots, \delta_m]$ to F such that $\delta_i \to \rho_i$ for $1 \leq i \leq m$. This K-homomorphism is a specialization of $K(D)$ in F; therefore, according to the preceding theorem, it is extendible to a place u of $K(D)$ in F.

Let A denote the domain of definition of u. We contend that $A = K(D)$. To see this, assume that $A \subset K(D)$; it then follows from

$$K[\delta_1, \delta_2, \ldots, \delta_m] \subseteq A \quad \text{and} \quad K[D] = K(D)$$

that there exists an index k such that $m+1 \leq k \leq n$ and $\delta_k \notin A$. Consequently $\delta_k \neq 0$ and $1/\delta_k \in \mathrm{Ker}(u)$; but then, since

$$\sum_{j=0}^{d_k} \frac{h_{k,j}(\delta_1, \delta_2, \ldots, \delta_m)}{\delta_k^{d_k - j}} = \left(\frac{1}{\delta_k^{d_k}} \right) \left(\sum_{j=0}^{d_k} h_{k,j}(\delta_1, \delta_2, \ldots, \delta_m) \delta_k^j \right) = 0,$$

it follows that

$$u\left(h_{k,d_k}(\delta_1, \delta_2, \ldots, \delta_m) \right) = 0;$$

this implies that

$$h_{k,d_k}(\rho_1, \rho_2, \ldots, \rho_m) = h_{k,d_k}(u(\delta_1), u(\delta_2), \ldots, u(\delta_m))$$
$$= u\left(h_{k,d_k}(\delta_1, \delta_2, \ldots, \delta_m) \right) = 0,$$

which leads to the contradictory conclusion that

$$h(\rho_1, \rho_2, \ldots, \rho_m) = \prod_{i=m+1}^n h_{i,d_i}(\rho_1, \rho_2, \ldots, \rho_m) = 0.$$

This proves our claim that $A = K(D)$.

Thus, u is a K-homomorphism from $K(D)$ to F, and hence it is a K-monomorphism. This shows that $K(D)$ is K-embeddable in F; and since F is algebraic over K, we conclude that $K(D)$ also is algebraic over K, whence $D \subseteq A(L/K)$. This, however, is incompatible with our assumptions on D.   □

In order to state and prove the Nullstellensatz, it will be convenient to introduce some commonly used terminology. Let K be a field, and let L be an extension field of K. If n is a positive integer and $\mathscr{I}$ is an ideal of $K[X_1, X_2, \ldots, X_n]$, by a **zero of** $\mathscr{I}$ **in** $L^{(n)}$ we shall understand an element of $L^{(n)}$ that is a zero of every polynomial in $\mathscr{I}$, that is, an $(\alpha_1, \alpha_2, \ldots, \alpha_n) \in L^{(n)}$ such that $f(\alpha_1, \alpha_2, \ldots, \alpha_n) = 0$ for every $f \in \mathscr{I}$.

If K is a field and L is an algebraic closure of K, it is evident that every proper ideal of $K[X]$ possesses a zero in L: Indeed, since $K[X]$ is a principal ideal domain, such an ideal admits a nonconstant polynomial as a generator; and every zero in L of this generator is a zero of the ideal in question.

As shown by the next result, which can be viewed as a weak version of the Nullstellensatz, this remains true in the case of polynomials in several variables.

4.3.14. Theorem. *Let K be a field, and let L be an algebraic closure of K. If n is a positive integer, then every proper ideal of $K[X_1, X_2, \ldots, X_n]$ admits a zero in $L^{(n)}$.*

Proof. By 0.0.4, every proper ideal of a ring is contained in a maximal ideal, so that it suffices to consider a maximal ideal $\mathcal{M}$ of $K[X_1, X_2, \ldots, X_n]$.

Let us write $C = K[X_1, X_2, \ldots, X_n]$ and $F = C/\mathcal{M}$. Then F is a field; and the natural projection from C to F defines, by restriction, a monomorphism from K to F. In the remainder of the proof, we shall regard K as a subfield of F, the "identification" in question being made by means of this monomorphism. Finally, for $1 \le i \le n$, the value at X_i of the natural projection from C to F will be denoted by σ_i; in other words, σ_i is the coset $X_i + \mathcal{M}$.

With these understandings, it is now seen that the natural projection from C to F can be described as the mapping $f(X_1, X_2, \ldots, X_n) \to f(\sigma_1, \sigma_2, \ldots, \sigma_n)$ from C to F, and hence $F = K[\sigma_1, \sigma_2, \ldots, \sigma_n]$. Since F is a field, we then have

$$F = K[\sigma_1, \sigma_2, \ldots, \sigma_n] = K(\sigma_1, \sigma_2, \ldots, \sigma_n).$$

But this, by the preceding theorem, implies that F is algebraic over K.

Consequently, there exists a K-monomorphism u from F to L. To conclude, let $\alpha_i = u(\sigma_i)$ for $1 \le i \le n$. Then $(\alpha_1, \alpha_2, \ldots, \alpha_n) \in L^{(n)}$; and for every $f \in C$, we have

$$f(\alpha_1, \alpha_2, \ldots, \alpha_n) = f(u(\sigma_1), u(\sigma_2), \ldots, u(\sigma_n))$$
$$= u(f(\sigma_1, \sigma_2, \ldots, \sigma_n)),$$

so that $f(\alpha_1, \alpha_2, \ldots, \alpha_n) = 0$ if and only if $f(\sigma_1, \sigma_2, \ldots, \sigma_n) = 0$. This means that $(\alpha_1, \alpha_2, \ldots, \alpha_n)$ is a zero in $L^{(n)}$ of a polynomial f in C if and only if f belongs to the kernel of the natural projection from C to F. Since this kernel is $\mathcal{M}$, it follows that $(\alpha_1, \alpha_2, \ldots, \alpha_n)$ is indeed a zero of $\mathcal{M}$ in $L^{(n)}$. □

With the help of a simple trick due to Rabinowitsch, we can now obtain what is known as the strong version of the Nullstellensatz.

4.3.15. Theorem (Hilbert Nullstellensatz). *Let K be a field, and let L be an algebraic closure of K. If n is a positive integer, if $\mathcal{J}$ is an ideal of $K[X_1, X_2, \ldots, X_n]$ and $f \in K[X_1, X_2, \ldots, X_n]$, and if every zero of $\mathcal{J}$ in $L^{(n)}$ is a zero of f, then there exists a positive integer k for which $f^k \in \mathcal{J}$.*

Proof. We may assume that $f \neq 0$. Let $\mathscr{C}$ denote the ideal of $K[X_1, X_2, \ldots, X_n, Y]$ generated by $\mathscr{J} \cup \{1 - Yf\}$. We contend that $\mathscr{C}$ admits no zeros in $L^{(n+1)}$: For if $(\alpha_1, \alpha_2, \ldots, \alpha_n, \beta)$ were such a zero, it would follow that

$$1 - \beta f(\alpha_1, \alpha_2, \ldots, \alpha_n) = (1 - Yf)(\alpha_1, \alpha_2, \ldots, \alpha_n, \beta) = 0;$$

moreover, $(\alpha_1, \alpha_2, \ldots, \alpha_n)$ would be a zero of $\mathscr{J}$ in $L^{(n)}$, and hence by hypothesis also of f, so that

$$1 - \beta f(\alpha_1, \alpha_2, \ldots, \alpha_n) = 1,$$

which contradicts the previous equality.

By the preceding theorem, we now have $\mathscr{C} = K[X_1, X_2, \ldots, X_n, Y]$, whence there obtains an equality of the form

$$\sum_{i \in I} g_i h_i + (1 - Yf)h = 1,$$

where I is a finite set, $g_i \in \mathscr{J}$ and $h_i \in K[X_1, X_2, \ldots, X_n, Y]$ for every $i \in I$, and $h \in K[X_1, X_2, \ldots, X_n, Y]$. Next, note that the element $(X_1, X_2, \ldots, X_n, 1/f)$ of $K(X_1, X_2, \ldots, X_n)^{(n+1)}$ is a zero of $1 - Yf$; therefore, if we evaluate the polynomials appearing in our equality at $(X_1, X_2, \ldots, X_n, 1/f)$, we see that the equality

$$\sum_{i \in I} g_i h_i(X_1, X_2, \ldots, X_n, 1/f) = 1$$

holds in $K(X_1, X_2, \ldots, X_n)$. Thus, for every positive integer k, the equality

$$f^k = \sum_{i \in I} g_i f^k h_i(X_1, X_2, \ldots, X_n, 1/f)$$

holds in $K(X_1, X_2, \ldots, X_n)$. If we now choose k so that

$$k > \deg_Y(h_i(X_1, X_2, \ldots, X_n, Y))$$

for every $i \in I$, we see that

$$f^k h_i(X_1, X_2, \ldots, X_n, 1/f) \in K[X_1, X_2, \ldots, X_n]$$

for every $i \in I$; but this, in view of the last equality, implies that $f^k \in \mathscr{J}$. $\square$

PROBLEMS

1. Let K be a field, and let A be a subdomain of K. Prove that if $\mathscr{P}$ and $\mathscr{Q}$ are prime ideals of A, then $\mathscr{P} \subseteq \mathscr{Q}$ if and only if $A_{\mathscr{P}} \supseteq A_{\mathscr{Q}}$.
2. Let K be a field, and let A and B be valuation domains in K. Show that $A \subseteq B$ if and only if $\mathrm{m}(A) \supseteq \mathrm{m}(B)$, in which case $B = A_{\mathrm{m}(B)}$.
3. Let K be a field, and let A be a valuation domain in K. Show that the set of all intermediate domains between A and K is a chain.

4. Let K be a field, and let A be a valuation domain in K. Prove that every element of K that is a zero of a monic polynomial in $A[X]$ belongs to A.

5. Let A and B be local rings such that A is a subring of B. Show that the following conditions are equivalent:
   a. $m(A) = A \cap m(B)$.
   b. $m(A) \subseteq m(B)$.
   c. $m(A)B \subset B$.
      When these conditions are satisfied, it is said that B **dominates** A.

6. Let K be a field, and let A be a local subdomain of K. Prove the following assertions:
   a. There exists a valuation domain in K that dominates A.
   b. A is a valuation domain in K if and only if A is the only intermediate local domain between A and K that dominates A.

7. Let K be a field, and let A be a subdomain of K. Prove that an element of K is a zero of a monic polynomial in $A[X]$ if and only if it belongs to every valuation domain in K containing A. (This result is known as **Krull's theorem.**)

8. Let A be a factorial domain, and let K be a field of fractions of A. Show that a proper subdomain of K containing A is a valuation domain in K if and only if it is of the form $A_{\pi A}$ for some irreducible element π in A.

9. Show that the valuation domains in $\mathbf{Q}$ different from $\mathbf{Q}$ are the subdomains of $\mathbf{Q}$ of the form $\mathbf{Z}_{p\mathbf{Z}}$ for some prime p.

10. Let K be a field, and let A be a valuation domain in $K(X)$ such that $K \subset A \subset K(X)$. Prove the following assertions:
    a. If $X \in A$, then $A = K[X]_{fK[X]}$ for some irreducible polynomial f in $K[X]$.
    b. If $X \notin A$, then $A = K[1/X]_{(1/X)K[1/X]}$, so that A consists of all rational functions in $K(X)$ of the form $g(X)/h(X)$ with $g, h \in K[X]$, $h \neq 0$, and $\deg(g) \leq \deg(h)$.

11. Let K be a field, let n be a positive integer, and let $(\alpha_1, \alpha_2, \ldots, \alpha_n) \in K^{(n)}$. Verify the following assertions:
    a. If $f \in K[X_1, X_2, \ldots, X_n]$, then

    $$f(X_1, X_2, \ldots, X_n) - f(\alpha_1, \alpha_2, \ldots, \alpha_n)$$

    $$\in \sum_{i=1}^{n} (X_i - \alpha_i) K[X_1, X_2, \ldots, X_n].$$

    b. $\sum_{i=1}^{n}(X_i - \alpha_i)K[X_1, X_2, \ldots, X_n]$ consists of the polynomials in $K[X_1, X_2, \ldots, X_n]$ admitting $(\alpha_1, \alpha_2, \ldots, \alpha_n)$ as a zero.
    c. $\sum_{i=1}^{n}(X_i - \alpha_i)K[X_1, X_2, \ldots, X_n]$ is a maximal ideal of $K[X_1, X_2, \ldots, X_n]$.

12. Let K be a field, let L be an algebraic closure of K, and let n be a positive integer. Prove the following assertions:
    a. An ideal of $K[X_1, X_2,\ldots,X_n]$ is maximal if and only if there exists an $(\alpha_1, \alpha_2,\ldots,\alpha_n) \in L^{(n)}$ such that it consists of the polynomials in $K[X_1, X_2,\ldots,X_n]$ admitting $(\alpha_1, \alpha_2,\ldots,\alpha_n)$ as a zero.
    b. If $(\alpha_1, \alpha_2,\ldots,\alpha_n)$, $(\beta_1, \beta_2,\ldots,\beta_n) \in L^{(n)}$, and if $\mathscr{A}$ and $\mathscr{B}$ denote, respectively, the maximal ideals of $K[X_1, X_2,\ldots,X_n]$ consisting of the polynomials in $K[X_1, X_2,\ldots,X_n]$ admitting $(\alpha_1, \alpha_2,\ldots,\alpha_n)$ and $(\beta_1, \beta_2,\ldots,\beta_n)$ as a zero, then $\mathscr{A} = \mathscr{B}$ if and only if there exists a K-automorphism of L such that $\alpha_i \to \beta_i$ for $1 \le i \le n$.

13. Show that if K is an algebraically closed field and n is a positive integer, then every maximal ideal of $K[X_1, X_2,\ldots,X_n]$ is of the form $\sum_{i=1}^{n}(X_i - \alpha_i)K[X_1, X_2,\ldots,X_n]$ for some $(\alpha_1, \alpha_2,\ldots,\alpha_n) \in K^{(n)}$.

14. Let K be a field, and let L be an algebraic closure of K. Prove that if n is a positive integer and $\mathscr{P}$ is a prime ideal of $K[X_1, X_2,\ldots,X_n]$, and if $f \in K[X_1, X_2,\ldots,X_n] - \mathscr{P}$, then there exists a zero of $\mathscr{P}$ in $L^{(n)}$ that is not a zero of f.

   Several authors give an alternative definition of a place in which a "point at infinity" is introduced. This is particularly suitable in certain applications, for example in classical algebraic geometry and in complex analysis. It will be seen in the next two problems that this definition is equivalent to the one adopted in the present section. Before stating these problems, we shall define the process of adjoining a "point at infinity" to a field.

   Given a field F, the symbol $\tilde{F}$ will denote the set obtained by adjoining to F one new element, usually denoted by the symbol ∞. The addition and multiplication on F are then extended to partially defined addition and multiplication on $\tilde{F}$ by the additional requirements that

$$\sigma + \infty = \infty = \infty + \sigma \quad \text{for every } \sigma \in F$$

and

$$\sigma\infty = \infty = \infty\sigma \quad \text{for every } \sigma \in F^* \cup \{\infty\}.$$

   Consequently, the extended addition is undefined only at (∞, ∞), whereas the extended multiplication is undefined at $(0, \infty)$ and $(\infty, 0)$.

15. Let K and F be fields, and let u be a mapping from K to $\tilde{F}$ satisfying the following three conditions: (i) $u(1) = 1$; (ii) if $\alpha, \beta \in K$ and $u(\alpha) + u(\beta)$ is defined, then $u(\alpha + \beta) = u(\alpha) + u(\beta)$; and (iii) if $\alpha, \beta \in K$ and $u(\alpha)u(\beta)$ is defined, then $u(\alpha\beta) = u(\alpha)u(\beta)$. Prove that $u^{-1}(F)$ is a valuation domain in K, and that the mapping from $u^{-1}(F)$ to F defined by restriction of u is a place of K in F.

16. Let K and F be fields, and let u be a place of K in F. Denote by $\tilde{u}$ the mapping from K to $\tilde{F}$ extending u and such that $\alpha \to \infty$ for every $\alpha \in K$

not belonging to the domain of definition of u. Show that $\tilde{u}$ satisfies the three conditions for mappings from K to $\tilde{F}$ stated in problem 15.

4.4. SEPARABLE EXTENSIONS

The concept of separability, which has played an important role in the preceding parts of the book, has been defined exclusively for algebraic extensions. Our main objective in this section is to extend it to the case of arbitrary field extensions.

We shall begin by studying the auxiliary notion of linear disjointness. Although this can be done in the general context of algebras over fields, we shall restrict ourselves to the particular case needed for our purposes.

Let K be a field, let F be an extension field of K, and let A and B be intermediate domains between K and F. We say that A and B are **linearly disjoint over** K when the following condition is satisfied: If $(\alpha_i)_{i \in I}$ and $(\beta_j)_{j \in J}$ are, respectively, families of elements of A and B that are linearly independent over K, then the family $(\alpha_i \beta_j)_{(i, j) \in I \times J}$ is linearly independent over K.

Let K be a field, and let F be an extension field of K. It is seen that if A is an intermediate domain between K and F, then K and A are linearly disjoint over K. Also, if $\bar{A}, A, \bar{B}, B$ are intermediate domains between K and F such that A and B are linearly disjoint over K and such that $\bar{A} \subseteq A$ and $\bar{B} \subseteq B$, then $\bar{A}$ and $\bar{B}$ are linearly disjoint over K.

4.4.1. Proposition. *Let K be a field, let F be an extension field of K, and let A and B be intermediate domains between K and F that are linearly disjoint over K.*

(i) *If $(\alpha_i)_{i \in I}$ and $(\beta_j)_{j \in J}$ are, respectively, bases of the K-spaces A and B, then $(\alpha_i \beta_j)_{(i, j) \in I \times J}$ is a base of the K-space AB.*

(ii) $[AB : K] = [A : K][B : K]$.

(iii) $A \cap B = K$.

(iv) *The fields of fractions of A and B in F are linearly disjoint over K.*

Proof. It is clear that (ii) follows from (i). To verify (i), note that the hypothesis implies that $(\alpha_i \beta_j)_{(i, j) \in I \times J}$ is linearly independent over K; and since this family is a generating system of the K-space AB, the desired conclusion follows at once.

To prove (iii), assume that $K \subset A \cap B$, and choose a $\gamma \in (A \cap B) - K$. Then the sequence $(1, \gamma)$ is linearly independent over K; and since it consists of elements of $A \cap B$, it follows from the hypothesis that the sequence $(1, \gamma, \gamma, \gamma^2)$ is linearly independent over K. Since the latter contains a repeated term, this is a contradiction.

To conclude, we now prove (iv). Let L and M denote, respectively, the fields of fractions of A and B in F, and assume that L and M are not linearly disjoint over K. Then there exist, respectively, finite families $(\alpha_i)_{i \in I}$ and $(\beta_j)_{j \in J}$ of elements of L and M that are linearly independent over K, but such that $(\alpha_i \beta_j)_{(i,j) \in I \times J}$ is linearly dependent over K. Therefore, we can choose a family $(\lambda_{ij})_{(i,j) \in I \times J}$ of elements of K, not all equal to 0, so that $\sum_{(i,j) \in I \times J} \lambda_{ij} \alpha_i \beta_j = 0$. Also, we can choose an $\alpha \in A - \{0\}$ and a $\beta \in B - \{0\}$ so that $\alpha\alpha_i \in A$ for every $i \in I$ and $\beta\beta_j \in B$ for every $j \in J$. Since

$$\sum_{(i,j) \in I \times J} \lambda_{ij}(\alpha\alpha_i)(\beta\beta_j) = 0,$$

it follows that $((\alpha\alpha_i)(\beta\beta_j))_{(i,j) \in I \times J}$ is linearly dependent over K. But this contradicts the linear disjointness of A and B over K, because the families $(\alpha\alpha_i)_{i \in I}$ and $(\beta\beta_j)_{j \in J}$ are linearly independent over K. □

4.4.2. Proposition. *Let K be a field, let F be an extension field of K, and let A and M be, respectively, an intermediate domain and an intermediate field between K and F that are linearly disjoint over K.*

(i) *Every family of elements of A that is linearly independent over K is linearly independent over M.*

(ii) *Every base of the K-space A is a base of the M-space AM.*

(iii) $[A : K] = [AM : M]$.

Proof. To prove (i), let $(\alpha_i)_{i \in I}$ be a family of elements of A that is linearly independent over K. To show that $(\alpha_i)_{i \in I}$ remains linearly independent over M, suppose that we have a relation of the form $\sum_{i \in I} \alpha_i \xi_i = 0$, where $(\xi_i)_{i \in I}$ is a family of elements of M such that $\xi_i = 0$ for almost every $i \in I$. Choose a linear base $(\beta_j)_{j \in J}$ of M over K; and for each $i \in I$, write $\xi_i = \sum_{j \in J} \gamma_{ij} \beta_j$, where $(\gamma_{ij})_{j \in J}$ is a family of elements of K such that $\gamma_{ij} = 0$ for almost every $j \in J$. It then follows that $\gamma_{ij} = 0$ for almost every $(i,j) \in I \times J$ and $\sum_{(i,j) \in I \times J} \gamma_{ij} \alpha_i \beta_j = 0$. The linear disjointness of A and M over K now implies that $\gamma_{ij} = 0$ for every $(i,j) \in I \times J$, whence $\beta_j = 0$ for every $j \in J$.

Since every generating system of the K-space A is a generating system of the M-space AM, we see that (ii) follows from (i). To conclude, note that (iii) is an immediate consequence of (ii). □

4.4.3. Proposition. *Let K be a field, let F be an extension field of K, and let A and B be intermediate domains between K and F. If there exist, respectively, bases $(\alpha_i)_{i \in I}$ and $(\beta_j)_{j \in J}$ of the K-spaces A and B such that $(\alpha_i \beta_j)_{(i,j) \in I \times J}$ is linearly independent over K, then A and B are linearly disjoint over K.*

Proof. In order to verify the condition of linear disjointness, suppose that $(\eta_h)_{1 \leq h \leq m}$ and $(\xi_k)_{1 \leq k \leq n}$ are, respectively, finite sequences of elements of A and B that are linearly independent over K. We have to show that $(\eta_h \xi_k)_{1 \leq h \leq m, 1 \leq k \leq n}$ is linearly independent over K.

It is seen, first, that there exist, respectively, finite subsets $\bar{I}$ and $\bar{J}$ of I and J such that $\eta_1, \eta_2, \ldots, \eta_m \in \sum_{i \in \bar{I}} K\alpha_i$ and $\xi_1, \xi_2, \ldots, \xi_n \in \sum_{j \in \bar{J}} K\beta_j$. Now put $r = \text{Card}(\bar{I})$ and $s = \text{Card}(\bar{J})$, so that

$$\left[\sum_{i \in \bar{I}} K\alpha_i : K \right] = r \quad \text{and} \quad \left[\sum_{j \in \bar{J}} K\beta_j : K \right] = s;$$

then $m \leq r$ and $n \leq s$, and elements $\eta_{m+1}, \eta_{m+2}, \ldots, \eta_r$ of A and $\xi_{n+1}, \xi_{n+2}, \ldots, \xi_s$ of B can be chosen so that $(\eta_h)_{1 \leq h \leq r}$ and $(\xi_k)_{1 \leq k \leq s}$ are, respectively, bases of the K-spaces $\sum_{i \in \bar{I}} K\alpha_i$ and $\sum_{j \in \bar{J}} K\beta_j$. Since

$$\sum_{1 \leq h \leq r} K\eta_h = \sum_{i \in \bar{I}} K\alpha_i \quad \text{and} \quad \sum_{1 \leq k \leq s} K\xi_k = \sum_{j \in \bar{J}} K\beta_j,$$

we get

$$\sum_{1 \leq h \leq r, 1 \leq k \leq s} K\eta_h \xi_k = \sum_{(i, j) \in \bar{I} \times \bar{J}} K\alpha_i \beta_j;$$

therefore $(\eta_h \xi_k)_{1 \leq h \leq r, 1 \leq k \leq s}$ is a generating system of the K-space $\sum_{(i, j) \in \bar{I} \times \bar{J}} K\alpha_i \beta_j$. Since our assumption implies that the latter is rs-dimensional, we conclude that it admits $(\eta_h \xi_k)_{1 \leq h \leq r, 1 \leq k \leq s}$ as a base; and this implies that $(\eta_h \xi_k)_{1 \leq h \leq m, 1 \leq k \leq n}$ is linearly independent over K. $\qquad \square$

4.4.4. Proposition. *Let K be a field, let F be an extension field of K, and let A and M be, respectively, an intermediate domain and an intermediate field between K and F. If there exists a base of the K-space A that is linearly independent over M, then A and M are linearly disjoint over K.*

Proof. Let $(\alpha_i)_{i \in I}$ be a base of the K-space A that remains linearly independent over M, and let $(\beta_j)_{j \in J}$ be a linear base of M over K. Since $(\alpha_i)_{i \in I}$ is a generating system of the M-space AM, we see that it is a base of the latter. It then follows from 0.0.2 that $(\alpha_i \beta_j)_{(i, j) \in I \times J}$ is a base of the K-space AM, and hence is linearly independent over K. The desired conclusion now follows at once from the preceding proposition. $\qquad \square$

The foregoing discussion contains the basic generalities on linear disjointness that will be needed in our study of separability. Before taking up this topic, it will be well to prove two special results on linear disjointness that will be useful in the sequel.

4.4.5. Proposition. *Let K be a field, let L be an extension field of K, and let M be an extension field of L. If $S \subseteq M$ and S is algebraically independent over L, then $K(S)$ and L are linearly disjoint over K.*

Proof. First consider a nonempty finite subset D of S. Put $n = \text{Card}(D)$, and write $D = \{\alpha_1, \alpha_2, \ldots, \alpha_n\}$. Since D is algebraically independent over L, the family $(\prod_{i=1}^{n} \alpha_i^{k_i})_{k_1, k_2, \ldots, k_n \in \mathbf{N}}$ is linearly independent over L, and hence it is a base of the K-space $K[D]$ that remains linearly independent over L. It then follows from 4.4.4 that $K[D]$ and L are linearly disjoint over K.

To conclude, let Ω denote the set of all finite subsets of S. Then $(K[D])_{D \in \Omega}$ is filtered and $K[S] = \cup_{D \in \Omega} K[D]$; therefore, by virtue of what was shown in the preceding paragraph, $K[S]$ and L are linearly disjoint over K. Since $K(S)$ is the field of fractions of $K[S]$ in M, the desired conclusion now follows immediately from 4.4.1. $\square$

As stated in the remarks following Proposition 1.4.6, the latter does not contain the full result on the transitivity property for finitely generated extensions. We shall now digress briefly in order to show that the complementary part is an easy consequence of the proposition just proved.

4.4.6. Corollary. *Let K be a field, let L be an extension field of K, and let M be an extension field of L. If M is finitely generated over K, then L is finitely generated over K.*

Proof. Let S and T denote, respectively, transcendence bases of L over K and of M over L. Then $S \cup T$ is a transcendence base of M over K; and our hypothesis implies that S and T are finite, and that M is finite over $K(S \cup T)$.

Since $K(S \cup T) = K(S)(T)$ and T is algebraically independent over L, the proposition shows that $K(S \cup T)$ and L are linearly disjoint over $K(S)$. Therefore, according to 4.4.2, every linear base of L over $K(S)$ is linearly independent over $K(S \cup T)$. Since M is finite over $K(S \cup T)$, we conclude that every linear base of L over $K(S)$ is finite, which means that L is finite over $K(S)$. And since S is finite, this implies that L is finitely generated over K. $\square$

4.4.7. Proposition. *Let K be a field of prime characteristic p, and let L be an extension field of K. If K and L^p are linearly disjoint over K^p, if $(\alpha_i)_{i \in I}$ is a family of elements of L that is linearly independent over K, and if n is a nonnegative integer, then $(\alpha_i^{p^n})_{i \in I}$ also is linearly independent over K. Conversely, if there exists a linear base $(\alpha_i)_{i \in I}$ of L over K such that $(\alpha_i^p)_{i \in I}$ is linearly independent over K, then K and L^p are linearly disjoint over K^p.*

Proof. It is clear that if $(\xi_j)_{j \in J}$ is a family of elements of L, then $(\xi_j)_{j \in I}$ is linearly independent over K if and only if $(\xi_j^p)_{j \in J}$ is linearly independent over K^p; and $(\xi_j)_{j \in J}$ is a linear base of L over K if and only if $(\xi_j^p)_{j \in J}$ is a linear base of L^p over K^p. Thus, the first assertion for $n = 1$ and the second assertion follow, respectively, from 4.4.2 and 4.4.4; and the general case of the first assertion is easily proved by induction. $\square$

Now we are in a position to proceed to the main topic of this section. In order to generalize the concept of separability, we shall first establish the connection between separability for algebraic extensions and linear disjointness.

4.4.8. Proposition. *Let K be a field of prime characteristic p, and let L be an algebraic extension of K. Then L is separable over K if and only if K and L^p are linearly disjoint over K^p.*

Proof. In view of the equivalence between the linear disjointness of K and L^p over K^p and the condition given in 4.4.7, it will suffice to repeat the argument used to prove 2.5.6.

Indeed, it was seen in the last part of the proof of 2.5.6 how the separability of L over K follows from that condition.

Conversely, suppose that L is separable over K. To show that the condition in question is satisfied, let $(\alpha_i)_{1 \leq i \leq n}$ be a finite sequence of elements of L that is linearly independent over K. If we put $F = K(\alpha_1, \alpha_2, \ldots, \alpha_n)$, then F is separable over K, and 2.5.6 shows that $F = K \vee F^p$; but F is also finite over K, and the argument given in the second paragraph of the proof of 2.5.6 then shows that $(\alpha_i^p)_{1 \leq i \leq n}$ is linearly independent over K, which is what was needed. $\square$

The preceding proposition and the fact that every algebraic extension of a field of characteristic 0 is separable now allow us to extend the notion of separability as follows.

Let K be a field, and let L be an extension field of K. We say that L is **separable over** K, or that L is a **separable extension** of K, either when K has characteristic 0; or when K has prime characteristic p, with K and L^p linearly disjoint over K^p.

According to this definition, in characteristic 0 all field extensions are separable. Therefore, this general notion of separability is of interest only for fields of prime characteristic.

The next result states necessary and sufficient conditions for separability in the case of prime characteristic. These conditions are used by several authors to define separability.

4.4.9. Proposition. *Let K be a field of prime characteristic, and let L be an extension field of K. If L is separable over K, and if M is an extension field of L, then L and $P(M/K)$ are linearly disjoint over K. Conversely, if there exists a perfect extension field M of L such that L and $P_1(M/K)$ are linearly disjoint over K, then L is separable over K.*

Proof. Let $p = \mathrm{Char}(K)$.

Suppose that L is separable over K, and let M be an extension field of L. By 4.4.4, in order to show that L and $P(M/K)$ are linearly disjoint over K, it suffices to verify that every nonempty finite family of elements of L that is linearly independent over K remains linearly independent over $P(M/K)$. Thus, let $(\alpha_i)_{i \in I}$ be such a family of elements of L, and assume that $\sum_{i \in I} \alpha_i \beta_i = 0$, where $\beta_i \in P(M/K)$ for every $i \in I$. If $i \in I$, there exists a positive integer n_i for which $\beta_i^{p^{n_i}} \in K$; putting $n = \sup_{i \in I} n_i$, we then have $\beta_i^{p^n} \in K$ for every $i \in I$. By 4.4.7, we know that $(\alpha_i^{p^n})_{i \in I}$ is linearly independent over K. Since

$$\sum_{i \in I} \alpha_i^{p^n} \beta_i^{p^n} = \left(\sum_{i \in I} \alpha_i \beta_i \right)^{p^n} = 0,$$

it then follows that $\beta_i^{p^n} = 0$ for every $i \in I$; therefore $\beta_i = 0$ for every $i \in I$, as required.

On the other hand, if M is a perfect extension field of L, the Frobenius mapping of M is an automorphism by which K, L^p, and K^p are, respectively, the images of $P_1(M/K)$, L, and K. Therefore, for every such field M, the linear disjointness of K and L^p over K^p is equivalent to that of $P_1(M/K)$ and L over K. □

As the following proposition shows, a field admits separable extensions of every possible transcendence degree.

4.4.10. Proposition. *If K is a field, then every purely transcendental extension of K is separable over K.*

Proof. We only need to consider the case where K has prime characteristic p. Let L be a purely transcendental extension of K. Choose a pure base S of L over K, and denote by T the image of S by the Frobenius mapping of L. Then $L^p = K(S)^p = K^p(T)$, and T is algebraically independent over K. And this, according to 4.4.5, implies the linear disjointness of K and L^p over K^p. □

4.4.11. Example. Let us give an example of a transcendental extension that is not separable. Consider a field K of prime characteristic p, and a purely transcendental extension $K(\rho, \sigma, \tau)$ of K having transcendence degree 3 over K. Next choose a pth root ξ of $\rho \tau^p + \sigma$ in an extension field of $K(\rho, \sigma, \tau)$. Then $K(\rho, \sigma, \tau, \xi)$ is transcendental over $K(\rho, \sigma)$, but not separable over $K(\rho, \sigma)$.

Our hypothesis implies that the sequence (ρ, σ, τ) is algebraically independent over K, and hence τ is transcendental over $K(\rho, \sigma)$; therefore $K(\rho, \sigma, \tau, \xi)$ is transcendental over $K(\rho, \sigma)$.

Writing the relation $\xi^p = \rho \tau^p + \sigma$ in the form $\sigma - \rho \tau^p - \xi^p = 0$, we see that the sequence $(1, \tau^p, \xi^p)$ is linearly dependent over $K(\rho, \sigma)$. By 4.4.7, in order to show that $K(\rho, \sigma, \tau, \xi)$ is not separable over $K(\rho, \sigma)$, it will suffice to verify that the sequence $(1, \tau, \xi)$ is linearly independent over $K(\rho, \sigma)$.

Thus, suppose that we have a relation of the form

$$\alpha + \beta \tau + \gamma \xi = 0,$$

where $\alpha, \beta, \gamma \in K(\rho, \sigma)$. Since $K(\rho, \sigma)$ is a field of fractions of $K[\rho, \sigma]$, we can express α, β, γ as fractions with numerators and a common denominator in $K[\rho, \sigma]$. Upon multiplication by this common denominator, the preceding relation yields

$$f(\rho,\sigma)+g(\rho,\sigma)\tau+h(\rho,\sigma)\xi=0,$$

where $f, g, h \in K[X, Y]$ and $f(\rho,\sigma), g(\rho,\sigma), h(\rho,\sigma)$ are, respectively, the numerators of the fractions corresponding to α, β, γ. Denoting by $\bar{f}, \bar{g}, \bar{h}$ the values at f, g, h of the endomorphism of $K[X, Y]$ defined by the Frobenius mapping of K, and applying the Frobenius mapping of $K(\rho,\sigma,\tau,\xi)$ to the latter relation, we get

$$\bar{f}(\rho^p,\sigma^p)+\bar{g}(\rho^p,\sigma^p)\tau^p+\bar{h}(\rho^p,\sigma^p)\xi^p=0.$$

Since $\xi^p=\rho\tau^p+\sigma$, we have

$$(\bar{f}(\rho^p,\sigma^p)+\sigma\bar{h}(\rho^p,\sigma^p))+(\bar{g}(\rho^p,\sigma^p)+\rho\bar{h}(\rho^p,\sigma^p))\tau^p=0;$$

the algebraic independence of the sequence (ρ,σ,τ) over K now implies that in the polynomial domain $K[X, Y]$ the equalities

$$\bar{f}(X^p,Y^p)+Y\bar{h}(X^p,Y^p)=0$$

and

$$\bar{g}(X^p,Y^p)+X\bar{h}(X^p,Y^p)=0$$

obtain. If $h \neq 0$, these equalities would imply that

$$p\deg_Y(f(X,Y))=1+p\deg_Y(h(X,Y))$$

and

$$p\deg_X(g(X,Y))=1+p\deg_X(h(X,Y)),$$

an obvious contradiction. Consequently $h = 0$, and so

$$\bar{f}(X^p,Y^p)=0=\bar{g}(X^p,Y^p);$$

but this means that

$$f(X,Y)^p=0=g(X,Y)^p,$$

and hence $f = 0 = g$. It then follows that $\alpha = \beta = \gamma = 0$, and we conclude that the sequence $(1, \tau, \xi)$ is indeed linearly independent over $K(\rho,\sigma)$.   □

We saw in 2.5.8 that separable algebraic extensions enjoy the transitivity property. In the general case, we have a partial result.

4.4.12. Proposition.   *Let K be a field, let L be an extension field of K, and let M be an extension field of L.*

(i) If L is separable over K, and if M is separable over L, then M is separable over K.

(ii) If M is separable over K, then L is separable over K.

Proof.   It suffices to consider the case of prime characteristic p.

To prove (i), let $(\alpha_i)_{i \in I}$ and $(\beta_j)_{j \in J}$ be, respectively, linear bases of L over K and of M over L. According to 4.4.7, the hypothesis in (i) implies that $(\alpha_i^p)_{i \in I}$ and $(\beta_j^p)_{j \in J}$ are, respectively, linearly independent over K and L. By 0.0.2, we then know that $(\alpha_i \beta_j)_{(i,j) \in I \times J}$ is a linear base of M over K such that $(\alpha_i^p \beta_j^p)_{(i,j) \in I \times J}$ is linearly independent over K. Finally, 4.4.7 shows that the existence of such a linear base implies the separability of M over K.

To prove (ii), we need only note that, in view of the inclusions $K^p \subseteq L^p \subseteq M^p$, the linear disjointness over K^p of K and M^p implies that of K and L^p. □

It can be shown that in the situation described in the preceding proposition, the separability of M over L cannot be deduced from that of M over K.

Indeed, if K is a field of prime characteristic p, and if $K(\alpha)$ is a simple transcendental extension of K, then 4.4.10 shows that $K(\alpha)$ is separable over K; on the other hand, since $K(\alpha)$ is purely inseparable over $K(\alpha^p)$ and $K(\alpha^p) \subset K(\alpha)$, it is clear that $K(\alpha)$ is not separable over $K(\alpha^p)$.

4.4.13. Proposition. Let K be a field of prime characteristic p, and let L be a separable extension of K. If $(\alpha_i)_{i \in I}$ is a linear base of L over K, then $(\alpha_i^p)_{i \in I}$ is a base of the K-space KL^p.

Proof. The linear independence of $(\alpha_i^p)_{i \in I}$ over K follows from 4.4.7. Moreover, since $L = \sum_{i \in I} K\alpha_i$, we have $L^p = \sum_{i \in I} K^p \alpha_i^p$; consequently $KL^p = \sum_{i \in I} K\alpha_i^p$, which means that $(\alpha_i^p)_{i \in I}$ is also a generating system of the K-space KL^p. □

4.4.14. Corollary. Let K be a field of prime characteristic p, and let L be a separable algebraic extension of K. If $(\alpha_i)_{i \in I}$ is a linear base of L over K, and if n is a nonnegative integer, then $(\alpha_i^{p^n})_{i \in I}$ also is a linear base of L over K.

Proof. An easy induction shows that we only need to verify our assertion when $n = 1$. In this case, it is seen to follow from the proposition: for, by 2.5.6 and 2.1.12, the hypothesis implies that $L = K \vee L^p = KL^p$. □

Let us return to the transitivity property for separable extensions. As we have already seen, only the partial result 4.4.12 can be proved in general. We shall now show that the full result is valid under additional hypotheses.

4.4.15. Proposition. Let K be a field, let L be an algebraic extension of K, and let M be an extension field of L that is separable over K. Then M is separable over L.

Proof. We may assume that the fields under consideration have prime characteristic p.

Let $(\alpha_i)_{i \in I}$ and $(\beta_j)_{j \in J}$ be, respectively, linear bases of L over K and of M over L. We then know from 0.0.2 that $(\alpha_i \beta_j)_{(i,j) \in I \times J}$ is a linear base of M over K.

By 4.4.7, the separability of M over K implies that $(\alpha_i^p \beta_j^p)_{(i,j) \in I \times J}$ is linearly independent over K; and by 4.4.12, it also implies that L is separable over K. Since L is separable and algebraic over K, it follows from 4.4.14 that $(\alpha_i^p)_{i \in L}$ is a linear base of L over K.

According to 4.4.7, the separability of M over L will be established if we show that $(\beta_j^p)_{j \in J}$ is linearly independent over L. To do this, suppose that $\sum_{j \in J} \gamma_j \beta_j^p = 0$, where $(\gamma_j)_{j \in J}$ is a family of elements of L such that $\gamma_j = 0$ for almost every $j \in J$. Since $(\alpha_i^p)_{i \in I}$ is a linear base of L over K, for each $j \in J$ we can write $\gamma_j = \sum_{i \in I} \lambda_{ij} \alpha_i^p$, where $(\lambda_{ij})_{i \in I}$ is a family of elements of K such that $\lambda_{ij} = 0$ for almost every $i \in I$. It then follows that $\lambda_{ij} = 0$ for almost every $(i, j) \in I \times J$ and

$$\sum_{(i,j) \in I \times J} \lambda_{ij} \alpha_i^p \beta_j^p = \sum_{j \in J} \gamma_j \beta_j^p = 0;$$

the linear independence of $(\alpha_i^p \beta_j^p)_{(i,j) \in I \times J}$ over K now implies that $\lambda_{ij} = 0$ for every $(i, j) \in I \times J$. Therefore $\gamma_j = 0$ for every $j \in J$, as required.   □

To conclude this section, we shall now show that perfect fields do not allow inseparability.

4.4.16. Proposition. *If K is a perfect field, then every extension field of K is separable over K.*

Proof. Supposing that K has prime characteristic p, we have $K^p = K$. For every extension field L of K, it is then seen that K and L^p are linearly disjoint over K^p.   □

PROBLEMS

1. Let K be a field, let L be an extension field of K, and let M be an extension field of L. Prove that if A and B are intermediate domains between K and M, and if $L \subseteq B$, then A and B are linearly disjoint over K if and only if A and L are linearly disjoint over K and AL and B are linearly disjoint over L.

2. Let K be a field, let N be an extension field of K, and let L and M be intermediate fields between K and N. Verify the following assertions:
   (i) If L is finite over K, then L and M are linearly disjoint over K if and only if $[L \vee M : M] = [L : K]$.
   (ii) If L and M are finite over K, then L and M are linearly disjoint over K if and only if $[L \vee M : K] = [L : K][M : K]$.

3. Let K be a field, and let N be an extension field of K. Show that if L and M are intermediate fields between K and N that are, respectively, separable and purely inseparable over K, then L and M are linearly disjoint over K.

4. Show that if F is a field and Γ is a group of automorphisms of F, then F is separable over $\mathrm{Inv}(\Gamma)$.

5. Let K be a field, let L be an extension field of K, and let M be an algebraically closed extension field of L. Prove that L is separable over K if and only if

$$\left[\mathrm{Res}_{\mathrm{Gal}(M/K)}(V, M): M\right] = [V: K]$$

   for every finite-dimensional K-subspace V of L.

6. Let K be a field of prime characteristic p, let L be an extension field of K, and let M be a separable extension of L. Show that if S and T are, respectively, p-bases of L over K and of M over L, then $S \cap T = \varnothing$ and $S \cup T$ is a p-base of M over K.

7. Let K be a field of prime characteristic, let L be an extension field of K having finite degree of imperfection over K, and let M be a finitely generated extension of L. Prove that the degree of imperfection of L over K is less than or equal to that of M over K.

8. Let K be a field of prime characteristic, let L be an extension field of K having degree of imperfection 0 over K, and let M be an extension field of L that is separable over K. Show that M is separable over L. (Note that by virtue of this result and of 2.5.6, we can now say that the full transitivity property for separable extensions is satisfied in a more general situation than that described in 4.4.15.)

9. Let K be a field of prime characteristic p, and let L be a separable extension of K. Prove the following assertions:
   a. Every subset of L that is p-independent in L over K is algebraically independent over K.
   b. The degree of imperfection of L over K is less than or equal to the transcendence degree of L over K.
   c. If S is a p-base of L over K, then L is separable over $K(S)$.
       Let K be a field, let N be an extension field of K, and let L and M be intermediate fields between K and N. We say that L and M are **algebraically disjoint over** K when the following condition is satisfied: If S and T are, respectively, subsets of L and M that are algebraically independent over K, then $S \cap T = \varnothing$ and $S \cup T$ is algebraically independent over K.

10. Let K be a field, let N be an extension field of K, and let L and M be intermediate fields between K and N that are algebraically disjoint over K. Prove the following assertions:
   a. If $\bar{L}$ and $\bar{M}$ are intermediate fields between K and N such that $\bar{L} \subseteq L$ and $\bar{M} \subseteq M$, then $\bar{L}$ and $\bar{M}$ are algebraically disjoint over K.
   b. $L \cap M$ is algebraic over K.
   c. If S and T are, respectively, transcendence bases of L and M over K, then $S \cup T$ is a transcendence base of $L \vee M$ over K.
   d. $\mathrm{tr.deg.}(L \vee M/K) = \mathrm{tr.deg.}(L/K) + \mathrm{tr.deg.}(M/K)$.

e. Every subset of L that is algebraically independent over K is algebraically independent over M.

f. Every transcendence base of L over K is a transcendence base of $L \vee M$ over M.

g. $\text{tr.deg.}(L \vee M/M) = \text{tr.deg.}(L/K)$.

h. If S and T are, respectively, subsets of L and M that are algebraically independent over K, then $K(S)$ and $K(T)$ are linearly disjoint over K.

11. Let K be a field, let N be an extension field of K, and let L and M be intermediate fields between K and N. Show that the following conditions are equivalent:

a. L and M are algebraically disjoint over K.

b. There exists a transcendence base of L over K that is algebraically independent over M.

c. There exist, respectively, transcendence bases S and T of L and M over K such that $S \cup T$ is algebraically independent over K and $S \cap T = \emptyset$.

d. $A(N/L)$ and $A(N/M)$ are algebraically disjoint over K.

e. There exist, respectively, transcendence bases S and T of L and M over K such that $K(S)$ and $K(T)$ are linearly disjoint over K.

12. Let K be a field, let N be an extension field of K, and let L and M be intermediate fields between K and N. Verify the following assertions:

a. If L has finite transcendence degree over K, then L and M are algebraically disjoint over K if and only if

$$\text{tr.deg.}(L \vee M/M) = \text{tr.deg.}(L/K).$$

b. If L and M have finite transcendence degree over K, then L and M are algebraically disjoint over K if and only if

$$\text{tr.deg.}(L \vee M/K) = \text{tr.deg.}(L/K) + \text{tr.deg.}(M/K).$$

13. Let K be a field, let N be an extension field of K, and let E, L, and M be intermediate fields between K and N such that $E \subseteq M$. Show that L and M are algebraically disjoint over K if and only if L and E are algebraically disjoint over K and $L \vee E$ and M are algebraically disjoint over E.

14. Let K be a field, let N be an extension field of K, and let L and M be intermediate fields between K and N. Verify the following assertions:

a. If either L or M is algebraic over K, then L and M are algebraically disjoint over K.

b. If L and M are linearly disjoint over K, then they are algebraically disjoint over K.

c. If L and M are algebraically disjoint over K, and if either L or M is purely transcendental over K, then L and M are linearly disjoint over K.

   d. If L and M are, respectively, algebraic and purely transcendental over K, then L and M are linearly disjoint over K.

   (Note that assertion d can be used to give a quick proof of the result stated in Section 4.2, Problem 15.)

4.5. DERIVATIONS OF FIELDS

The general theory of derivations and differentials is of considerable interest in commutative algebra and algebraic geometry. However, since our interest in derivations lies exclusively in their application to the study of separability for field extensions, we shall confine ourselves to the discussion of derivations of fields.

   Let F be a field, and let A be a subdomain of F. By a **derivation of A in F** we understand a mapping D from A to F such that

$$D(\alpha + \beta) = D(\alpha) + D(\beta) \quad \text{and} \quad D(\alpha\beta) = \beta D(\alpha) + \alpha D(\beta)$$

for all $\alpha, \beta \in A$.

   If F is a field, a derivation of F in F will simply be called a **derivation of F**.

4.5.1. Examples

   a. Let F be a field, and let A be a subdomain of F. Then the zero mapping from A to F is a derivation.

   b. Let K be a field. Then the mapping $f \to f'$ from $K[X]$ to $K(X)$ is a derivation.

   c. Let K be a field, and let I be a set. For each $k \in I$, the mapping $f \to \partial f / \partial X_k$ from $K[X_i]_{i \in I}$ to $K(X_i)_{i \in I}$ is a derivation. □

   The next three propositions state the basic properties of derivations that will be needed in the sequel.

   4.5.2. Proposition. *Let F be a field, let A be a subdomain of F, and let D be a derivation of A in F.*
   (i) $D(1) = 0$.
   (ii) *If $\alpha \in A$ and n is a positive integer, then*

$$D(\alpha^n) = n\alpha^{n-1}D(\alpha).$$

   (iii) *If $\alpha \in A^*$ and n is an integer, then*

$$D(\alpha^n) = n\alpha^{n-1}D(\alpha).$$

   (iv) *If $\alpha \in A$ and $\beta \in A^*$, then*

$$D(\alpha/\beta) = (\beta D(\alpha) - \alpha D(\beta))/\beta^2.$$

Proof. To prove (i), note that

$$D(1) = D(1^2) = 1D(1) + 1D(1) = D(1) + D(1),$$

whence $D(1) = 0$.

It is easy to prove (ii) by induction on n. Indeed, the stated equality is evident for $n = 1$. If we now assume that it holds for some positive integer n, then

$$D(\alpha^{n+1}) = D(\alpha\alpha^n) = \alpha^n D(\alpha) + \alpha D(\alpha^n)$$
$$= \alpha^n D(\alpha) + \alpha(n\alpha^{n-1}D(\alpha)) = \alpha^n D(\alpha) + n\alpha^n D(\alpha)$$
$$= (n+1)\alpha^n D(\alpha),$$

which shows that it also holds for $n + 1$.

In view of (i) and (ii), it is sufficient to verify (iii) for $n < 0$. In this case, since $\alpha^n \alpha^{-n} = 1$ and $-n > 0$, we have

$$0 = D(1) = D(\alpha^n \alpha^{-n}) = \alpha^{-n}D(\alpha^n) + \alpha^n D(\alpha^{-n})$$
$$= \alpha^{-n}D(\alpha^n) + \alpha^n(-n\alpha^{-n-1}D(\alpha))$$
$$= \alpha^{-n}D(\alpha^n) - n\alpha^{-1}D(\alpha),$$

and hence

$$D(\alpha^n) = n\alpha^{n-1}D(\alpha).$$

To verify (iv), note that $\beta(\alpha/\beta) = \alpha$, so that

$$D(\alpha) = D(\beta(\alpha/\beta)) = (\alpha/\beta)D(\beta) + \beta D(\alpha/\beta),$$

and therefore

$$D(\alpha/\beta) = (\beta D(\alpha) - \alpha D(\beta))/\beta^2. \qquad \square$$

4.5.3. Proposition.   *Let A be a domain, let K be a field of fractions of A, and let L be an extension field of K. If D is a derivation of A in L, then there exists a mapping from K to L such that*

$$\alpha/\beta \to (\beta D(\alpha) - \alpha D(\beta))/\beta^2$$

whenever $\alpha, \beta \in A$ and $\beta \neq 0$; and this mapping is the only derivation of K in L extending D.

Proof. According to 4.5.2, for every derivation E of K in L extending D we have

$$E(\alpha/\beta) = (\beta E(\alpha) - \alpha E(\beta))/\beta^2 = (\beta D(\alpha) - \alpha D(\beta))/\beta^2$$

when $\alpha, \beta \in A$ and $\beta \neq 0$. This proves that there exists at most one derivation of K in L extending D, and suggests how to define the required mapping from K to L.

If $\alpha, \beta, \gamma, \delta \in A$ and $\beta \neq 0 \neq \delta$, and if $\alpha/\beta = \gamma/\delta$, then $\alpha\delta = \beta\gamma$, and hence

$$\delta D(\alpha) + \alpha D(\delta) = D(\alpha\delta) = D(\beta\gamma) = \gamma D(\beta) + \beta D(\gamma);$$

we then have

$$\delta D(\alpha) - \gamma D(\beta) = \beta D(\gamma) - \alpha D(\delta),$$

which can be written as

$$\delta D(\alpha) - (\alpha\delta/\beta) D(\beta) = \beta D(\gamma) - (\beta\gamma/\delta) D(\delta);$$

dividing by $\beta\delta$, we now obtain

$$(\beta D(\alpha) - \alpha D(\beta))/\beta^2 = (\delta D(\gamma) - \gamma D(\delta))/\delta^2.$$

It then follows that there exists a mapping from K to L such that

$$\alpha/\beta \to (\beta D(\alpha) - \alpha D(\beta))/\beta^2$$

when $\alpha, \beta \in A$ and $\beta \neq 0$. To complete the proof, it only remains to verify that this mapping is a derivation of K in L extending D. The rather tedious details involved in this verification are straightforward, and will be left to the reader.   □

4.5.4. Proposition.   *Let K be a field, let L be an extension field of K, and let D and E be derivations of K in L.*

   (i) *The $\alpha \in K$ for which $D(\alpha) = E(\alpha)$ form a subfield of K.*

   (ii) *If $S \subseteq K$, and if D and E agree on S, then D and E agree on the subfield of K generated by S.*

   (iii) *D and E agree on the prime subfield of K.*

Proof.   The first assertion follows directly from the definition of derivation and from 4.5.2; the second is a consequence of the first; and the third is a particular case of the second.   □

Let K be a field, and let L be an extension field of K. It is evident from the last proposition that if D is a derivation of K in L, then the $\alpha \in K$ for which $D(\alpha) = 0$ make up a subfield of K; and also that if K is a prime field, then the only derivation of K in L is the zero derivation.

In studying derivations of fields, we shall frequently find it convenient to restrict our considerations to derivations of a field that vanish on a prescribed subfield. For this reason, it will be well to introduce the following terminology.

Let K be a field, let L be an extension field of K, and let M be an extension field of L. A derivation D of L in M is said to be a K-**derivation**, or a **derivation over** K, when $D(\alpha) = 0$ for every $\alpha \in K$.

4.5.5. Examples

As consequences of 4.5.1 and 4.5.3, we have:

   a. Let K be a field. Then there exists a K-derivation of $K(X)$ such that

$$f/g \to (gf' - fg')/g^2$$

whenever $f, g \in K[X]$ and $g \neq 0$; and it is the only derivation of $K(X)$ extending the derivation $f \to f'$ of $K[X]$ in $K(X)$.

b. Let K be a field, and let I be a set. For each $k \in I$, there exists a K-derivation of $K(X_i)_{i \in I}$ such that

$$f/g \to (g\partial f/\partial X_k - f\partial g/\partial X_k)/g^2$$

when $f, g \in K[X_i]_{i \in I}$ and $g \neq 0$; and it is the only derivation of $K(X_i)_{i \in I}$ extending the derivation $f \to \partial f/\partial X_k$ of $K[X_i]_{i \in I}$ in $K(X_i)_{i \in I}$. $\qquad\square$

4.5.6. Proposition. *Let K be a field, and let L be an extension field of K.*

(i) *If P denotes the prime subfield of K, then every derivation of K in L is a P-derivation.*

(ii) *If K has prime characteristic p, then every derivation of K in L is a K^p-derivation.*

Proof. Let D be a derivation of K in L. Since the prime subfield P of K is contained in the subfield of K consisting of the $\alpha \in K$ for which $D(\alpha) = 0$, it follows that D is a P-derivation. And if K has prime characteristic p, then

$$D(\alpha^p) = p\alpha^{p-1}D(\alpha) = 0$$

for every $\alpha \in K$, which shows that D is a K^p-derivation. $\qquad\square$

4.5.7. Proposition. *Let K be a field, let L be an extension field of K, and let M be an extension field of L.*

(i) *A derivation of L in M is a K-derivation if and only if it is K-linear.*

(ii) *If K has prime characteristic p, then every K-derivation of L in M is a $K \vee L^p$-derivation.*

Proof. Let D be a derivation of L in M.

If D is K-linear, for every $\alpha \in K$ we have

$$D(\alpha) = D(\alpha 1) = \alpha D(1) = 0;$$

consequently, D is a K-derivation. And conversely, if D is a K-derivation, for all $\alpha \in K$ and $\beta \in L$ we have

$$D(\alpha\beta) = \beta D(\alpha) + \alpha D(\beta) = \alpha D(\beta),$$

which shows that D is K-linear.

To conclude, suppose that K has prime characteristic p and that D is a K-derivation. According to 4.5.6, D is also an L^p-derivation. Therefore $D(\alpha) = 0$ for every $\alpha \in K \cup L^p$; since $K \vee L^p$ is the subfield of L generated by $K \cup L^p$, it now follows from 4.5.4 that $D(\alpha) = 0$ for every $\alpha \in K \vee L^p$, which is what we wanted. $\qquad\square$

We now turn to the principal theme of the present section, the extendibility of derivations of fields. We shall begin by establishing a basic criterion in the case of simple field extensions. This criterion, together with

Zorn's lemma, will then be used to derive the main results for more general field extensions.

Let K be a field, let L be an extension field of K, and let D be a derivation of K in L. For every polynomial f in $D[X]$, the symbol f^D will denote the polynomial in $L[X]$ obtained by applying D to the coefficients of f; in other words, if $f = \sum_{i=0}^{n} \gamma_i X^i$, where n is a nonnegative integer and $\gamma_0, \gamma_1, \ldots, \gamma_n \in K$, then $f^D = \sum_{i=0}^{n} D(\gamma_i) X^i$.

The following elementary result will be needed in our discussion.

4.5.8. **Proposition.** *Let K be a field, let L be an extension field of K, and let D be a derivation of K in L.*

(i) *The mapping $f \to f^D$ from $K[X]$ to $L(X)$ is a derivation.*

(ii) *If $\alpha \in K$ and $f \in K[X]$, then*

$$D(f(\alpha)) = f^D(\alpha) + f'(\alpha)D(\alpha).$$

Proof. The first assertion is readily verified by direct computation. As to the second, let $\alpha \in K$ and $f \in K[X]$, and write $f = \sum_{i=0}^{n} \gamma_i X^i$, where n is a nonnegative integer and $\gamma_0, \gamma_1, \ldots, \gamma_n \in K$; then

$$D(f(\alpha)) = D\left(\sum_{i=0}^{n} \gamma_i \alpha^i \right) = \sum_{i=0}^{n} D(\gamma_i \alpha^i)$$

$$= \sum_{i=0}^{n} D(\gamma_i)\alpha^i + \sum_{i=0}^{n} \gamma_i D(\alpha^i)$$

$$= \sum_{i=0}^{n} D(\gamma_i)\alpha^i + \sum_{i=1}^{n} i\gamma_i \alpha^{i-1} D(\alpha)$$

$$= f^D(\alpha) + f'(\alpha)D(\alpha),$$

as required. $\square$

We now are in a position to give, in the case of simple field extensions, a necessary and sufficient condition for the extendibility of a derivation.

4.5.9. **Proposition.** *Let K be a field, let L be an extension field of K, and let D be a derivation of K in L. Let $\alpha, \beta \in L$, and let $\mathscr{I}$ denote the ideal of algebraic relations of α over K. If D is extendible to a derivation of $K(\alpha)$ in L such that $\alpha \to \beta$, then $f^D(\alpha) + f'(\alpha)\beta = 0$ for every $f \in \mathscr{I}$. Conversely, if there exists a generator f of $\mathscr{I}$ for which $f^D(\alpha) + f'(\alpha)\beta = 0$, then D is uniquely extendible to a derivation of $K(\alpha)$ in L such that $\alpha \to \beta$.*

Proof. First suppose that there exists a derivation E of $K(\alpha)$ in L extending D and such that $E(\alpha) = \beta$. If $f \in \mathscr{I}$, we have $f(\alpha) = 0$ and $f^E = f^D$, and the preceding proposition shows that

$$f^D(\alpha) + f'(\alpha)\beta = f^E(\alpha) + f'(\alpha)E(\alpha) = E(f(\alpha)) = 0.$$

Now suppose that there exists a generator f of $\mathscr{J}$ for which $f^D(\alpha)+f'(\alpha)\beta = 0$. Note that the uniqueness of a derivation of $K(\alpha)$ in L extending D and such that $\alpha \to \beta$ follows immediately from 4.5.4: Every two such derivations agree on $K \cup \{\alpha\}$, and hence also on $K(\alpha)$. It remains, therefore, to establish the existence of a derivation of $K(\alpha)$ in L enjoying the required properties.

To do this, we begin by showing that there exists a mapping from the domain $K[\alpha]$ to L such that $g(\alpha) \to g^D(\alpha)+g'(\alpha)\beta$ for every $g \in K[X]$. Indeed, if $r, s \in K[X]$ and $r(\alpha)=s(\alpha)$, then $r - s \in \mathscr{J}$; since $\mathscr{J} = fK[X]$, we can write $r - s = ft$ with $t \in K[X]$; since $f(\alpha)=0$, it now follows that

$$r^D(\alpha)-s^D(\alpha) = (r-s)^D(\alpha) = (ft)^D(\alpha)$$
$$= t(\alpha)f^D(\alpha)+f(\alpha)t^D(\alpha) = t(\alpha)f^D(\alpha)$$

and similarly

$$r'(\alpha)-s'(\alpha) = (r-s)'(\alpha) = (ft)'(\alpha)$$
$$= t(\alpha)f'(\alpha)+f(\alpha)t'(\alpha) = t(\alpha)f'(\alpha),$$

whence

$$(r^D(\alpha)+r'(\alpha)\beta)-(s^D(\alpha)+s'(\alpha)\beta)$$
$$= (r^D(\alpha)-s^D(\alpha))+(r'(\alpha)-s'(\alpha))\beta$$
$$= t(\alpha)f^D(\alpha)+t(\alpha)f'(\alpha)\beta = t(\alpha)(f^D(\alpha)+f'(\alpha)\beta) = 0,$$

which is what was needed.

Let us denote this mapping from $K[\alpha]$ to L by E, so that $E(g(\alpha)) = g^D(\alpha)+g'(\alpha)\beta$ for every $g \in K[X]$. If we specialize this condition to the case of constant polynomials in $K[X]$, we find that E is an extension of D; and if we specialize it to the case where $g(X) = X$, we find that $E(\alpha)=\beta$. Finally, an elementary calculation shows that E is a derivation.

Thus, E is a derivation of $K[\alpha]$ in L extending D and such that $\alpha \to \beta$. Since $K(\alpha)$ is the field of fractions of $K[\alpha]$ in L, we know from 4.5.3 that E is extendible to a derivation of $K(\alpha)$ in L; and every such derivation is a derivation of $K(\alpha)$ in L extending D and such that $\alpha \to \beta$. $\square$

Applying the criterion just established to various types of simple field extensions, we obtain the following useful corollary.

4.5.10. Corollary. *Let K be a field, and let L be an extension field of K. Let D be a derivation of K in L, and let $\alpha \in L$.*

(i) If α is transcendental over K, and if $\beta \in L$, then D is uniquely extendible to a derivation of $K(\alpha)$ in L such that $\alpha \to \beta$.

(ii) If α is algebraic and separable over K, then D is uniquely extendible to a derivation of $K(\alpha)$ in L. Moreover, if f denotes the minimal polynomial of α over K, then the derivation of $K(\alpha)$ in L extending D sends α

to $- f^D(\alpha)/f'(\alpha)$.

(iii) *If K is of prime characteristic p, if α is purely inseparable over K and $\alpha \notin K$, if e denotes the smallest element of the set of all positive integers i for which $\alpha^{p^i} \in K$, and if $D(\alpha^{p^e}) = 0$ and $\beta \in L$, then D is uniquely extendible to a derivation of $K(\alpha)$ in L such that $\alpha \to \beta$.*

Proof. All assertions will follow from the criterion given in the proposition. Let $\mathscr{I}$ denote the ideal of algebraic relations of α over K.

To prove (i), suppose that α is transcendental over K. Then $\mathscr{I}$ is the null ideal of $K[X]$, which is generated by the zero polynomial. And it is clear that the condition in our criterion is satisfied by this polynomial and by every $\beta \in L$.

Now suppose that α is algebraic over K, and let f denote the minimal polynomial of α over K. Since f is the monic generator of $\mathscr{I}$, our concern lies in determining the $\beta \in L$ satisfying the condition $f^D(\alpha) + f'(\alpha)\beta = 0$.

To prove (ii), suppose that α is separable over K. We then have $f'(\alpha) \neq 0$, and so the only $\beta \in L$ satisfying the required condition is $\beta = - f^D(\alpha)/f'(\alpha)$.

Finally, in the situation described in (iii), we have $f(X) = X^{p^e} - \alpha^{p^e}$, and hence

$$f^D = - D(\alpha^{p^e}) = 0 \quad \text{and} \quad f' = p^e X^{p^e - 1} = 0.$$

Consequently, the condition in question is satisfied by every $\beta \in L$. $\square$

We shall now proceed to the general results on the extendibility of derivations.

4.5.11. Theorem. *Let K be a field, let L be a purely transcendental extension of K, and let M be an extension field of L. If S is a pure base of L over K, and if $(\beta_\alpha)_{\alpha \in S}$ is a family of elements of M indexed by S, then every derivation of K in M is uniquely extendible to a derivation of L in M such that $\alpha \to \beta_\alpha$ for every $\alpha \in S$. In particular, every derivation of K in M is extendible to a derivation of L in M.*

Proof. Let D be a derivation of K in M, and let Ω denote the set of all subsets A of S such that D is extendible to a derivation of $K(A)$ in M such that $\alpha \to \beta_\alpha$ for every $\alpha \in A$. It then follows from 4.5.4 that for each $A \in \Omega$, there exists a unique derivation of $K(A)$ in M extending D and such that $\alpha \to \beta_\alpha$ for every $\alpha \in A$.

To prove the theorem, it suffices to verify that $S \in \Omega$. To do this, we order Ω by the inclusion relation. Note that $\varnothing \in \Omega$, so that Ω is nonempty. Moreover, taking into account the last assertion of the preceding paragraph, it is seen that Ω is inductive. By Zorn's lemma, there exists a maximal element T of Ω.

To conclude, we shall show that $T = S$. Indeed, assume that $T \subset S$, and choose an $\omega \in S - T$. The algebraic independence of S over K implies that ω is transcendental over $K(S - \{\omega\})$; and this, in view of the inclusion $T \subseteq S - \{\omega\}$, implies that ω is transcendental over $K(T)$. Since $T \in \Omega$, we also know that there exists a derivation E of $K(T)$ in M extending D and such that $\alpha \to \beta_\alpha$ for every $\alpha \in T$. It now follows from 4.5.10 and the equality $K(T \cup \{\omega\}) = K(T)(\omega)$ that E is extendible to a derivation of $K(T \cup \{\omega\})$ in M such that $\omega \to \beta_\omega$. And since the latter is a derivation extending D and such that $\alpha \to \beta_\alpha$ for every $\alpha \in T \cup \{\omega\}$, this shows that $T \cup \{\omega\} \in \Omega$. But this is a contradiction, because $T \subset T \cup \{\omega\}$ and T is a maximal element of Ω.   □

Let K be a field. According to the theorem just proved, the derivation discussed in example 4.5.5a can be described as the K-derivation of $K(X)$ such that $X \to 1$; and if I is a set and $k \in I$, then the derivation discussed in 4.5.5b can be described as the K-derivation of $K(X_i)_{i \in I}$ such that $X_k \to 1$ and such that $X_j \to 0$ for every $j \in I - \{k\}$.

4.5.12. Theorem. *Let K be a field, let L be a separable algebraic extension of K, and let M be an extension field of L. Then every derivation of K in M is uniquely extendible to a derivation of L in M. In particular, the only K-derivation of L in M is the zero derivation.*

Proof. Let D be a derivation of K in M. If A is an intermediate field between K and L, and if $\alpha \in A$, it follows from 4.5.10 that a derivation of A in M that extends D has to send α to $-f^D(\alpha)/f'(\alpha)$, where f denotes the minimal polynomial of α over K. Thus, for every intermediate field A between K and L, there exists at most one derivation of A in M extending D.

The proof will be complete if we show that D is extendible to a derivation of L in M, which we shall do by imitating the argument in the corresponding part of the preceding proof. We denote by Ω the set of all intermediate fields A between K and L such that D is extendible to a derivation of A in M; and we order Ω by the inclusion relation. We then have $K \in \Omega$, which implies that Ω is nonempty. Furthermore, it follows from what was said in the preceding paragraph that Ω is inductive. Therefore, Zorn's lemma can be invoked to obtain a maximal element T of Ω.

It is readily verified that $T = L$. For, assume that $T \subset L$, and choose an $\omega \in L - T$. Since ω is algebraic and separable over K, it is algebraic and separable over T. Also, since $T \in \Omega$, we know that there exists a derivation E of T in M extending D. It then follows from 4.5.10 that E is extendible to a derivation of $T(\omega)$ in M; and such a derivation obviously extends D. We then have $T(\omega) \in \Omega$ and $T \subset T(\omega)$, which is incompatible with the maximality of T in Ω.

We conclude that $L \in \Omega$, and hence that D is extendible to a derivation of L in M.   □

In the next section, we shall see that for finitely generated field extensions, the property expressed by the last assertion in the statement of the theorem just proved characterizes separable algebraic extensions. However, it can be shown that such a characterization is not valid for arbitrary field extensions (problem 3).

4.5.13. Theorem. *Let K be a field of prime characteristic p, let L be an extension field of K such that $L^p \subseteq K$, and let M be an extension field of L. Then a derivation of K in M is extendible to a derivation of L in M if and only if it is an L^p-derivation.*

Proof. Let D be a derivation of K in M. If there exists a derivation E of L in M extending D, then

$$D(\alpha^p) = E(\alpha^p) = p\alpha^{p-1}E(\alpha) = 0$$

for every $\alpha \in L$, and hence D is an $L^p =$ derivation.

Conversely, suppose that D is an L^p-derivation. To show that D is extendible to a derivation of L in M, we cannot proceed exactly as in the previous two proofs, because the uniqueness of the possible extensions of D to intermediate fields between K and L is not implied in the present situation. It will be presently seen, however, that a minor variation yields a proof of the theorem.

Let Ω denote the set of all pairs (A, U) consisting of an intermediate field A between K and L and a derivation U of A in M extending D. We now order Ω in the natural way: If $(A, U), (B, V) \in \Omega$, then $(A, U) \le (B, V)$ when $A \subseteq B$ and V extends U. Since $(K, D) \in \Omega$, it is seen that Ω is nonempty; and it is readily checked that Ω is inductive. By Zorn's lemma, we then know that there exists a maximal element (T, E) of Ω.

To conclude, it suffices to show that $T = L$, since this will prove that E is a derivation of L in M extending D. Assume that $T \subset L$, and choose an $\omega \in L - T$. Since $\omega^p \in L^p$ and $L^p \subseteq K \subseteq T$, we have $\omega^p \in T$ and $\omega \notin T$; and since D is an L^p-derivation, we also have $E(\omega^p) = D(\omega^p) = 0$. By 4.5.10, we then know that E is extendible to a derivation V of $T(\omega)$ in M. It follows that $(T(\omega), V) \in \Omega$ and $(T, E) < (T(\omega), V)$, which contradicts the maximality of (T, E) in Ω. $\square$

4.5.14. Corollary. *Let K be a field of prime characteristic p, let L be an extension field of K, and let M be an extension field of L. Then $K \vee L^p$ consists of the $\alpha \in L$ such that $D(\alpha) = 0$ for every K-derivation D of L in M.*

Proof. We have already shown in 4.5.7 that every K-derivation of L in M is a $K \vee L^p$-derivation. Therefore, if $\alpha \in K \vee L^p$, then $D(\alpha) = 0$ for every K-derivation D of L in M.

Now let $\alpha \in L - K \vee L^p$. We have to show that there exists a K-derivation of L in M that does not vanish at α. To do this, note that

$\alpha^p \in L^p$, so that $\alpha^p \in K \vee L^p$ and $\alpha \notin K \vee L^p$; by 4.5.10, the zero derivation of $K \vee L^p$ in M is extendible to a derivation U of $(K \vee L^p)(\alpha)$ in M
such that $U(\alpha) = 1$. Furthermore, since L is an extension field of $K \vee L^p$
and $L^p \subseteq K \vee L^p$, the theorem implies the existence of a derivation D of L
in M extending U. Finally, since U clearly is a K-derivation, we see that D is
a K-derivation; and since $D(\alpha) = U(\alpha) = 1$, we also have $D(\alpha) \neq 0$.   □

4.5.15. Corollary. *Let K be a field of prime characteristic p, and let
L be an extension field of K. Then K^p consists of the $\alpha \in K$ such that $D(\alpha) = 0$
for every derivation D of K in L.*

Proof. This is an immediate consequence of 4.5.6 and the preceding
corollary.   □

4.5.16. Corollary. *Let K be a field of prime characteristic, and let L
be an extension field of K. Then K is perfect if and only if the only derivation of
K in L is the zero derivation.*

Proof. Let $p = \mathrm{Char}(K)$. Since to say that K is perfect means that
$K^p = K$, the conclusion follows at once from 4.5.6 and the preceding
corollary.   □

In regard to the last corollary, note that a field of characteristic 0 is
perfect, but may admit nonzero derivations. For example, the **Q**-derivation
of $\mathbf{Q}(X)$ described in example 4.5.5a sends X to 1, and therefore is different
from the zero derivation.

The preceding results will now be applied in order to obtain an
important characterization of separability in terms of the extendibility of
derivations.

4.5.17. Theorem. *Let K be a field, let L be an extension field of K,
and let M be an extension field of L. Then L is separable over K if and only if
every derivation of K in M is extendible to a derivation of L in M.*

Proof. It is easy to prove the theorem for fields of characteristic 0.
For in this case, all field extensions are separable; and since L is an
algebraic extension of a purely transcendental extension of K, it is seen
from 4.5.11 and 4.5.12 that every derivation of K in an extension field M of
L is extendible to a derivation of L in M.

Thus, in the remainder of the proof, we shall assume that p is a prime
and that all fields under consideration have characteristic p.

To prove the direct implication, let us assume first that L is separable
over K, and let D be a derivation of K in M. We have to show that D is
extendible to a derivation of L in M. To do this, we begin by selecting a
linear base $(\xi_i)_{i \in I}$ of L^p over K^p containing 1, that is, such that $\xi_r = 1$ for
one index $r \in I$. For all $j, k \in I$, we have $\xi_j \xi_k \in L^p$, and hence we can write

$\xi_j \xi_k = \sum_{i \in I} \gamma_{ijk} \xi_i$, where $(\gamma_{ijk})_{i \in I}$ is a family of elements of K^p such that $\gamma_{ijk} = 0$ for almost every $i \in I$.

According to 4.4.13, the separability of L over K implies that $(\xi_i)_{i \in I}$ is a base of the K-space KL^p. It is meaningful, therefore, to speak of the mapping from KL^p to M such that $\sum_{i \in I} \lambda_i \xi_i \to \sum_{i \in I} D(\lambda_i) \xi_i$ whenever $(\lambda_i)_{i \in I}$ is a family of elements of K such that $\lambda_i = 0$ for almost every $i \in I$. Let U denote this mapping from KL^p to M.

Since $(\xi_i)_{i \in I}$ contains 1, it follows that U extends D. Also, note that $U(\alpha) = 0$ for every $\alpha \in L^p$: Indeed, we can write each $\alpha \in L^p$ in the form $\alpha = \sum_{i \in I} \alpha_i \xi_i$, where $(\alpha_i)_{i \in I}$ is a family of elements of K^p such that $\alpha_i = 0$ for almost every $i \in I$; but, by 4.5.6, we know that D is a K^p-derivation, so that $D(\alpha_i) = 0$ for every $i \in I$, and this implies that $U(\alpha) = 0$.

We shall now verify that U is a derivation of KL^p in M. Let $\alpha, \beta \in KL^p$, and write

$$\alpha = \sum_{i \in I} \alpha_i \xi_i \quad \text{and} \quad \beta = \sum_{i \in I} \beta_i \xi_i,$$

where $(\alpha_i)_{i \in I}$ and $(\beta_i)_{i \in I}$ are families of elements of K such that $\alpha_i = 0 = \beta_i$ for almost every $i \in I$. Then

$$\alpha + \beta = \sum_{i \in I} \alpha_i \xi_i + \sum_{i \in I} \beta_i \xi_i = \sum_{i \in I} (\alpha_i + \beta_i) \xi_i,$$

so that

$$U(\alpha + \beta) = \sum_{i \in I} D(\alpha_i + \beta_i) \xi_i$$

$$= \sum_{i \in I} D(\alpha_i) \xi_i + \sum_{i \in I} D(\beta_i) \xi_i = U(\alpha) + U(\beta).$$

Also, since

$$\alpha \beta = \left(\sum_{j \in I} \alpha_j \xi_j \right) \left(\sum_{k \in I} \beta_k \xi_k \right) = \sum_{j,k \in I} \alpha_j \beta_k \xi_j \xi_k$$

$$= \sum_{i,j,k \in I} \alpha_j \beta_k \gamma_{ijk} \xi_i,$$

we get

$$U(\alpha \beta) = \sum_{i,j,k \in I} D(\alpha_j \beta_k \gamma_{ijk}) \xi_i.$$

We know from 4.5.6 and 4.5.7 that D is K^p-linear; and since $\gamma_{ijk} \in K^p$ for all $i, j, k \in I$, it then follows that

$$U(\alpha \beta) = \sum_{i,j,k \in I} D(\alpha_j \beta_k) \gamma_{ijk} \xi_i$$

$$= \sum_{j,k \in I} D(\alpha_j \beta_k) \xi_j \xi_k = \sum_{j,k \in I} (\beta_k D(\alpha_j) + \alpha_j D(\beta_k)) \xi_j \xi_k$$

$$= \sum_{j,k \in I} \beta_k D(\alpha_j) \xi_j \xi_k + \sum_{j,k \in I} \alpha_j D(\beta_k) \xi_j \xi_k$$

$$= \left(\sum_{k \in I} \beta_k \xi_k \right) \left(\sum_{j \in I} D(\alpha_j) \xi_j \right) + \left(\sum_{j \in I} \alpha_j \xi_j \right) \left(\sum_{k \in I} D(\beta_k) \xi_k \right)$$

$$= \beta U(\alpha) + \alpha U(\beta).$$

Consequently, U is indeed a derivation of KL^p in M.

Since $KL^p = K[L^p]$ and $K \vee L^p = K(L^p)$, we see that $K \vee L^p$ is the field of fractions of KL^p in L. By 4.5.3, we then know that U can be extended to a derivation V of $K \vee L^p$ in M; and furthermore, the remarks above imply that V is an L^p-derivation extending D. To conclude, we need only note that 4.5.13 can now be applied in order to obtain a derivation of L in M extending V, because every such derivation is an extension of D.

In order to prove the opposite implication, let us now suppose that every derivation of K in M is extendible to a derivation of L in M. We have to show that L is separable over K.

Assume, on the contrary, that K and L^p are not linearly disjoint over K^p. Then there exist finite subsets of L^p that are linearly independent over K^p but linearly dependent over K. From among such finite subsets of L^p, select one with the smallest possible cardinality, and denote it by S.

Let $n = \mathrm{Card}(S)$, and write $S = \{\alpha_1, \alpha_2, \ldots, \alpha_n\}$. It follows that $n > 1$ and that there exist elements $\beta_1, \beta_2, \ldots, \beta_n$ of K, at least one of which is equal to 1, such that $\sum_{i=1}^n \alpha_i \beta_i = 0$. And furthermore, the notation can be chosen so that $\beta_n = 1$.

It is seen next that $D(\beta_1) = D(\beta_2) = \cdots = D(\beta_{n-1}) = 0$ for every derivation D of K: Indeed, if D is a derivation of K, our assumption implies the existence of a derivation E of L in M extending D; since $\beta_n = 1$ and $\alpha_1, \alpha_2, \ldots, \alpha_n \in L^p$, it now follows from 4.5.2, 4.5.6, and 4.5.7 that

$$\sum_{i=1}^{n-1} \alpha_i D(\beta_i) = \sum_{i=1}^n \alpha_i D(\beta_i) = \sum_{i=1}^n \alpha_i E(\beta_i)$$

$$= E\left(\sum_{i=1}^n \alpha_i \beta_i \right) = 0;$$

finally, since $D(\beta_i) \in K$ for $1 \leq i \leq n-1$, the relation $\sum_{i=1}^{n-1} \alpha_i D(\beta_i) = 0$ implies that $D(\beta_i) = 0$ for $1 \leq i \leq n-1$, for otherwise $\{\alpha_1, \alpha_2, \ldots, \alpha_{n-1}\}$ would be a subset of L^p of cardinality $n-1$ that is linearly independent over K^p and linearly dependent over K, and this is impossible.

By 4.5.15, what we have just shown implies that $\beta_1, \beta_2, \ldots, \beta_n \in K^p$; in view of the relation $\sum_{i=1}^n \alpha_i \beta_i = 0$, this leads to the contradictory conclusion that S is linearly dependent over K^p. $\qquad\square$

We shall now define some vector spaces of derivations of fields. Here we shall confine ourselves to the required basic generalities on these vector spaces; the special type of field extension for which interesting results can be obtained will be considered in the next section.

Let F be a field, and let A be a subdomain of F. It is evident that the derivations of A in F make up an F-subspace of the F-space $\mathrm{Map}(A, F)$ of all mappings from A to F. We shall denote this F-subspace of $\mathrm{Map}(A, F)$ by $\mathrm{Der}(A, F)$.

Next consider a field K, an extension field L of K, and an extension field M of L. It is also seen that the K-derivations of L in M form an M-subspace of $\mathrm{Der}(L, M)$. The symbol $\mathrm{Der}_K(L, M)$ will be used to denote this M-subspace of $\mathrm{Der}(L, M)$.

We have shown in 4.5.12 that if L is separable and algebraic over K, then the only K-derivation of L in M is the zero derivation, which is equivalent to saying that the M-space $\mathrm{Der}_K(L, M)$ is null. We shall show in the next section that, conversely, under the additional hypothesis of finite generation of L over K, the vanishing of $\mathrm{Der}_K(L, M)$ implies that L is separable and algebraic over K.

To close this section, we now state and prove a useful technicality.

4.5.18. Proposition. *Let K be a field, let L be an extension field of K, and let M be an extension field of L. If S is a finite subset of L such that L is separable and algebraic over $K(S)$, then*

$$\left[\mathrm{Der}_K(L, M) : M\right] \le \mathrm{Card}(S).$$

Proof. If $S = \emptyset$, then L is separable and algebraic over K; therefore, as noted in the preceding remarks, $\mathrm{Der}_K(L, M)$ is null, and so there is nothing to prove.

Now assume that $S \ne \emptyset$. Write

$$n = \mathrm{Card}(S) \quad \text{and} \quad S = \{\alpha_1, \alpha_2, \ldots, \alpha_n\}.$$

Note then that the mapping $D \to (D(\alpha_1), D(\alpha_2), \ldots, D(\alpha_n))$ from $\mathrm{Der}_K(L, M)$ to $M^{(n)}$ is M-linear. If D belongs to its kernel, then D is a K-derivation of L in M satisfying $D(\alpha_i) = 0$ for $1 \le i \le n$, whence it is a $K(S)$-derivation; but since L is separable and algebraic over $K(S)$, it then follows from 4.5.12 that $D = 0$. Thus, our M-linear mapping from $\mathrm{Der}_K(L, M)$ to $M^{(n)}$ is injective; and hence

$$\left[\mathrm{Der}_K(L, M) : M\right] \le [M^{(n)} : M] = n = \mathrm{Card}(S),$$

as required. □

PROBLEMS

1. Let K be a field, and let L be a finite separable extension of K. Show that $D \circ T_{L/K} = T_{L/K} \circ D$ for every derivation D of L.

2. Let K be a field, let L be an extension field of K, and let D be a derivation of K in L. Suppose that $\alpha \in L$ and α is algebraic but not separable over K, and denote by f its minimal polynomial over K. Prove that if $f^D(\alpha) = 0$ and $\beta \in L$, then D is uniquely extendible to a derivation of $K(\alpha)$ in L such that $\alpha \to \beta$.

3. Let K be a field of prime characteristic, and let F be an algebraic closure of K. Write $L = P(F/K)$, and let M be an extension field of L. Show that if K is not perfect, then L is an infinite purely inseparable extension of K such that the only K-derivation of L in M is the zero derivation.

4. Let K be a field, let L be an extension field of K, and let M be an extension field of L. Prove that if $\mathrm{Der}_K(L, L)$ is null, then $\mathrm{Der}_K(L, M)$ is also null.

5. Let K be a field, let L be a purely transcendental extension of K, and let M be an extension field of L. Verify the following assertions:

   a. If S is a pure base of L over K, and if for each $\alpha \in S$ the symbol D_α denotes the K-derivation of L in M such that $\alpha \to 1$ and such that $\beta \to 0$ for every $\beta \in S - \{\alpha\}$, then $(D_\alpha)_{\alpha \in S}$ is a linearly independent family of elements of $\mathrm{Der}_K(L, M)$; and if S is finite, then $(D_\alpha)_{\alpha \in S}$ is a base of $\mathrm{Der}_K(L, M)$.

   b. $\mathrm{Der}_K(L, M)$ is finite-dimensional if and only if L has finite transcendence degree over K, in which case

   $$[\mathrm{Der}_K(L, M) : M] = \mathrm{tr.deg.}(L/K).$$

   (These results apply in an obvious way to the case where $L = K(X_i)_{i \in I}$ for some set I. Note that in assertion a, for each $k \in I$, the symbol D_k then represents the K-derivation of $K(X_i)_{i \in I}$ in M such that $f \to \partial f / \partial X_k$ for every $f \in K[X_i]_{i \in I}$. Also, note that assertion b states that $\mathrm{Der}_K(K(X_i)_{i \in I}, M)$ is finite-dimensional if and only if I is finite, in which case

   $$[\mathrm{Der}_K(K(X_i)_{i \in I}, M) : M] = \mathrm{Card}(I).)$$

The results contained in the next two problems are the natural generalizations of 4.5.8 and 4.5.9 to the case of polynomials in several variables.

Before stating these, we shall extend the pertinent notation to this more general situation. If K is a field and L is an extension field of K, if D is a derivation of K in L, and if n is a positive integer and $f \in K[X_1, X_2, \ldots, X_n]$, the symbol f^D will be used to denote the polynomial in $L[X_1, X_2, \ldots, X_n]$ obtained when D is applied to the coefficients of f. Thus, if $(\gamma_{i_1 i_2 \cdots i_n})_{i_1, i_2, \ldots, i_n \in \mathbf{N}}$ is the family of elements of K such that $\gamma_{i_1, i_2, \ldots, i_n} = 0$ for almost every $(i_1, i_2, \ldots, i_n) \in \mathbf{N}^{(n)}$ and for which

$$f = \sum_{i_1, i_2, \ldots, i_n \in \mathbf{N}} \gamma_{i_1 i_2 \cdots i_n} X_1^{i_1} X_2^{i_2} \cdots X_n^{i_n},$$

then

$$f^D = \sum_{i_1, i_2, \ldots, i_n \in N} D(\gamma_{i_1 i_2 \cdots i_n}) X_1^{i_1} X_2^{i_2} \cdots X_n^{i_n}.$$

6. Let K be a field, and let L be an extension field of K. Let D be a derivation of K in L, and let n be a positive integer. Verify the following assertions:

    a. The mapping $f \rightarrow f^D$ from $K[X_1, X_2, \ldots, X_n]$ to $L(X_1, X_2, \ldots, X_n)$ is a derivation.

    b. If $\alpha_1, \alpha_2, \ldots, \alpha_n \in K$ and $f \in K[X_1, X_2, \ldots, X_n]$, then

    $$D(f(\alpha_1, \alpha_2, \ldots, \alpha_n)) = f^D(\alpha_1, \alpha_2, \ldots, \alpha_n)$$

    $$+ \sum_{i=1}^{n} \frac{\partial f}{\partial X_i}(\alpha_1, \alpha_2, \ldots, \alpha_n) D(\alpha_i).$$

7. Let K be a field, let L be an extension field of K, and let D be a derivation of K in L. Let n be a positive integer and $\alpha_1, \alpha_2, \ldots, \alpha_n, \beta_1, \beta_2, \ldots, \beta_n \in L$, and denote by $\mathscr{I}$ the ideal of algebraic relations of $(\alpha_1, \alpha_2, \ldots, \alpha_n)$ over K. Prove that if D is extendible to a derivation of $K(\alpha_1, \alpha_2, \ldots, \alpha_n)$ in L such that $\alpha_i \rightarrow \beta_i$ for $1 \le i \le n$, then

    $$f^D(\alpha_1, \alpha_2, \ldots, \alpha_n) + \sum_{i=1}^{n} \frac{\partial f}{\partial X_i}(\alpha_1, \alpha_2, \ldots, \alpha_n) \beta_i = 0$$

    for every $f \in \mathscr{I}$. Prove also that, conversely, if there exists a generating set W of $\mathscr{I}$ such that

    $$f^D(\alpha_1, \alpha_2, \ldots, \alpha_n) + \sum_{i=1}^{n} \frac{\partial f}{\partial X_i}(\alpha_1, \alpha_2, \ldots, \alpha_n) \beta_i = 0$$

    for every $f \in W$, then D is uniquely extendible to a derivation of $K(\alpha_1, \alpha_2, \ldots, \alpha_n)$ in L such that $\alpha_i \rightarrow \beta_i$ for $1 \le i \le n$.

8. Let K be a field of prime characteristic p, let L be an extension field of K, and let M be an extension field of L. Prove the following assertions:

    a. If $S \subseteq L$, then S is p-independent in L over K if and only if there exists a family $(D_\alpha)_{\alpha \in S}$ of K-derivations of L in M such that $D_\alpha(\alpha) = 1$ for every $\alpha \in S$ and such that $D_\alpha(\beta) = 0$ for all $\alpha, \beta \in S$ with $\alpha \ne \beta$; in which case every such $(D_\alpha)_{\alpha \in S}$ is a linearly independent family of elements of $\mathrm{Der}_K(L, M)$.

    b. If S is a p-base of L over K, and if $\alpha \in S$, then there exists a unique $K(S - \{\alpha\})$-derivation of L in M such that $\alpha \rightarrow 1$.

    c. If S is a p-base of L over K, and if for each $\alpha \in S$ the $K(S - \{\alpha\})$-derivation of L in M such that $\alpha \rightarrow 1$ is denoted by D_α, then $(D_\alpha)_{\alpha \in S}$ is a linearly independent family of elements of $\mathrm{Der}_K(L, M)$; and it is a base of $\mathrm{Der}_K(L, M)$ if S is finite.

    d. L has finite degree of imperfection over K if and only if $\mathrm{Der}_K(L, M)$ is finite-dimensional, in which case $[\mathrm{Der}_K(L, M): M]$ is the degree of imperfection of L over K.

e. If S is a p-base of L over K, then every mapping from S to M is uniquely extendible to a K-derivation of L in M.

f. If S is a p-base of L over K, and if for each mapping u from S to M the K-derivation of L in M extending u is denoted by D_u, then the mapping $u \to D_u$ from $\mathrm{Map}(S, M)$ to $\mathrm{Der}_K(L, M)$ is an M-linear isomorphism.

9. Let K be a field of prime characteristic p, let N be an extension field of K, and let L and M be intermediate fields between K and N such that $L \subseteq M$. Prove that if every K-derivation of L in N is extendible to a K-derivation of M in N, then every subset of L that is p-independent in L over K is p-independent in M over K. Prove also that, conversely, if there exists a p-base of L over K that is p-independent in M over K, then every K-derivation of L in N is extendible to a K-derivation of M in N.

10. Let K be a field of prime characteristic p, let L be an extension field of K, and let P denote the prime subfield of K. Show that if L is separable over K, then every subset of K that is p-independent in K over P is p-independent in L over P. Show also that, conversely, if there exists a p-base of K over P that is p-independent in L over P, then L is separable over K.

The remaining problems are devoted to a sequence of results leading to a correspondence, due to Jacobson, between two classes of objects associated with a given field F of prime characteristic. One of these classes consists of subfields of F, and the other of spaces of derivations of F.

Let F be a field. The F-space $\mathrm{Der}(F, F)$ of derivations of F will be denoted more simply by $\mathrm{Der}(F)$; and for every subfield P of F, the F-space $\mathrm{Der}_P(F, F)$ of P-derivations of F will be denoted by $\mathrm{Der}_P(F)$.

It is easily verified that if $D, E \in \mathrm{Der}(F)$ and $\alpha, \beta \in F$, then

$$E(D(\alpha\beta)) = \beta E(D(\alpha)) + E(\alpha)D(\beta) + E(\beta)D(\alpha) + \alpha E(D(\beta)).$$

This equality shows that the composition of two derivations of F need not be a derivation. It suggests, however, how to use composition of mappings in order to define an operation on $\mathrm{Der}(F)$.

11. Let F be a field, and let $D, E \in \mathrm{Der}(F)$. We then write $[D, E] = D \circ E - E \circ D$, and say that $[D, E]$ is the **Lie commutator of D and E**. Show that $[D, E] \in \mathrm{Der}(F)$.

12. Let F be a field, and let $D \in \mathrm{Der}(F)$. We define a sequence $(D^k)_{k \geq 0}$ of mappings from F to F inductively as follows: $D^0 = i_F$; and $D^{k+1} = D \circ D^k$ for every $k \geq 0$. Verify the following assertions:

a. If k is a nonnegative integer and $\alpha, \beta \in F$, then

$$D^k(\alpha\beta) = \sum_{j=0}^{k} \binom{k}{j} D^j(\alpha) D^{k-j}(\beta).$$

b. If F has prime characteristic p, then $D^p \in \mathrm{Der}(F)$.
(Assertion a is known as **Leibniz's rule** for derivations.)

13. Let F be a field of prime characteristic p. By a **restricted Lie algebra in** $\mathrm{Der}(F)$ we shall understand an F-subspace Ω of $\mathrm{Der}(F)$ with the following two properties: (i) If $D, E \in \Omega$, then $[D, E] \in \Omega$; and (ii) if $D \in \Omega$, then $D^p \in \Omega$. Show that if P is a subfield of F, then $\mathrm{Der}_P(F)$ is a restricted Lie algebra in $\mathrm{Der}(F)$.

14. Let F be a field of prime characteristic p, and let Ω be a restricted Lie algebra in $\mathrm{Der}(F)$. We then denote by $C(\Omega)$ the set of all $\alpha \in F$ such that $D(\alpha) = 0$ for every $D \in \Omega$. Prove that $C(\Omega)$ is the largest element of the set of all subfields P of F such that every element of Ω is a P-derivation, and hence that $F^p \subseteq C(\Omega)$.

15. Let F be a field of prime characteristic p, and let P be a subfield of F such that $F^p \subseteq P$. Prove the following assertions:
    a. $C(\mathrm{Der}_P(F)) = P$.
    b. F is finite over P if and only if $\mathrm{Der}_P(F)$ is finite-dimensional, in which case

$$[F:P] = p^{[\mathrm{Der}_P(F):F]}.$$

16. Let F be a field of prime characteristic p, let n be a positive integer, and let Ω be an n-dimensional restricted Lie algebra in $\mathrm{Der}(F)$. Let $(D_1, D_2, \ldots, D_n)$ be a base of Ω, and let Φ be the F-subspace of $\mathrm{Map}(F, F)$ generated by the family

$$\left(D_1^{k_1} \circ D_2^{k_2} \circ \cdots \circ D_n^{k_n} \right)_{0 \le k_1, k_2, \ldots, k_n \le p-1}.$$

Prove the following assertions:
    a. Φ is an F-algebra.
    b. $R(\Phi) = C(\Omega)$.
    c. $[\Phi : F] = p^n$.

17. Let F be a field of prime characteristic p, and let Ω be a restricted Lie algebra in $\mathrm{Der}(F)$. Show that Ω is finite-dimensional if and only if F is finite over $C(\Omega)$, in which case

$$[F : C(\Omega)] = p^{[\Omega : F]}.$$

18. Let F be a field of prime characteristic p. In view of the foregoing results, it is meaningful to speak of the mapping $P \rightarrow \mathrm{Der}_P(F)$ from the set of all subfields of F containing F^p and over which F is finite to the set of all finite-dimensional restricted Lie algebras in $\mathrm{Der}(F)$, and of the mapping $\Omega \rightarrow C(\Omega)$ from the set of all finite-dimensional restricted Lie algebras in $\mathrm{Der}(F)$ to the set of all subfields of F containing F^p and over which F is finite. Prove that these mappings are mutually inverse inclusion-reversing bijections. (These bijections constitute the **Jacobson differential correspondence** defined by the field F.)

4.6. DERIVATIONS OF ALGEBRAIC FUNCTION FIELDS

The primary purpose of this section is to prove some classical results on the separability of finitely generated extensions. Our approach will be based on rather technical considerations involving transcendence degrees, the dimensions of various vector spaces of derivations, and the indices of certain linear mappings between these vector spaces.

Some elementary auxiliary facts on linear mappings of finite index, which have been discussed in 0.0.8, will be requited for our presentation. Let us recall that if F is a field, an F-linear mapping is said to have finite index when its kernel and cokernel are finite-dimensional F-spaces; and that for every such mapping u, the index is the integer $[u:F]$ defined by

$$[u:F] = [\text{Ker}(u):F] - [\text{Coker}(u):F].$$

We shall need only two properties of F-linear mappings of finite index. First, every F-linear isomorphism has finite index 0. Second, if u and v are F-linear mappings of finite index such that $v \circ u$ is defined, then $v \circ u$ has finite index, and

$$[v \circ u:F] = [u:F] + [v:F].$$

Let K be a field, let L be an extension field of K, and let M be an extension field of L. If a mapping from K to M is obtained by restriction of a derivation of L in M, then it is a derivation of K in M. We shall denote by $r_{K,L,M}$ the mapping from $\text{Der}(L, M)$ to $\text{Der}(K, M)$ assigning to each derivation of L in M the derivation of K in M that it defines by restriction.

It is evident that $r_{K,L,M}$ is an M-linear mapping. Also, we have $\text{Ker}(r_{K,L,M}) = \text{Der}_K(L, M)$, so that $\text{Ker}(r_{K,L,M})$ is null if and only if $\text{Der}_K(L, M)$ is null. Finally, $\text{Im}(r_{K,L,M})$ consists of the derivations of K in M that are extendible to derivations of L in M, and hence $\text{Coker}(r_{K,L,M})$ is null if and only if every derivation of K in M is extendible to a derivation of L in M.

Generally speaking, the "restriction mappings" just defined do not have finite index. However, as the next result shows, interesting things happen when suitable assumptions of finiteness are made.

4.6.1. Theorem. *Let K be a field, let L be a finitely generated extension of K, and let M be an extension field of L. Then $r_{K,L,M}$ is an M-linear mapping of finite index, and*

$$[r_{K,L,M}:M] = \text{tr.deg.}(L/K).$$

Proof. We begin by showing that it suffices to prove a particular case of the theorem.

Lemma. *The validity of the theorem follows from its validity in the case where the words "finitely generated" are replaced by the word "simple".*

Proof. It will be convenient to consider a field M as given once and for all, and to view the theorem as an assertion on finitely generated extensions consisting of subfields of M.

We shall proceed by induction on the number of generators for the field extensions under consideration. The case of one generator is that whose validity we are assuming. Suppose next that n is a positive integer, and that the required conclusion holds in the case of n generators.

Consider then subfields A and B of M such that $B = A(\sigma_1, \sigma_2, \ldots, \sigma_{n+1})$ with $\sigma_1, \sigma_2, \ldots, \sigma_{n+1} \in M$. If we write $F = A(\sigma_1, \sigma_2, \ldots, \sigma_n)$, the induction hypothesis implies that $r_{A, F, M}$ has finite index and $[r_{A, F, M} : M] = $ tr.deg.(F/A). But $B = F(\sigma_{n+1})$, and our initial assumption implies that $r_{F, B, M}$ has finite index and $[r_{F, B, M} : M] = $ tr.deg.(B/F). In view of the relation $r_{A, B, M} = r_{A, F, M} \circ r_{F, B, M}$, it now follows that $r_{A, B, M}$ has finite index and

$$[r_{A, B, M} : M] = [r_{F, B, M} : M] + [r_{A, F, M} : M]$$
$$= \text{tr.deg.}(B/F) + \text{tr.deg.}(F/A) = \text{tr.deg.}(B/A),$$

which is what was needed. This completes the proof of the lemma.

Thus, we can now take K, L, and M as in the statement of the theorem, and impose the restriction that $L = K(\alpha)$ for some $\alpha \in L$. We can further assume that $\alpha \notin K$, because the theorem is trivial when $K = L$. We shall consider four cases.

Case 1. α *is transcendental over* K.

In this case, we have to prove that $r_{K, L, M}$ has finite index 1. Note first that, according to 4.5.10, every derivation of K in M is extendible to a derivation of L in M; as observed earlier, this amounts to saying that Coker$(r_{K, L, M})$ is null.

To reach the desired conclusion, we must now verify that $[\text{Ker}(r_{K, L, M}) : M] = 1$. This is also seen to follow from 4.5.10. Indeed, there exists a K-derivation $\bar{D}$ of L in M such that $\bar{D}(\alpha) = 1$. Then $\text{Der}_K(L, M) = M\bar{D}$: For let $D \in \text{Der}_K(L, M)$; if $\mu = D(\alpha)$, then $D(\gamma) = 0 = \mu\bar{D}(\gamma)$ for every $\gamma \in K$ and $D(\alpha) = \mu = \mu\bar{D}(\alpha)$, which implies that $D = \mu\bar{D}$. Consequently

$$[\text{Der}_K(L, M) : M] = [M\bar{D} : M] = 1,$$

which is what we wanted.

Case 2. α *is algebraic and separable over* K.

This time we have to prove that $r_{K, L, M}$ has finite index 0. This is a consequence of 4.5.10, which shows that the only K-derivation of L in M is the zero derivation, and that every derivation of K in L is extendible to a derivation of L in M; and this means that $\text{Ker}(r_{K, L, M})$ and $\text{Coker}(r_{K, L, M})$ are null, which implies the required conclusion.

Case 3. K has prime characteristic; and α is purely inseparable over K.

Again, we have to show that $r_{K,L,M}$ has finite index 0. Let $p = \text{Char}(K)$, and let e denote the smallest element of the set of all positive integers i for which $\alpha^{p^i} \in K$. By 4.5.10, there exists a K-derivation of L in M such that $\alpha \to 1$. Exactly as in case 1, the latter generates the M-space $\text{Der}_K(L, M)$, which implies that $[\text{Ker}(r_{K,L,M}): M] = 1$.

To conclude, it remains to verify that $[\text{Coker}(r_{K,L,M}): M] = 1$. To this end, note that the definition of e implies that $\alpha^{p^e} \notin K^p$; therefore, by 4.5.15, a derivation $\overline{D}$ of K in M can be chosen so that $\overline{D}(\alpha^{p^e}) = 1$. We now contend that $\text{Der}(K, M)$ decomposes as the internal direct sum of its M-subspaces $\text{Im}(r_{K,L,M})$ and $M\overline{D}$; this will give us what we need, since it implies that $\text{Coker}(r_{K,L,M})$ and $M\overline{D}$ are isomorphic M-spaces.

To establish our claim, suppose first that $D \in \text{Im}(r_{K,L,M}) \cap M\overline{D}$. Then D is extendible to a derivation E of L in M, whence

$$D(\alpha^{p^e}) = E(\alpha^{p^e}) = p^e \alpha^{p^e-1} E(\alpha) = 0;$$

also, $D = \mu \overline{D}$ for some $\mu \in M$, so that

$$\mu = \mu \overline{D}(\alpha^{p^e}) = D(\alpha^{p^e}) = 0,$$

which implies that $D = \mu \overline{D} = 0$.

Finally, suppose that $D \in \text{Der}(K, M)$. If we let $\mu = D(\alpha^{p^e})$, we get

$$(D - \mu \overline{D})(\alpha^{p^e}) = D(\alpha^{p^e}) - \mu \overline{D}(\alpha^{p^e}) = \mu - \mu = 0.$$

By 4.5.10, it now follows that $D - \mu \overline{D}$ is extendible to a derivation of L in M, which means that $D - \mu \overline{D} \in \text{Im}(r_{K,L,M})$. Therefore

$$D = (D - \mu \overline{D}) + \mu \overline{D} \in \text{Im}(r_{K,L,M}) + M\overline{D}.$$

Case 4. K has prime characteristic; and α is algebraic over K, but not separable over K.

Write $S = S(L/K)$. Then $L = S(\alpha), \alpha \notin S$, and α is purely insepara-ble over S; and so it follows from case 3 that $r_{S,L,M}$ has finite index 0. Moreover, since S is separable and algebraic over K, the argument given in Case 2—invoking 4.5.12 instead of 4.5.10—shows that $r_{K,S,M}$ has finite index 0. Finally, note that $r_{K,L,M} = r_{K,S,M} \circ r_{S,L,M}$, so that $r_{K,L,M}$ has finite index and

$$\text{ind}(r_{K,L,M}) = \text{ind}(r_{S,L,M}) + \text{ind}(r_{K,S,M}) = 0;$$

but this implies the required conclusion, because tr.deg.$(L/K) = 0$. $\square$

4.6.2. Corollary. *Let K be a field, let L be a finitely generated extension of K, and let M be an extension field of L. Then $\text{Der}_K(L, M)$ is finite-dimensional, and*

$$\text{tr.deg.}(L/K) \leq [\text{Der}_K(L, M): M].$$

We can now proceed to the study of separability for finitely generated extensions. We begin by showing that the finitely generated extensions for which the inequality appearing in the preceding corollary becomes an equality are precisely the separable ones.

4.6.3. Theorem. *Let K be a field, let L be a finitely generated extension of K, and let M be an extension field of L. Then L is separable over K if and only if*

$$\text{tr.deg.}(L/K) = [\text{Der}_K(L, M): M].$$

Proof. It follows from 4.6.1 that the stated equality holds if and only if $\text{Coker}(r_{K, L, M})$ is null, and hence if and only if every derivation of K in M is extendible to a derivation of L in M. But according to 4.5.17, the latter assertion means that L is separable over K, and our conclusion follows at once. □

We have already observed that if K is a field and L is a separable algebraic extension of K, and if M is an extension field of L, then $\text{Der}_K(L, M)$ is null. For finitely generated extensions, we can now prove the following converse.

4.6.4. Theorem. *Let K be a field, let L be a finitely generated extension of K, and let M be an extension field of L. Then L is separable and algebraic over K if and only if $\text{Der}_K(L, M)$ is null.*

Proof. We need only concern ourselves with the opposite implication. Assume, therefore, that $\text{Der}_K(L, M)$ is null. By 4.6.1, this implies that

$$\text{tr.deg.}(L/K) = -[\text{Coker}(r_{K, L, M}): M],$$

which in turn implies the equalities

$$\text{tr.deg.}(L/K) = 0 = [\text{Coker}(r_{K, L, M}): M].$$

The first of these shows that L is algebraic over K. And the second shows that $\text{Coker}(r_{K, L, M})$ is null, so that every derivation of K in M is extendible to a derivation of L in M; and by 4.5.17, this proves the separability of L over K. □

4.6.5. Corollary. *Let K be a field, let L be a finitely generated extension of K, and let M be an extension field of L. Suppose that $\text{Der}_K(L, M)$ is nonnull, write $n = [\text{Der}_K(L, M): M]$, and let $\alpha_1, \alpha_2, \ldots, \alpha_n \in L$. If there exists a base $(D_1, D_2, \ldots, D_n)$ of $\text{Der}_K(L, M)$ such that the matrix $[D_i(\alpha_j)]_{1 \le i, j \le n}$ in $\text{Mat}_n(M)$ is nonsingular, then L is separable and algebraic over $K(\alpha_1, \alpha_2, \ldots, \alpha_n)$.*

Proof. Let $P = K(\alpha_1, \alpha_2, \ldots, \alpha_n)$. Then L is finitely generated over P; and, by the theorem, the conclusion will follow if we show that $\text{Der}_P(L, M)$ is null.

Thus, let $D \in \text{Der}_P(L, M)$. Then $D \in \text{Der}_K(L, M)$, and we can write $D = \sum_{i=1}^n \mu_i D_i$ with $\mu_1, \mu_2, \ldots, \mu_n \in M$. Since $\alpha_1, \alpha_2, \ldots, \alpha_n \in P$, for $1 \le j \le n$ we have the equalities

$$\sum_{i=1}^n \mu_i D_i(\alpha_j) = D(\alpha_j) = 0.$$

But by hypothesis, the n rows of the matrix $[D_i(\alpha_j)]_{1 \le i, j \le n}$ are linearly independent vectors in $M^{(n)}$; therefore, these equalities imply that $\mu_i = 0$ for $1 \le i \le n$, and hence $D = 0$.   □

In each of the equalities stated in 4.6.1 and 4.6.3, an integer depending on the three fields K, L, and M is related to tr.deg.(L/K), which is independent of M. In this sense, when L is finitely generated over K, it can be said that the index $[r_{K, L, M} : M]$ is independent of M; and that if L is further assumed to be separable over K, the dimension $[\text{Der}_K(L, M) : M]$ is also independent of M.

We shall now show that even without assuming the separability of L over K, we can describe the dimension $[\text{Der}_K(L, M) : M]$ independently of M.

4.6.6. Theorem. *Let K be a field, let L be a finitely generated extension of K, and let M be an extension field of L. Then $[\text{Der}_K(L, M) : M]$ is the smallest element of the set of all nonnegative integers k for which there exists a subset S of L with $\text{Card}(S) = k$ and such that L is separable and algebraic over $K(S)$. Moreover, if $T \subseteq L$ and $L = K(T)$, then there exists a subset S of T with $\text{Card}(S) = [\text{Der}_K(L, M) : M]$ and such that L is separable and algebraic over $K(S)$.*

Proof. We have already seen in 4.5.18 that

$$[\text{Der}_K(L, M) : M] \le \text{Card}(S)$$

whenever S is a finite subset of L such that L is separable and algebraic over $K(S)$. Consequently, in order to prove the theorem, it suffices to verify the last assertion in the conclusion.

This presents no difficulty if $\text{Der}_K(L, M)$ is null; for in this case, it follows from 4.6.4 that L is separable and algebraic over K, and hence we can take $S = \varnothing$.

Now assume that $\text{Der}_K(L, M)$ is nonnull. Then $K \subset L$, and 4.2.4 shows that there exist a positive integer r and elements $\alpha_1, \alpha_2, \ldots, \alpha_r$ of T such that $L = K(\alpha_1, \alpha_2, \ldots, \alpha_r)$. The mapping $D \to (D(\alpha_1), D(\alpha_2), \ldots, D(\alpha_r))$ from $\text{Der}_K(L, M)$ to $M^{(r)}$ is M-linear; and furthermore, it is seen to be injective: If D belongs to its kernel, then D is a K-derivation of $K(\alpha_1, \alpha_2, \ldots, \alpha_r)$ in M such that $D(\alpha_j) = 0$ for $1 \le j \le r$, whence $D = 0$.

Finally, write $n = [\mathrm{Der}_K(L, M) : M]$ and choose a base $(D_1, D_2, \ldots, D_n)$ of $\mathrm{Der}_K(L, M)$. For $1 \le i \le n$, define a vector v_i in $M^{(r)}$ by

$$v_i = (D_i(\alpha_1), D_i(\alpha_2), \ldots, D_i(\alpha_r)),$$

so that $D_i \to v_i$ by the linear mapping defined above. The injectivity of the latter implies that $(v_1, v_2, \ldots, v_n)$ is a linearly independent sequence of vectors in $M^{(r)}$; therefore $n \le r$, and the matrix $[D_i(\alpha_j)]_{1 \le i \le n, 1 \le j \le r}$ in $\mathrm{Mat}_{n \times r}(M)$ has rank n. Thus, indices $k_1, k_2, \ldots, k_n$ can be chosen so that $1 \le k_1 < k_2 < \cdots < k_n \le r$ and so that the matrix $[D_i(\alpha_{k_j})]_{1 \le i, j \le n}$ in $\mathrm{Mat}_n(M)$ is nonsingular. According to 4.6.5, L is seen to be separable and algebraic over $K(\alpha_{k_1}, \alpha_{k_2}, \ldots, \alpha_{k_n})$. To conclude, we now need only take $S = \{\alpha_{k_1}, \alpha_{k_2}, \ldots, \alpha_{k_n}\}$. $\qquad\square$

We have thus found that in the determination of the dimension of $\mathrm{Der}_K(L, M)$, a natural role is played by the subsets S of L having the property that L is separable and algebraic over $K(S)$.

If S is a transcendence base of L over K, then L is of course algebraic over $K(S)$; but L may fail to be separable over $K(S)$.

We shall presently see that, under the usual hypothesis of finite generation underlying our discussion, separability implies the existence of a transcendence base having this additional property. In order to establish this, we shall first introduce some traditional terminology.

Let K be a field, and let L be an extension field of K. By a **separating transcendence base of L over K** we understand a transcendence base S of L over K such that L is separable over $K(S)$. We say that L is **separably generated over K**, or that L is a **separably generated extension of K**, when L admits a separating transcendence base over K.

The concepts just defined are of no interest for fields of characteristic 0: Every transcendence base is then separating, and hence every field extension is separably generated. It is also evident that every pure base is separating, so that every purely transcendental extension is separably generated; on the other hand, it should be noted that in prime characteristic, every purely transcendental extension admits a nonseparating transcendence base (problem 5).

It follows from 4.4.10 and 4.4.12 that every separably generated extension is separable. The converse of this statement is not difficult to disprove; in fact, we can show that every field of prime characteristic admits a separable extension that is not separably generated (problem 6). For finitely generated extensions, however, separability implies separable generation; indeed, the preceding results now allow us to give a quick proof of the following more precise statement.

4.6.7. Theorem (MacLane). *Let K be a field. If L is a finitely generated separable extension of K, then every subset of L that generates L over K contains a separating transcendence base of L over K. In particular, every finitely generated separable extension of K is separably generated over K.*

Proof. Let $T \subseteq L$ and $L = K(T)$. Combining 4.6.6 and 4.6.3, we see that there exists a subset S of T such that

$$\text{Card}(S) = \text{tr.deg.}(L/K)$$

and such that L is separable and algebraic over $K(S)$. To conclude, we need only observe that S is also a transcendence base of L over K, which is evident from the preceding conditions and the finiteness of tr.deg.(L/K). $\square$

4.6.8. **Corollary** (Schmidt). *Let K be a perfect field. If L is a finitely generated extension of K, then every subset of L that generates L over K contains a separating transcendence base of L over K. In particular, every finitely generated extension of K is separably generated over K.*

PROBLEMS

1. Let K be a field, and let L be a finitely generated extension of K. Write $L = K(\alpha_1, \alpha_2, \ldots, \alpha_n)$, where n is a positive integer and $\alpha_1, \alpha_2, \ldots, \alpha_n \in L$, and denote by $\mathscr{J}$ the ideal of algebraic relations of $(\alpha_1, \alpha_2, \ldots, \alpha_n)$ over K. Prove the following assertions:
   a. If there exists polynomials $f_1, f_2, \ldots, f_n$ in $\mathscr{J}$ such that the matrix $[(\partial f_i/\partial X_j)(\alpha_1, \alpha_2, \ldots, \alpha_n)]_{1 \le i, j \le n}$ in $\text{Mat}_n(L)$ is nonsingular, then L is separable and algebraic over K.
   b. If for every $f \in \mathscr{J}$ the L-linear form

   $$(\rho_1, \rho_2, \ldots, \rho_n) \to \sum_{i=1}^n \frac{\partial f}{\partial X_i}(\alpha_1, \alpha_2, \ldots, \alpha_n)\rho_i$$

   on $L^{(n)}$ is denoted by d_f, then the L-subspace of the dual $\widehat{L^{(n)}}$ generated by $(d_f)_{f \in \mathscr{J}}$ has dimension $n - [\text{Der}_K(L): L]$.
2. Let K be a field, let N be an extension field of K, and let L and M be intermediate fields between K and N that are finitely generated over K and such that $L \subseteq M$. Show that

   $$[\text{Der}_K(L, N): N] \le [\text{Der}_K(M, N): N].$$

   (This result is due to Faith.)
3. Let K be a field of prime characteristic, and let L be a finitely generated extension of K. Prove that the transcendence degree of L over K is less than or equal to the degree of imperfection of L over K.
4. Let K be a field of prime characteristic p, let L be a finitely generated extension of K, and let M be an extension field of L. Suppose that $\text{Der}_K(L, M)$ is nonnull, write $n = [\text{Der}_K(L, M): M]$, and let $(\alpha_1, \alpha_2, \ldots, \alpha_n)$ and $(D_1, D_2, \ldots, D_n)$ be, respectively, sequences of elements of L and $\text{Der}_K(L, M)$. Prove the following assertions:
   a. If $(D_1, D_2, \ldots, D_n)$ is a base of $\text{Der}_K(L, M)$, then $\{\alpha_1, \alpha_2, \ldots, \alpha_n\}$ is a p-base of L over K if and only if the matrix $[D_i(\alpha_j)]_{1 \le i, j \le n}$ in $\text{Mat}_n(M)$ is nonsingular.
   b. If $\{\alpha_1, \alpha_2, \ldots, \alpha_n\}$ is a p-base of L over K, then $(D_1, D_2, \ldots, D_n)$ is a

    base of $\text{Der}_K(L, M)$ if and only if the matrix $[D_i(\alpha_j)]_{1 \le i, j \le n}$ in Mat$_n(M)$ is nonsingular.

5. Let K be a field of prime characteristic, and let L be a purely transcendental extension of K. Show that if S is a pure base of L over K, then the image of S by the Frobenius mapping of L is a nonseparating transcendence base of L over K.

6. Let K be a field of prime characteristic p, let $K(\alpha)$ be a simple transcendental extension of K, and let F be an algebraic closure of $K(\alpha)$. For every nonnegative integer n, let β_n denote the p^nth root of α in F. Verify the following assertions:

    a. $(K(\beta_n))_{n \ge 0}$ is a strictly increasing sequence of intermediate fields between $K(\alpha)$ and F.

    b. $\cup_{n=0}^{\infty} K(\beta_n)$ is separable but not separably generated over K.

7. Let K be a field, let N be an extension field of K, and let L and M be intermediate fields between K and N. Prove the following assertions:

    a. If L and M are algebraically disjoint over K, and if L is separable over K, then $L \vee M$ is separable over M.

    b. If L and M are algebraically disjoint over K, and if L and M are separable over K, then $L \vee M$ is separable over K.

    c. If L and M are linearly disjoint over K, then L is separable over K if and only if $L \vee M$ is separable over M.

8. Let K be a field of prime characteristic p, and let L be a separable extension of K having finite transcendence degree over K. Prove that L is separably generated over K if and only if the transcendence degree of L over K is equal to the degree of imperfection of L over K, in which case the separating transcendence bases of L over K are the p-bases of L over K.

    Apply this result to prove MacLane's theorem 4.6.7 without using 4.6.6.

9. Let K be a field, and let L be an extension field of K. For each $\alpha \in L$, the mapping $D \to D(\alpha)$ from $\text{Der}_K(L)$ to L obviously is an L-linear form on $\text{Der}_K(L)$. The mapping from L to the dual $\widehat{\text{Der}_K(L)}$ assigning to each $\alpha \in L$ the L-linear form $D \to D(\alpha)$ on $\text{Der}_K(L)$ will be denoted by $d_{L/K}$. Verify the following assertions:

    a. $d_{L/K}$ is a K-linear mapping from L to $\widehat{\text{Der}_K(L)}$.

    b. If $\alpha \in K$, then $d_{L/K}(\alpha) = 0$.

    c. If $\alpha, \beta \in L$, then

$$d_{L/K}(\alpha\beta) = \beta d_{L/K}(\alpha) + \alpha d_{L/K}(\beta).$$

10. Let K be a field, let L be a finitely generated extension of K, and let $S \subseteq L$ and $\text{Card}(S) = \text{tr.deg.}(L/K)$. Show that S is a separating transcendence base of L over K if and only if $(d_{L/K}(\alpha))_{\alpha \in S}$ is a base of $\widehat{\text{Der}_K(L)}$.

    Show that Maclane's theorem 4.6.7 can be deduced from this result and 4.6.5, without using any other result in this section.

NOTES

Section 4.1.   The dimension theory presented in this section is given in Bourbaki [3: sec. 5.5] as an exercise (note that the third condition in the definition of a dimensional operator given here does not appear in this exercise, since it is superfluous in the presence of the other conditions). For a dimension theory closer in spirit to that in Steinitz [1: §22], see Jacobson [4: sec. 4.3; 5: sec. 3.6].

The notions of *p*-independence, *p*-base, and degree of imperfection discussed in problems 4–13 were introduced by Teichmüller [1].

Section 4.2.   The concepts of purely transcendental extension, transcendence base, and transcendence degree were introduced by Steinitz [1: §§22, 23]; some of his proofs are by transfinite induction, and so are quite different from those commonly found in the more recent expositions on transcendental extensions. The discussion on the extendibility of field monomorphisms given here follows Bourbaki [3: sec. 5.6].

As regards the questions on intermediate fields of a purely transcendental extension, the positive results stated at the end of the section were given in the works of Lüroth [1], Weber [2: II, §124], Igusa [1], Castelnuovo [1], and Zariski [3]. On the negative side, it was shown by Segre [1] that for a purely transcendental extension of **Q**, an intermediate field with transcendence degree 2 over **Q** need not be purely transcendental over **Q**. And, with the help of highly sophisticated tools from algebraic geometry, it has been shown recently by Iskovskih and Manin [1], Artin and Mumford [1], and Clemens and Griffiths [1] that for a purely transcendental extension of **C**, an intermediate field of transcendence degree 3 over **C** need not be purely transcendental over **C**. These results are discussed briefly in Deligne [1].

The first example implying a negative answer to Noether's problem was given in Swan [1]. Simpler examples were given subsequently in Lenstra [1].

Section 4.3.   The notions of local ring and valuation domain were first studied systematically by Krull [2; 3]. The lemma in the proof of 4.3.10 is contained in the proof of 4.3.12 given in Chevalley [2: sec. 1.4]. The extension theorem (4.3.11) was first proved by van der Waerden [1]; it was applied by Zariski [1] in the study of birational correspondences. The proof used to derive Zariski's theorem (4.3.13) from the extension theorem is a minor variation of an argument in Jacobson [4: sec. 5.12]. The original proof of 4.3.13 was published by Zariski [2], who in fact used this theorem together with the result in Rabinowitsch [1] in order to derive the Nullstellensatz. The original proof of the Nullstellensatz was given in Hilbert [2: §3].

Much of the material in this section was initially developed in response to the needs of algebraic geometry, and is treated in texts on commutative algebra. However, the presentations in these texts differ from the one given here in that the discussion on valuation domains and places is preceded by a study of integral dependence (see Bourbaki [1], Atiyah and Macdonald [1], Kaplansky [2], Nagata [1], and Zariski and Samuel [1]).

As shown by Flanders [1], the properties of the norm mapping can be combined with the most elementary considerations on integral elements and integrally closed domains to give a simple proof of Zariski's theorem (4.3.13).

Section 4.4. The generalization of separability to transcendental extensions was first given in MacLane [1]; in this work, essential use is made of the properties of p-independent sets and p-bases in section 4.1, problems 4–13. In his original article, MacLane introduced the term "preservation of p-independence" for the general property, and reserved the term "separability" exclusively for algebraic extensions.

The expositions on separability given in most texts on field extensions—including ours—usually follow that given in Weil [1], where linear disjointness was first discussed. An alternative approach to the study of separability, based on Artin's theorem (section 3.2, problem 12) and problem 5, is used in Bourbaki [3: sec. 5.7.2].

The results on algebraic disjointness stated in problems 10–14 can be found in Bourbaki [3: sec. 5.5.4].

Section 4.5. The differentiation of rational functions of one variable was treated in Steinitz [1: §9]. The use of derivations in the study of separability for algebraic function fields was initiated in Weil [1]. We have departed here from the standard presentations in that the case of algebraic function fields is not emphasized from the outset; in this section, we have dealt almost exclusively with those questions in which finiteness conditions are not relevant. Alternative expositions on derivations of fields can be found in Bourbaki [3: sec. 5.9], Jacobson [4: sec. 4.7], and Zariski and Samuel [1: I, sec. 2.17].

The results stated in problems 8–10 are given in Jacobson [4: sec. 4.7]; problem 10 is the statement that separability, as defined here, is equivalent to the property called "preservation of p-independence" by MacLane [1].

The Jacobson differential correspondence was first given in Jacobson [1]; the outline given in problems 11–18 is based on Jacobson [4: sec. 4.8]. An alternative treatment of this topic can be found in Winter [1], where additional results on restricted Lie algebras are given.

Section 4.6. The presentation in this section is based on the technical result 4.6.1, which is due to the present author [1]. The combination of this result with others from the preceding sections yields simple proofs for some of the classical theorems on separability for algebraic function fields.

These results are also discussed in the sources quoted in the notes to section 4.5.

The notion of a separating transcendence base was defined in MacLane [1]; in this work, MacLane also proved theorem 4.6.7, which is a generalization of a classical result of Schmidt (4.6.8). For an excellent exposition that requires minimal background and explains how separably generated extensions arise in the study of algebraic and geometric problems, we refer the reader to MacLane [2].

The results stated in problems 9 and 10 on the "differential mapping" defined by a field extension can be found in Lang [4: sec. 4.10].

Suggestions for further reading

1. The arithmetic theory of algebraic function fields was initiated in the works of Dedekind and Weber [1] and Hensel and Landsberg [1]. The extension to algebraic functions of two variables was carried out by Jung [1]. For more recent expositions of the theory of algebraic functions of one variable, the reader is referred to the books of Artin [5], Chevalley [2], Deuring [2], Eichler [1], and Lang [2]. The arithmetic theories of algebraic number fields and of algebraic function fields of one variable can be treated to a large extent in a unified manner, as is done in several of the books on algebraic number theory referred to in the notes to the preceding chapter.

2. After reading the section on transcendence bases and transcendence degree, the reader may undertake the study of homogeneous polynomial equations in several variables. A suitable reference for this is the monograph by Greenberg [1], in which interesting results on p-adic fields are proved and connections with mathematical logic are discussed. An introductory treatment on homogeneous polynomials in several variables, quadratic forms, and formally real fields can be found in Jacobson [5: II, sec. 11.5, 11.6].

3. Our discussion on valuation domains and places was limited to the minimum required for the proof of the Nullstellensatz. The reader interested in pursuing the study of valuation domains and places may consult Bourbaki [1: ch. 6] and Zariski and Samuel [1: II, ch. 6].

4. Transcendental extensions play an important role in the study of algebraic varieties. The elementary treatments in Walker [1] and Jenner [1], which require minimal prerequisites, can be used as an introduction to algebraic curves and algebraic varieties. The works of Artin [2], Fogarty [1], Fulton [1], Lang [1], and Northcott [1] are accessible to any reader who has studied transcendental extensions and has a modest background in commutative algebra. For higher-level presentations of various topics in algebraic geometry, the reader may consult the books by Borel [1], Humphreys [1], Lang [2], Mumford [1], Šafarevič [1], and Weil [1]. However, a more extensive background is required for the appreciation of these works.

References and Selected Bibliography

ABEL, N. H.
1 *Oeuvres Complètes*, 2 volumes, Christiania: Grondahl and Sons.
2 Démonstration de l'impossibilité de la résolution algébrique des équations générales qui passent le quatrième degré, *J. Reine Angew. Math.* 1(1826), 65–84.
   See also Abel [1: I, 66–87].
3 Mémoir sur une classe particulière d'équations résolubles algébriquement, *J. Reine Angew. Math.* 4(1829), 131–156.
   See also Abel [1: I, 478–507].

ALBERT, A. A.
1 Cyclic fields of degree p^n over F of characteristic p, *Bull. Amer. Math. Soc.* 40(1934), 625–631.

ARTIN, E.
1 Über die Zerlegung definiter Funktionen in Quadrate, *Abh. Math. Sem. Univ. Hamburg* 5(1927), 100–115.
2 *Elements of Algebraic Geometry*, New York: (Lecture Notes), New York University, 1955.
3 *Galois Theory* (Notre Dame Mathematical Lectures No. 2), Notre Dame, IN: Notre Dame University Press, 1959.
4 *Theory of Algebraic Numbers*, Göttingen: George Striker, Schildweg 12, 1959.
5 *Algebraic Numbers and Algebraic Functions*, New York: Gordon & Breach, 1967.

ARTIN, E., AND SCHREIER, O.
1 Algebraische Konstruktion reeller Körper, *Abh. Math. Sem. Univ. Hamburg* 5(1927), 85–99.

2   Eine Kennzeichnung der reell absgeschlossenen Körper, *Abh. Math. Sem. Univ. Hamburg* **5**(1927), 225–231.

ARTIN, E., AND TATE, J. T.
1   *Class Field Theory* (Lecture Notes), Cambridge, MA: Harvard University Press, 1961.

ARTIN, M., AND MUMFORD, D.
1   Some elementary examples of unirational varieties which are non-rational, *Proc. London Math. Soc.* **25**(1972), 75–95.

ATIYAH, M. F., AND MACDONALD, I. G.
1   *Introduction to Commutative Algebra*, Reading, MA: Addison-Wesley, 1969.

BASTIDA, J. R.
1   On derivations of fields of algebraic functions, *Proc. Edinburgh Math. Soc.* (1980), 239–241.

BERGER, T. R., AND REINER, I.
1   A proof of the normal basis theorem, *Amer. Math. Monthly* **82**(1975), 915–918.

BIEBERBACH, L.
1   *Theorie der Geometrischen Konstruktionen*, Basel: Birkhäuser Verlag, 1952.

BOREL, A.
1   *Linear Algebraic Groups*, Reading, MA: W. A. Benjamin, 1969.

BOREVIČ, Z. I., AND ŠAFAREVIČ, I. R.
1   *Number Theory*, New York: Academic Press, 1966.

BOURBAKI, N.
1   *Algèbre Commutative (Ch. 5, 6)*, Paris: Hermann, 1964.
2   *Algèbre (Ch. 6, 7)*, Paris: Hermann, 1964.
3   *Algèbre (Ch. 4, 5)*, Paris: Hermann, 1967.

CAPELLI, A.
1   Sulla riduttibilità delle equazioni algebriche I, *Rend. Accad. Sci. Fis. Mat. Napoli* **3**(1897), 243–252.
2   Sulla riduttibilità delle equazioni algebriche II, *Rend, Accad. Sci. Fis. Mat. Napoli* **4**(1898), 84–90.

CARDANO, G.
1   *Ars Magna*, Nürnberg, 1545. (English translation: *The Great Art or the Rules of Algebra*, Cambridge, MA: M.I.T. Press, 1968.)

CARTAN, H.
1   Théorie de Galois pour les corps non commutatifs, *Ann. Sci. École Norm. Sup.* **64**(1947), 59–77.

CASSELS, J. W. S., AND FRÖHLICH, A. (EDS.)
1   *Algebraic Number Theory*, New York: Academic Press, 1967.

CASTELNUOVO, G.
1   Sulla razionalità delle involuzioni piane, *Math. Ann.* **44**(1894), 125–155.

CAUCHY, A.-L.
1 *Oeuvres Complètes*, 12 volumes, Paris: Gauthier-Villars.
2 Mémoire sur une nouvelle théorie des imaginaires, et sur les racines symboliques des équations et des équivalences, *C. R. Acad. Sci. Paris* 24(1847), 1120–1131.
See also Cauchy [1: X, 312–323].
3 Mémoire sur l'application de la nouvelle théorie des imaginaires aux diverses branches des sciences mathématiques, *C. R. Acad. Sci. Paris* 25(1847), 129–132.
See also Cauchy [1: X, 351–354].

ČEBOTAREV, N. (TSCHEBOTARÖW, N.)
1 *Grundzüge der Galois'schen Theorie*, Groningen: P. Noordhoff, 1950.

CHASE, S. U., HARRISON, D. K., AND ROSENBERG, A.
1 Galois theory and Galois cohomology of commutative rings, *Mem. Amer. Math. Soc.* 52(1968), 1–19.

CHEVALLEY, C.
1 Démonstration d'une hypothèse de M. Artin, *Abh. Math. Sem. Univ. Hamburg* 11(1936), 73–75.
2 *Introduction to the Theory of Algebraic Functions of One Variable*, Providence, RI: American Mathematical Society, 1951.

CLEMENS, C. H., AND GRIFFITHS, P. A.
1 The intermediate Jacobian of the cubic threefold, *Ann. of Math.* 92(1972), 281–356.

DEDEKIND, R.
1 *Gesammelte Mathematische Werke*, 3 Bände, Braunschweig: Vieweg. Reprinted by Chelsea, New York, 1969.
2 Abriss einer Theorie der höheren Kongruenzen in bezug auf einen reellen Primzahl-Modulus, *J. Reine Angew. Math.* 54(1857), 1–26.
See also Dedekind [1: I, 40–66].
3 Beweis für die Irreduktibilität der Kreisteilungs-Gleichungen, *J. Reine Angew. Math.* 54(1857), 27–30.
See also Dedekind [1: I, 68–71].
4 Über die Theorie der ganzen algebraischen Zahlen. In Dirichlet [1: supp. XI].
See also Dedekind [1: III, 1–222]. (Originally, this source appeared as supp. X to the 1871 edition of Dirichlet's book.)

DEDEKIND, R., AND WEBER, H.
1 Theorie der algebraischen Funktionen einer Veränderlichen, *J. Reine Angew. Math.* 92(1882), 181–290.
See also Dedekind [1: I, 238–349].

DEHN, E.
1 *Algebraic Equations*, New York: Columbia University Press, 1930.

DELIGNE, P.
1 Variétés unirationnelles non-rationnelles, *Sém. Bourbaki 1971/72, Exp. 402* (Lecture Notes in Math. No. 317), Berlin: Springer-Verlag, 1973.

DEURING, M.
1 Galoissche Theorie und Darstellungstheorie, *Math. Ann.* **107**(1932), 140–144.
2 *Lectures on the Theory of Algebraic Functions of One Variable* (Lecture Notes in Math. No. 314), New York: Springer-Verlag, 1973.

DIRICHLET, P. G. L.
1 *Vorlesungen über Zahlentheorie*, Braunschweig: Vieweg, 1894.

EICHLER, M.
1 *Einführung in die Theorie der Algebraischen Zahlen und Funktionen*, Basel: Birkhäuser Verlag, 1963. (English translation: *Introduction to the Theory of Algebraic Numbers and Functions*, New York: Academic Press, 1966.)

FLANDERS, H.
1 A remark on Hilbert's Nullstellensatz, *J. Math. Soc. Japan* **6**(1954), 160–161.

FOGARTY, J.
1 *Invariant Theory*, Reading, MA: W. A. Benjamin, 1969.

FRÖHLICH, A. (ED.)
1 *Algebraic Number Fields*, London: Academic Press, 1977.

FULTON, W.
1 *Algebraic Curves*, Reading, MA: W. A. Benjamin, 1969.

GALOIS, É.
1 *Oeuvres Mathématiques*, Paris: Gauthier-Villars.
2 Sur la théorie des nombres, *Bull. Sci. Math. Férussac* **13**(1830), 428–435.
   See also, Galois [1: 15–23].

GAUSS, C. F.
1 *Werke*, 14 Bände, Göttingen: Königliche Gesellschaft der Wissenschaften.
2 *Disquisitiones Arithmeticae*, Leipzig: Gehr. Fleischer, Jun., 1801. (English translation: *Disquisitiones Arithmeticae*, New Haven, CT: Yale University Press, 1966.)
   See also Gauss [1: XIII].

GILMER, R.
1 A note on the algebraic closure of a field, *Amer. Math. Monthly* **75**(1968), 1101–1102.

GOLDSTEIN, L. J.
1 *Analytic Number Theory*, Englewood Cliffs, NJ: Prentice-Hall, 1971.

GREENBERG, M. J.
1 *Lectures on Forms in Many Variables*, Reading, MA: W. A. Benjamin, 1969.

HADLOCK, C. R.
1 *Field Theory and Its Classical Problems*, Buffalo: Mathematical Association of America, 1978.

HENSEL, K.
1 *Theorie der Algebraischen Zahlen*, Leipzig: Teubner Verlag, 1908.

HENSEL, K., AND LANDSBERG, G.
1 *Theorie der Algebraischen Funktionen einer Variabeln*, Leipzig: Teubner Verlag, 1902. Reprinted by Chelsea, New York, 1965.

HERMITE, C.
1 *Oeuvres de Charles Hermite*, 3 volumes, Paris: Gauthier-Villars.
2 Sur la fonction exponentielle, *C. R. Acad. Sci.* **77**(1873), 18–24, 74–79, 226–233, 285–293.
   See also Hermite [1: III, 150–181].

HILBERT, D.
1 *Gesammelte Abhandlungen*, 3 Bände, Berlin: Springer-Verlag. Reprinted by Chelsea, New York, 1965.
2 Über die vollen Invariantensysteme, *Math. Ann.* **42**(1893), 313–373.
3 Die Theorie der algebraischen Zahlkörper, *Jahrber. Deutsch. MathVerein.* **4**(1897), 175–546.

HUMPHREYS, J. E.
1 *Linear Algebraic Groups*, New York: Springer-Verlag, 1975.

HUNGERFORD, T. W.
1 *Algebra*, New York: Springer-Verlag, 1980.

IGUSA, J.
1 On a theorem of Lüroth, *Mem. Coll. Sci. Univ. Kyoto* **26**(1951), 251–253.

ISAACS, I. M.
1 Roots of polynomials in algebraic extensions of fields, *Amer. Math. Monthly* **87**(1980), 543–544.

ISKOVSKIH, V. A., AND MANIN, YU. I.
1 Three-dimensional quartics and counterexamples to the Lüroth problem, *Mat. Sb.* (N. S.) **86**(1971), 140–166.

IYANAGA, S. (ED.)
1 *The Theory of Numbers*, Amsterdam: North-Holland, 1975.

JACOBSON, N.
1 Galois theory of purely inseparable fields of exponent one, *Amer. J. Math.* **66**(1944), 645–648.
2 A note on division rings, *Amer. J. Math.* **69**(1947), 27–36.
3 *Structure of Rings*, Providence, RI: American Mathematical Society, 1964.
4 *Lectures in Abstract Algebra III, Theory of Fields and Galois Theory*, New York: Springer-Verlag, 1964.
5 *Basic Algebra I, II*, San Francisco: W. H. Freeman, 1974, 1980.

JANUSZ, G. J.
1 *Algebraic Number Fields*, New York: Academic Press, 1973.

JENNER, W. E.
1 *Rudiments of Algebraic Geometry*, New York: Oxford University Press, 1963.

JUNG, H. W. E.
1 *Einführung in die Theorie der Algebraischen Funktionen zweier Veränderlicher*, Berlin: Akademie-Verlag, 1951.

KAPLANSKY, I.
1 *An Introduction to Differential Algebra*, Paris: Hermann, 1957.
2 *Commutative Rings*, Boston: Allyn & Bacon, 1970.
3 *Fields and Rings*, Chicago: University of Chicago Press, 1972.

KOCH, H.
1 *Galoissche Theorie der p-Erweiterungen*, Berlin: VEB Deutscher Verlag der Wissenschaften, 1970.

KRONECKER, L.
1 *Werke*, 6 Bände, Leipzig: Teubner Verlag.
2 Grundzüge einer arithmetischen Theorie der algebraischen Grössen, *J. Reine Angew. Math.* **92**(1882), 1–122.
   See also Kronecker [1: II, 245–387].
3 Ein Fundamentalsatz der allgemeinen Arithmetik, *J. Reine Angew. Math.* **100**(1887), 490–510.
   See also Kronecker [1: III, 209–240].

KRULL, W.
1 Galoissche Theorie der unendlichen algebraischen Erweiterungen, *Math. Ann.* **100**(1928), 687–698.
2 Allgemeine Bewertungstheorie, *J. Reine Angew. Math.* **167**(1931), 160–196.
3 Dimensionstheorie in Stellenringe, *J. Reine Angew. Math.* **179**(1938), 204–226.

LAGRANGE, J.-L.
1 *Oeuvres*, 14 volumes, Paris: Gauthier-Villars.
2 Réflexions sur la résolution algébrique des équations, *Mém. Acad. Berlin*, 1770–1771.
   See also Lagrange [1: III, 205–421].
3 Sur la forme des racines imaginaires des équations, *Mém. Acad. Berlin*, 1772.
   See also Lagrange [1: III, 479–516].

LANG, S.
1 *Introduction to Algebraic Geometry*, New York: Wiley (Interscience), 1959.
2 *Diophantine Geometry*, New York: Wiley (Interscience), 1962.
3 *Algebraic Number Theory*, Reading, MA: Addison-Wesley, 1970.
4 *Algebra*, Reading, MA: Addison-Wesley, 1971.
5 *Introduction to Algebraic and Abelian Functions*, Reading, MA: Addison-Wesley, 1972.

LENSTRA, H. W.
1 Rational functions invariant under a finite Abelian group, *Invent. Math.* **25**(1974), 299–325.

LIDL, R., AND NIEDERREITER, H.
1 *Finite Fields* (Encyclopedia of Mathematics and Its Applications, No. 20), Reading, MA: Addison-Wesley, 1983.

LINDEMANN, C. L. F.
1 Über die Zahl π, *Math. Ann.* **20**(1882), 213–225.

LIOUVILLE, J.
1 Sur des classes très étendues de quantités dont la valeur n'est ni algébrique, ni même réductible à des irrationnelles algébriques, *J. Math. Pures Appl.* **16**(1851), 133–142.

LONG, R. L.
1 *Algebraic Number Theory*, New York: Marcel Dekker, 1977.

LÜROTH, J.
1 Beweis eines Satzes über rationalen Kurven, *Math. Ann.* **9**(1876), 163–165.

MACLANE, S.
1 Modular fields: I, Separating transcendence bases, *Duke Math. J.* **5**(1939), 372–393.
2 Modular fields, *Amer. Math. Monthly* **47**(1940), 259–274.

MILLER, M. D., AND GURALNICK, R. M.
1 Subfields of algebraically closed fields, *Math. Mag.* **50**(1977), 260–261.

MOORE, E. H.
1 A doubly infinite system of simple groups, Chicago: Papers read at the Congress of Mathematics (1896), 208–242.

MUMFORD, D.
1 *Algebraic Geometry I, Complex Projective Varieties*, Berlin: Springer-Verlag, 1976.

NAGATA, M.
1 *Local Rings*, New York: Wiley (Interscience), 1962.
2 *Field Theory*, New York: Marcel Dekker, 1977.

NOETHER, E.
1 Normalbasis bei Körpern ohne höhere Verzweigung, *J. Reine Angew. Math.* **167**(1931), 147–152.

NORTHCOTT, D. G.
1 *Affine Sets and Affine Groups*, Cambridge: Cambridge University Press, 1980.

O'MEARA, O. T.
1 *Introduction to Quadratic Forms*, Berlin: Springer-Verlag, 1963.

RABINOWITSCH, J. L.
1 Zum Hilbertschen Nullstellensatz, *Math. Ann.* **102**(1929), 520.

RÉDEI, L.
1 *Algebra I, II*, Leipzig: Akademische Verlaggesellschaft, 1959.

RIBENBOIM, P.
1 *Algebraic Numbers*, New York: Wiley, 1972.
2 *L'Arithmétique des Corps*, Paris: Hermann, 1972.

ROTHMAN, T.
1   Genius and biographers: The fictionalization of Évariste Galois, *Amer. Math. Monthly* **89**(1982), 84–106.

RUFFINI, P.
1   *Opere Matematiche*, 3 tomi, Roma: Edizioni Cremonese.

ŠAFAREVIČ, I. R.
1   *Basic Algebraic Geometry*, Berlin: Springer-Verlag, 1977.

SALTMAN, D. J.
1   Generic Galois extensions, *Proc. Nat. Acad. Sci. U.S.A.* **77**(1980), 1250–1251.

SAMUEL, P.
1   *Théorie Algébrique des Nombres*, Paris: Hermann, 1967. (English translation: *Algebraic Theory of Numbers*, Boston: Houghton Mifflin, 1970.)

SCHUR, I.
1   *Gesammelte Abhandlungen*, 3 Bände, Berlin: Springer-Verlag.

SEGRE, B.
1   Sull'esistenza, sia nel campo razionale che nel campo reale, di involunzioni piane non-birazionali, *Rend. Accad. Naz.* **40**(5), 8(1951), 564–570.

SERRE, J.-P.
1   *Corps Locaux*, Paris: Hermann, 1962. (English translation: *Local Fields*, New York: Springer-Verlag, 1979.)
2   *Cohomologie Galoisienne* (Lecture Notes in Math. No. 5), Berlin: Springer-Verlag, 1964.

SERRET, J.-A.
1   *Cours d'Algèbre Supérieure*, Paris: Mallet-Bachelier, 1854.

SHATZ, S. S.
1   *Profinite Groups, Arithmetic, and Geometry*, Princeton, NJ: Princeton University Press, 1972.

STEINITZ, E.
1   Algebraische Theorie der Körper, *J. Reine Angew. Math.* **137**(1910), 167–309. Reprinted by Chelsea, New York, 1950.

STEWART, I.
1   *Galois Theory*, London: Chapman & Hall, 1973.

SWAN, R. G.
1   Invariant rational functions and a problem of Steenrod, *Invent. Math.* **7**(1969), 148–158.

TEICHMÜLLER, O.
1   *p*-Algebren, *Deutsch. Math.* **1**(1936), 362–388.

TOMINAGA, H., AND NAGAHARA, T.
1   *Galois Theory of Simple Rings*, Okayama: Okayama University, 1970.

VAHLEN, K. TH.
1 Über reductible Binome, *Acta Math.* **19**(1895), 195–198.

VAN DER WAERDEN, B. L.
1 *Einführung in die Algebraische Geometrie*, Berlin: Springer-Verlag, 1939.
2 *Algebra I, II*, Berlin: Springer-Verlag, 1960, 1959. (English translation: *Algebra I, II*, New York: Frederick Ungar 1970.)

WALKER, R. J.
1 *Algebraic Curves*, New York: Dover, 1950.

WANTZEL, P. L.
1 Recherches sur les moyens de reconnaître si un problème de géométrie peut se résoudre avec la règle et le compas, *J. Math. Pures Appl.* **2**(1837), 366–372.

WARNING, E.
1 Bemerkung zur vorstehenden Arbeit von Herrn Chevalley, *Abh. Math. Sem. Univ. Hamburg* **11**(1936), 76–83.

WATERHOUSE, W. C.
1 The normal basis theorem, *Amer. Math. Monthly* **86**(1979), 212.

WEBER, H.
1 Untersuchungen über die allgemeinen Grundlagen der Galois'schen Gleichungstheorie, *Math. Ann.* **43**(1893), 521–549.
2 *Lehrbuch der Algebra I, II, III*, Braunschweig: Vieweg, 1894, 1899, 1908. Reprinted by Chelsea, New York, 1963.

WEIERSTRASS, K.
1 Zu Lindemann's Abhandlung "Über die Ludolph'sche Zahl", *Sitzber. Berliner Akad.*, 1885.

WEIL, A.
1 *Foundations of Algebraic Geometry*, Providence, RI: American Mathematical Society, 1962.
2 *Basic Number Theory*, New York: Springer-Verlag, 1973.

WEISS, E.
1 *Algebraic Number Theory*, New York: McGraw-Hill, 1963.

WINTER, D. J.
1 *Structure of Fields*, New York: Springer-Verlag, 1974.

WITT, E.
1 Zyklische Körper und Algebren der Charakteristik p vom Grad p^n, *J. Reine Angew. Math.* **176**(1936), 126–140.

ZARISKI, O.
1 Foundations of a general theory of birational correspondences, *Trans. Amer. Math. Soc.* **53**(1943), 490–542.
2 A new proof of Hilbert's Nullstellensatz, *Bull. Amer. Math. Soc.* **53**(1947), 362–368.

3   On Castelnuovo's criterion of rationality $p_a = p_2 = 0$ of an algebraic surface, *Illinois J. Math.* **2**(1958), 305–315.

ZARISKI, O., AND SAMUEL, P.
1   *Commutative Algebra I, II*, Princeton, NJ: Van Nostrand, 1958, 1960.

Bibliography

Books that contain expositions on field extensions and Galois theory.

Adamson, I. T. *Introduction to Field Theory*, New York: Wiley, 1964.
Albert, A. A. *Modern Higher Algebra*, Chicago: University of Chicago Press, 1937.
_____. *Fundamental Concepts of Higher Algebra*, Chicago: University of Chicago Press, 1956.
Cohn, P. M. *Algebra I, II*, New York: Wiley, 1974, 1977.
Gaal, L. *Classical Galois Theory with Examples*, New York: Chelsea, 1973.
Goldhaber, J. K., and Ehrlich, G. *Algebra*, New York: Macmillan, 1970.
Hasse, H. *Höhere Algebra I, II*, Berlin: Walter de Gruyter, 1951. (English translation: *Higher Algebra I, II*, New York: Frederick Ungar, 1954.)
Herstein, I. N. *Topics in Algebra*, New York: Wiley, 1975.
McCarthy, P. J. *Algebraic Extensions of Fields*, Waltham, MA: Blaisdell, 1966.
Postnikov, M. M. *Foundations of Galois Theory*, New York: Macmillan, 1962.

Of historical interest

Bell, E. T. *Men of Mathematics*, New York: Simon & Schuster, 1937.
Bourgne, R., and Azra, J.-P. (eds.) *Écrits et Mémoires Mathématiques d'Évariste Galois*, Paris: Gauthier-Villars, 1976.
Boyer, C. B. *A History of Mathematics*, New York: Wiley, 1968.
Dupuy, P. La vie d'Évariste Galois, *Ann. École Norm.* **13**(1896), 197–266.
Hoyle, F. *Ten Faces of the Universe*, San Francisco: W. H. Freeman, 1977.
Infeld, L. *Whom the Gods Love: The Story of Évariste Galois*, New York: Whittlesey House, 1948.
Kiernan, M. B. The development of Galois theory from Lagrange to Artin, *Arch. Hist. Exact Sci.* **8**(1971), 40–154.

Index